Ecosystem Observation, Simulation and Assessment

Ecosystem Observation, Simulation and Assessment

Editors

Peng Hou
Weiguo Jiang
Wei Li
Li Zhang

MDPI • Basel • Beijing • Wuhan • Barcelona • Belgrade • Manchester • Tokyo • Cluj • Tianjin

Editors
Peng Hou
Ministry of Ecology and Environment
China

Weiguo Jiang
Beijing Normal University
China

Wei Li
Chinese Academy of Forestry
China

Li Zhang
Chinese Academy of Sciences
China

Editorial Office
MDPI
St. Alban-Anlage 66
4052 Basel, Switzerland

This is a reprint of articles from the Special Issue published online in the open access journal *Diversity* (ISSN 1424-2818) (available at: https://www.mdpi.com/journal/diversity/special_issues/Ecosystem_Observation).

For citation purposes, cite each article independently as indicated on the article page online and as indicated below:

LastName, A.A.; LastName, B.B.; LastName, C.C. Article Title. *Journal Name* **Year**, *Volume Number*, Page Range.

ISBN 978-3-0365-6774-7 (Hbk)
ISBN 978-3-0365-6775-4 (PDF)

Cover image courtesy of Peng Hou

Contents

About the Editors . . . . **vii**

Peng Hou
Ecosystem Observation, Simulation and Assessment: Progress and Challenges
Reprinted from: *Diversity* **2023**, *15*, 255, doi:10.3390/d15020255 . . . . **1**

Wenchang Zhou, Shanshan Xiang, Yuhu Shi, Xiuhuan Xu, Huicui Lu, Wenhui Ou and Jiawei Yang
Invasive Water Hyacinth (*Eichhornia crassipes*) Increases Methane Emissions from a Subtropical Lake in the Yangtze River in China
Reprinted from: *Diversity* **2022**, *14*, 1036, doi:10.3390/d14121036 . . . . **7**

Yi Wu, Xufeng Mao, Liang Xia, Hongyan Yu, Yao Yu, Wenjia Tang, et al.
Microbial Community Abundance Affects the Methane Ebullition Flux in Dahejia Reservoir of the Yellow River in the Warm Season
Reprinted from: *Diversity* **2023**, *15*, 154, doi:10.3390/d15020154 . . . . **21**

Cun Du, Shahbaz Khan, Yinghai Ke and Demin Zhou
Assessment of Spatiotemporal Dynamics of Mangrove in Five Typical Mangrove Reserve Wetlands in Asia, Africa and Oceania
Reprinted from: *Diversity* **2023**, *15*, 148, doi:10.3390/d15020148 . . . . **37**

Peng Hou, Jun Zhai, Dian-Dian Jin, Yan Zhou, Yan Chen and Hai-Feng Gao
Assessment of Changes in Key Ecosystem Factors and Water Conservation with Remote Sensing in the Zoige
Reprinted from: *Diversity* **2022**, *14*, 552, doi:10.3390/d14070552 . . . . **59**

Jiarong Ma, Zhuowei Hu, Quanqin Shao, Yongcai Wang, Yanqiong Zhou, Jiayan Liu and Shuchao Liu
Detection of Large Herbivores in UAV Images: A New Method for Small Target Recognition in Large-Scale Images
Reprinted from: *Diversity* **2022**, *14*, 624, doi:10.3390/d14080624 . . . . **73**

Leichao Nie, Zhiguo Dou, Lijuan Cui, Xiying Tang, Xiajie Zhai, Xinsheng Zhao, et al.
Hyperspectral Inversion of Soil Carbon and Nutrient Contents in the Yellow River Delta Wetland
Reprinted from: *Diversity* **2022**, *14*, 862, doi:10.3390/d14100862 . . . . **95**

Ziyan Ling, Weiguo Jiang, Chaoming Liao, Yanshun Li, Yurong Ling, Kaifeng Peng and Yawen Deng
Evaluation of Production–Living–Ecological Functions in Support of SDG Target 11.a: Case Study of the Guangxi Beibu Gulf Urban Agglomeration, China
Reprinted from: *Diversity* **2022**, *14*, 469, doi:10.3390/d14060469 . . . . **113**

Qi Ma, Mengquan Zhou, Jiechao Liu, Jing Zhao and Meijun Xi
Coupling Relationship between Ecosystem Service Value and Socioeconomic Development in the Qinba Mountains, China
Reprinted from: *Diversity* **2022**, *14*, 1105, doi:10.3390/d14121105 . . . . **133**

Chenyan Sha, Jian Wu, Jianqiang Wu, Chunmei Ye, Cheng Shen, Jinghua Su and Min Wang
Effects of Different Fertilizers on Soil Microbial Diversity during Long-Term Fertilization of a Corn Field in Shanghai, China
Reprinted from: *Diversity* **2023**, *15*, 78, doi:10.3390/d15010078 . . . . **153**

Yuqing Xu, Fengjin Xiao and Yaoming Liao
Assessment of Grassland Ecosystem Service Value in Response to Climate Change in China
Reprinted from: *Diversity* **2022**, *14*, 160, doi:10.3390/d14030160 . . . . **169**

Li Yu, Botao Zhou, Yuqing Xu, Yongxiang Zhang and Fengxue Gu
Projections of the Net Primary Production of Terrestrial Ecosystem and Spatiotemporal Responses to Climate Change in the Yangtze River Economic Belt
Reprinted from: *Diversity* **2022**, *14*, 327, doi:10.3390/d14050327 . . . . **187**

Junjun Bai, Peng Hou, Diandian Jin, Jun Zhai, Yutiao Ma and Jiajun Zhao
Habitat Suitability Assessment of Black-Necked Crane (*Grus nigricollis*) in the Zoige Grassland Wetland Ecological Function Zone on the Eastern Tibetan Plateau
Reprinted from: *Diversity* **2022**, *14*, 579, doi:10.3390/d14070579 . . . . **203**

Dehao Li, Shiying Zhu, Jin Gao, Haibo Jiang, Guangyi Deng, Lianxi Sheng, et al.
The Influence of Ecological Engineering on Waterbird Diversity in Different Habitats within the Xianghai Nature Reserve
Reprinted from: *Diversity* **2022**, *14*, 1016, doi:10.3390/d14121016 . . . . **217**

Yan Wang, Naifeng Lin, Jixi Gao, Changxin Zou and Delin Xu
Impacts of China's Western Development and Protection Strategy: An Ecosystem Services Perspective of Western China
Reprinted from: *Diversity* **2022**, *14*, 863, doi:10.3390/d14100863 . . . . **229**

About the Editors

Peng Hou

Peng Hou, Ph.D., is Professor for the Center for Satellite Application on Ecology and Environment, Ministry of Ecology and Environment, China. He is interested in ecosystem monitoring and assessment, ecology and environment of remote sensing, ecosystem protection, nature reserves and national parks.

Weiguo Jiang

Weiguo Jiang, Ph.D., is currently the vice minister and professor of the Faculty of Geographical Science in Beijing Normal University. They are mainly engaged in the theory, method and application research of remote sensing science and technology in ecological hydrology, wetland environment, flood disaster and other fields. They presided over more than 10 national projects, such as projects of the National Natural Science Foundation of China and National Key Research and Development Program.

Wei Li

Wei Li, Ph.D., is Professor for wetlands ecology at the Institute of Wetland Research in Chinese Academy of Forestry, China. His research focuses on studying wetland restoration, wetland ecological processes and effects, and wetland vegetation monitoring by hyperspectral remote sensing. He obtained a PhD and master's degree in Chinese Academy of Forestry (2012 and 2009) and a Bachelor's degree in Shandong Agricultural University (2005). He is currently acting as a doctoral supervisor and the deputy director of the Institute of Wetland Research.

Li Zhang

Li Zhang, Ph.D., is an associate professor at Key Laboratory of Ecosystem Network Observation and Modeling, Institute of Geographic Sciences and Natural Resources Research, Chinese Academy of Sciences. Her research focuses on process-based ecosystem modelling, terrestrial carbon cycle model-data fusion, and assessment of ecosystem services and ecosystem quality. She obtained a PhD (2009) in Ecology from Graduate University of Chinese Academy of Sciences (Beijing, China). She teaches the theory and application of ecological informatics at the College of Resources and Environment, University of Chinese Academy of Sciences.

diversity MDPI

Editorial

Ecosystem Observation, Simulation and Assessment: Progress and Challenges

Peng Hou

Satellite Environment Application Center, Ministry of Ecology and Environment, Beijing 100094, China; houpcy@163.com

Ecosystems provide supply, regulation, culture and support services for human-being, and overall support human survival and sustainable development. As an open, dynamic and integrated system, the internal components of an ecosystem are constantly realizing the dynamic coordination of interaction to achieve a new balance through the process of material and energy exchange and actualizing the mutual adaptation and self-evolution of the ecosystem and the external environment. However, driven by multiple factors such as climate change, population growth, urbanization, and exploitation of mineral resources, global problems such as ecosystem degradation and biodiversity loss have affected the sustainable development of human beings. It has become a hot spot in ecology research to develop basic theories, model methods and technical means for ecosystem observation, simulation and evaluation, for the quantitative analysis of the structure, process and function of ecosystems, and for the improvement of the scientific understanding of the changing characteristics and evolution laws of natural ecosystems.

1. Ecosystem Observation Progress

According to the spatial distribution of ecosystem characteristics, the parameters of an ecosystem are measured on the spot by setting sample plots, quadrats, sample points and sample belts in the field investigation. Alternatively, quantitative analysis is carried out in the laboratory after ecological sample collection. By these representative investigation methods, the overall observation of a regional ecosystem can be realized. This field survey type exemplifies the earliest and most basic method of ecosystem observation. To obtain more continuous times ecosystem observation data so as to better understand the ecosystem process and its mechanism, the observation equipment is deployed at a fixed sample point to observe positioning or site. For the same ecological parameter, different observation methods can be selected according to different research purposes, such as the study of methane emission intensity in wetlands [1,2].

Due to the limitation of the continuity and representativeness of observation data, the long-term positioning observation data of a single site cannot reveal the universal law of ecology. To take a mountain ecosystem as an example, the observation data from a certain station on the sunny slope can be used to analyze the laws of the mountain's sunny slope ecosystem but cannot be used to analyze the laws of other mountain or shady slope ecosystems. If we aim to determine a general law of mountain ecosystems, we must establish up multiple observation stations across different mountains and ecosystems according to their distribution. Positioning observation has developed from a single position to a network, forming ecosystems observation networks on national, intercontinental and global scales. In particular, the International Biological Program (IBP) launched in the 1960s and the international symposium "Long-Term Ecological Research: Global Prospects" held in the 1980s played important roles in promoting the development of positioning observation worldwide. Over 40 years, a national-scale ecosystem observation network has formed which is represented by the United States, China, the United Kingdom, Japan, Canada and Australia. This is joined by intercontinental-scale ecosystem observation networks,

Citation: Hou, P. Ecosystem Observation, Simulation and Assessment: Progress and Challenges. *Diversity* **2023**, *15*, 255. https://doi.org/10.3390/d15020255

Received: 29 January 2023
Revised: 5 February 2023
Accepted: 6 February 2023
Published: 11 February 2023

represented by the Asian Alux Observation System and European Integrated Carbon Observation System, and series of globe-scale ecosystem observation networks represented by the International Long-Term Ecosystem Research Network, the Group on Earth Observations Bio-diversity Observation Network, and the Global Terrestrial Observing System.

In addition to ground observation, remote sensing observation has become an indispensable means of ecosystem observation. Since the launch of the first manmade satellite in 1957, Earth observation has changed. The era in which humans can only make local observations from the Earth's surface, which spanned millennia, has ended. Up to now, the spatial resolution of remote sensing observation includes kilometer-scale, meter-scale and sub-meter-scale resolutions, the spectral resolution of remote sensing observation includes multi-spectrum and hyperspectral resolutions, and the spectral resolution of remote sensing observation includes ultraviolet, visible light, infrared, microwave resolutions, etc. The increasing use of UAV (unmanned aerial vehicle) remote sensing platforms has significantly improved remote sensing observation ability. Remote sensing observation has the unique characteristics of large-scale synchronous observation, which significantly improves the capability of researchers to observe the ecosystem in a spatially continuous or synchronous manner. At the same time, based on historical images, we can carry out retrospective monitoring and assessments of regional ecological change characteristics. Remote sensing observation has been widely used in regional ecosystem change monitoring [3,4]. The use of high spatial resolution UAV remote sensing observation technology can better monitor and identify large animals, and significantly improve the efficiency and data accuracy of biodiversity monitoring and investigation [5]. For hyperspectral remote sensing data, refined spectral information can better identify ecosystem process indicators, such as soil carbon or nutrients [6].

2. Ecosystem Simulation Progress

Ecosystem simulation is mainly realized through the ecological principles model. Ecological model simulation is based on a good understanding of the ecological process, structure and function. Based on the basic elements and key processes of the ecosystem, an ecological model is put forward by the parameterized, digital and quantitative expression of the complex ecosystem. As a consequence, it is unrealistic to aspire to describe the ecosystem completely and accurately through model simulation. Simulation model construction is the trade-off between model accuracy and simulation efficiency. In general, the fewer the basic elements and key process nodes of the ecosystem, the higher the simulation efficiency of the model but the lower the accuracy of the model. On the contrary, the simulation efficiency of the model is low, but the model accuracy is high. If the ecosystem has more basic elements and key process nodes then the simulated model will be closer to the real and complex ecosystem.

Ecology is a wide range of contents, and there are many types of ecosystem models. Representative models include eco-geographic and biogeochemical models. Eco-geographic models mainly simulate the spatial distribution of terrestrial ecosystem and its relationship with geographical environment, as in a Holdridge life zone system model. Ecosystem biogeochemical models, such as the CASA vegetation net primary productivity model, CENTRY biogeochemical cycle model, or BIOME-BGC carbon and water flux model, mainly simulate ecosystem processes and functions including net primary productivity, carbon and nitrogen water cycle, nutrient circulation, etc. With the development of remote sensing and geographic information system technology, the spatial and geographic development trend of ecological models is obvious. At the same time, with the increasingly obvious impact of climate change and human disturbance on the ecosystem, the model simulation of the impact of climate change and human disturbance on the structure and function of the ecosystem has received more attention.

3. Ecosystem Assessment Progress

The main content of ecosystem assessment is the analysis of the spatio-temporal change process of the ecosystem and of the interaction between the ecosystem and human society, climate, hydrology, etc. The Millennium Ecosystem Assessment, carried out by the United Nations organization, has promoted the development of integrated ecosystem assessment in landmark fashion. Different organizations or countries have carried out many comprehensive ecosystem assessment practices at different scales, such as global, regional and national. The assessment framework models can be summarized into four categories: First is the "Ecological Pressure–Policy Response" ecosystem assessment framework, represented by the DPSIR (Drivers–Pressure–State–Impact–Response) framework adopted by the Global Environment Outlook of the United Nations Environment Programme. Second is the "Ecosystem Services–Human Welfare" ecosystem assessment framework, represented by the "Ecosystem Services–Material Supply Human Welfare-Change Driving Force" adopted by the United Nations Millennium Ecosystem Assessment. The third is the "Natural Benefit–Ecosystem Management" ecosystem assessment framework, represented by The Economics of Ecosystems and Biodiversity (TEEB) project plan promoted by the United Nations Environment Programme. The fourth is the "Comprehensive Status–Change Trend" ecosystem assessment framework, represented by China's regular national ecosystem survey and assessment.

In addition to the comprehensive assessment of ecosystems, there have been some new research hotspots in ecosystem assessment focusing on global challenges and sustainable development of human society. In terms of sustainable development, measured by progress towards the 2030 Sustainable Development Goals proposed by the United Nations [7], the coordinated development of ecological protection and social economy [8–10] has attracted much attention. In the realm of climate change challenges, some researchers have assessed ecosystem responses to climate change, especially the response of ecosystem services [11] and production functions [12] to climate change. In terms of the challenges of biodiversity and ecological degradation, the biodiversity conservation assessment [13,14] and implementation effectiveness assessment of ecological conservation and restoration policies or measures [15–17] are becoming popular.

4. Challenges in the Future

Although people have made significant progress in the research and cognition of ecosystems, the cognitive level remains very limited in the face of complex and comprehensive ecosystems. There is no doubt that scientific research of ecosystems is needed.

The observation of ecosystem structure, process and function cannot be realized by using any existing ecosystem observation means alone. In the future, it will be necessary to build an intelligent ecosystem observation network by integrating three observation methods with the help of networks, IOT (Internet of Things), improving ecosystem observation ability and realizing fine observation. At present, due to the relatively mature observation technologies and methods for the main ecological parameters including vegetation and hydrology, there are many observation and research methods for natural ecosystems such as forests, grasslands and wetlands. However, for ecosystems that are not dominated by vegetation, such as deserts and glaciers, the observation technologies and methods need to be strengthened to better serve the challenges of climate change.

The ecosystem parameter data obtained by multi-observation means are multi-source heterogeneous and differ in space–time scale. Developing a means to process data from different observation methods and achieve the matching of a space–time scale is the most important and basic objective in ecosystem research. The accuracy of spatio-temporal matching of observation data directly affects the reliability and accuracy of ecosystem research conclusions. However, due to the significant difference in data scales, such as the spatial matching between the ground survey data in 1 square meter samples and the satellite remote sensing data with tens of meters or kilometers of spatial resolution, the current models and methods for processing these basic data still have great limitations. Of course, the scale problem not only exists in data processing, but persists across wide areas of ecological research, such as evaluation and simulation, and represents a fundamental topic in ecological research. For the same ecological problem, with different space-time scales, the research conclusions may be inconsistent, and contradictory conclusions may even appear. Spatial scale includes both spatial scope and spatial granularity (or resolution), and time scale includes both time span and time frequency. A means of choosing and determining the best space–time scale is the basis and key of ecological research. Other scale concepts also constitute the object of ecological research: cells, tissues, organs, systems, individuals, populations, communities, ecosystems and biosphere.

Ecological research is not only a scientific problem, but also a management problem relating to the sustainable development of human beings. How to quantitatively analyze the interaction between ecosystems and human society, ecosystems and climate change, establish a universal "natural ecology human society" ecological assessment framework, and establish a high-accuracy "natural ecology climate change" ecosystem process model are the major challenge ecosystem assessment. In the face of major challenges such as current ecosystem degradation, biodiversity loss and climate change, many questions still require scientific explanation and accurate answers. Examples of such issues include developing a means to quantitatively understand and establish the driving effect of protection and restoration measures on ecosystem change, the prediction of impacts of human disturbance and destruction on ecosystems, the contribution of ecological protection and restoration to mitigating climate change, the impact of ecosystem change on natural disasters, the simulation and prediction of global large-scale ecological change, and the high-accuracy simulation and evaluation of local scale ecological processes. The in-depth study and solution of these problems are of great scientific significance to the sustainable use of nature, the optimization and adjustment of ecosystem protection, and to restoration strategies for human society.

Conflicts of Interest: The author declares no conflict of interest.

References

1. Zhou, W.; Xiang, S.; Shi, Y.; Xu, X.; Lu, H.; Ou, W.; Yang, J. Invasive Water Hyacinth (*Eichhornia crassipes*) Increases Methane Emissions from a Subtropical Lake in the Yangtze River in China. *Diversity* **2022**, *14*, 1036. [CrossRef]
2. Wu, Y.; Mao, X.; Xia, L.; Yu, H.; Yu, Y.; Tang, W.; Xiao, F.; Ji, H. Microbial Community Abundance Affects the Methane Ebullition Flux in Dahejia Reservoir of the Yellow River in the Warm Season. *Diversity* **2023**, *15*, 154. [CrossRef]
3. Du, C.; Khan, S.; Ke, Y.; Zhou, D. Assessment of Spatiotemporal Dynamics of Mangrove in Five Typical Mangrove Reserve Wetlands in Asia, Africa and Oceania. *Diversity* **2023**, *15*, 148. [CrossRef]
4. Hou, P.; Zhai, J.; Jin, D.-D.; Zhou, Y.; Chen, Y.; Gao, H.-F. Assessment of Changes in Key Ecosystem Factors and Water Conservation with Remote Sensing in the Zoige. *Diversity* **2022**, *14*, 552. [CrossRef]
5. Ma, J.; Hu, Z.; Shao, Q.; Wang, Y.; Zhou, Y.; Liu, J.; Liu, S. Detection of Large Herbivores in UAV Images: A New Method for Small Target Recognition in Large-Scale Images. *Diversity* **2022**, *14*, 624. [CrossRef]
6. Nie, L.; Dou, Z.; Cui, L.; Tang, X.; Zhai, X.; Zhao, X.; Lei, Y.; Li, J.; Wang, J.; Li, W. Hyperspectral Inversion of Soil Carbon and Nutrient Contents in the Yellow River Delta Wetland. *Diversity* **2022**, *14*, 862. [CrossRef]
7. Ling, Z.; Jiang, W.; Liao, C.; Li, Y.; Ling, Y.; Peng, K.; Deng, Y. Evaluation of Production–Living–Ecological Functions in Support of SDG Target 11.a: Case Study of the Guangxi Beibu Gulf Urban Agglomeration, China. *Diversity* **2022**, *14*, 469. [CrossRef]
8. Ma, Q.; Zhou, M.; Liu, J.; Zhao, J.; Xi, M. Coupling Relationship between Ecosystem Service Value and Socioeconomic Development in the Qinba Mountains, China. *Diversity* **2022**, *14*, 1105. [CrossRef]
9. Sha, C.; Wu, J.; Wu, J.; Ye, C.; Shen, C.; Su, J.; Wang, M. Effects of Different Fertilizers on Soil Microbial Diversity during Long-Term Fertilization of a Corn Field in Shanghai, China. *Diversity* **2023**, *15*, 78. [CrossRef]
10. Zhao, Y.; Hou, P.; Jiang, J.; Zhai, J.; Chen, Y.; Wang, Y.; Bai, J.; Zhang, B.; Xu, H. Coordination Study on Ecological and Economic Coupling of the Yellow River Basin. *Int. J. Environ. Res. Public Health* **2021**, *18*, 10664. [CrossRef]
11. Xu, Y.; Xiao, F.; Liao, Y. Assessment of Grassland Ecosystem Service Value in Response to Climate Change in China. *Diversity* **2022**, *14*, 160. [CrossRef]
12. Yu, L.; Zhou, B.; Xu, Y.; Zhang, Y.; Gu, F. Projections of the Net Primary Production of Terrestrial Ecosystem and Spatiotemporal Responses to Climate Change in the Yangtze River Economic Belt. *Diversity* **2022**, *14*, 327. [CrossRef]
13. Bai, J.; Hou, P.; Jin, D.; Zhai, J.; Ma, Y.; Zhao, J. Habitat Suitability Assessment of Black-Necked Crane (Grus nigricollis) in the Zoige Grassland Wetland Ecological Function Zone on the Eastern Tibetan Plateau. *Diversity* **2022**, *14*, 579. [CrossRef]

14. Hou, P.; Bai, J.; Chen, Y.; Hou, J.; Zhao, J.; Ma, Y.; Zhai, J. Analysis on the hotspot characteristics of bird diversity distribution along the continental coastline of China. *Front. Mar. Sci.* **2022**, *9*, 1007442. [CrossRef]
15. Hou, P.; Chen, Y.; Zhai, J.; Hou, J.; Yang, M.; Jin, D.-D.; Zhu, H.-S.; Gao, H.-F.; Wan, H.-W.; Fu, Z. Natural Ecological Background of Terrestrial Vegetation in China: Identification Methods and Spatial Differentiation Characteristics. *Sens. Mater.* **2021**, *33*, 4659–4680. [CrossRef]
16. Li, D.; Zhu, S.; Gao, J.; Jiang, H.; Deng, G.; Sheng, L.; Cao, Y.; Li, L.; Lin, B. The Influence of Ecological Engineering on Waterbird Diversity in Different Habitats within the Xianghai Nature Reserve. *Diversity* **2022**, *14*, 1016. [CrossRef]
17. Wang, Y.; Lin, N.; Gao, J.; Zou, C.; Xu, D. Impacts of China's Western Development and Protection Strategy: An Ecosystem Services Perspective of Western China. *Diversity* **2022**, *14*, 863. [CrossRef]

diversity

MDPI

Article

Invasive Water Hyacinth (*Eichhornia crassipes*) Increases Methane Emissions from a Subtropical Lake in the Yangtze River in China

Wenchang Zhou [1,2,*], Shanshan Xiang [1,2], Yuhu Shi [1,2,*], Xiuhuan Xu [1,2], Huicui Lu [3], Wenhui Ou [1,2] and Jiawei Yang [1]

1 Institute of Wetland Research, Hubei Academy of Forestry, Wuhan 430075, China
2 Hubei Honghu Wetland Ecosystem Research Station, Honghu 433200, China
3 Faculty of Forestry, Qingdao Agricultural University, Qingdao 266109, China
* Correspondence: zwclky@126.com (W.Z.); shiyuhu@126.com (Y.S.)

Abstract: Lakes represent an important source of atmospheric methane (CH_4); however, there are few studies on which lake-dwelling invasive aquatic plants generate CH_4. Therefore, in this study, CH_4 emissions were measured using a floating chamber and gas chromatography in a subtropical lake in China. We considered four community zones of invasive plants (*Eichhornia crassipes*), emergent vegetation (*Zizania latifolia*), floating-plant (*Trapa natans*) and open-water zones. The results indicate that the flux of CH_4 emissions varied between -5.38 and 102.68 mg m^{-2} h^{-1}. The higher emission values were attributed to lake eutrophication. Moreover, the flux of CH_4 emissions in the invasive plant zone was 140–220% higher than that in the open-water and the floating-plant zones. However, there was no significant difference in CH_4 emissions between the invasive plant and the emergent vegetation zones. This may be due to a higher production of plants, as well as the rapid reproductive rate of the invasive plants. Finally, CH_4 emissions were positively associated with the air and water temperature; however, the emissions were also negatively associated with water depth. Our results suggest that invasive plants enhance freshwater CH_4 emissions, thus contributing to global warming.

Keywords: methane emission; lakes; water hyacinth; climate change; greenhouse gases

Citation: Zhou, W.; Xiang, S.; Shi, Y.; Xu, X.; Lu, H.; Ou, W.; Yang, J. Invasive Water Hyacinth (*Eichhornia crassipes*) Increases Methane Emissions from a Subtropical Lake in the Yangtze River in China. *Diversity* **2022**, *14*, 1036. https://doi.org/10.3390/d14121036

Academic Editors: Peng Hou, Weiguo Jiang, Wei Li and Li Zhang

Received: 13 October 2022
Accepted: 14 November 2022
Published: 26 November 2022

1. Introduction

Despite the fact that lake ecosystems cover approximately 3.7% of the Earth's continental land area [1,2], they are believed to be a major source of the greenhouse gas (GHG) methane (CH_4) [3,4]. The greenhouse effect of CH_4 is approximately 28-fold higher than that of CO_2 for the century-long time scale, accounting for approximately 20% of total global warming [3,5]. The CH_4 concentration in the atmosphere, which is mainly caused by human activity, has increased by 150% since pre-industrial times and continues to increase. This may further enhance global warming to a greater extent [5,6].

Top-down and bottom-up estimates for global CH_4 emissions are 576 and 727 Tg yr^{-1}, respectively, of which CH_4 emissions from freshwater wetlands (including lakes and rivers) account for 308 Tg yr^{-1} [5]. CH_4 emissions between and within lakes exhibit high spatiotemporal variability [7,8], and it is estimated that CH_4 emissions from lakes account for 8–48 Tg yr^{-1}, with approximately 50% of the flux being attributed to tropical/subtropical regions [3,9]. Although several studies have determined the CH_4 emissions from these lakes are largely the result of a warming climate, invasive alien plants, pollution and enclosure aquaculture [10–13], there are few studies on the contribution of human activity (e.g., introduction of alien plants) to CH_4 emissions from lakes. In addition, these studies suggest that current and future increases in CH_4 emissions will intensify climate change [12–14]; therefore, it is necessary to further explore CH_4 emissions from lakes.

The free-floating water hyacinth (*Eichhornia crassipes*) is one of the world's most invasive aquatic plants. It causes significant ecological and socio-economic effects [15]. As an ornamental plant originating from tropical South America, this invasive water hyacinth weed was introduced into China in the 1900s [16], and it has subsequently been extensively cultivated as animal feed since the 1950s. It is distributed widely in the aquatic ecosystems of the Yangtze River in China [16]. The water hyacinth commonly forms dense, interlocking mat-forming floating aquatic plants on the water surface. This results from a rapid reproductive rate, complex root structure and a doubling of its biomass within five days [10,15,17,18]. The water hyacinth mats prevent the transfer of oxygen from the air to the water's surface and block the light required for photosynthesis by phytoplankton and submersed vegetation [15,18]. The water hyacinths on the water surface can prevent light penetration into the water column below [18], which decreases the temperature (water and sediment) [19,20]. Finally, changes in these factors affect the spatiotemporal variability of CH_4 production and emissions from the lake and impact whole-lake emission estimates on an annual basis [3,7,10]. Therefore, it is important to study the influence of the invasive water hyacinth weed on CH_4 emissions in these lakes.

Hong Lake is the seventh largest shallow lake in China and the largest natural lake in Hubei Province, which is located within the middle reaches of the Yangtze River [21]. Because of the abundant natural resources in Hong Lake, rapid socio-economic development has caused the lake to undergo a variety of environmental changes over the past few decades, including a shrinking water area, deterioration of water quality and a decline in biodiversity [21–23]. Several studies have confirmed that CH_4 emissions from the invasive *Spartina altrniflora* weed have significantly increased compared with the CH_4 emissions from the native plant community [24–26]. Banik et al. [27] also reported that the invasive water hyacinth has clearly increased CH_4 emissions from the freshwater ecosystems in India, which were estimated to reach 1.2 Tg yr^{-1}. Conversely, Attermeyer et al. [10] reported that CH_4 emissions from invasive water hyacinth zones were reduced compared with those from open-water zones. This was caused by the oxidation of CH_4 catalysed by methanotrophic bacteria. Therefore, the influence of invasive aquatic plants on CH_4 emission rates in freshwater ecosystems requires further examination.

In the present study, we focused on the effects of the invasive water hyacinth on CH_4 emissions in a shallow lake in subtropical China. CH_4 emissions in the shallow lake were measured with floating chambers and a gas chromatography method. In addition, to reveal the scope of its influence, we analysed the relationship between ecological factors (water depth, temperature and dissolved oxygen from water) and CH_4 emissions.

2. Materials and Methods

2.1. Study area Description

The study was conducted at the Hong Lake Natural Reserve (113°12′–113°26′ N, 29°40′–29°58′ E) towards the middle reaches of the Yangtze River. Hong Lake has a surface area of 344 km^2 with an open-water area of 308 km^2, a littoral area of 36 km^2 and a mean water depth of 1.5 m [21,28]. Hong Lake was listed as an internationally important wetland in the Ramsar convention in 2008 and was entered into the China Wetland Ecosystem Research Network in 2014. The region is characterised by a north subtropical humid monsoon climate, with a mean annual temperature of 15.9 °C–16.6 °C. The minimum and maximum mean monthly temperatures were 3.8 °C in January and 28.9 °C in July and/or August, respectively. The annual average evaporation is 1000–1300 mm and the mean annual precipitation is 1174 mm, 74% of which occurs between April and October [21].

According to the change in water depth and the vegetation type from the littoral zone to the open water, four zones in the study region were selected to monitor CH_4 emission flux between April and October of 2021 (Figure 1). The first site (OPs) was the open water of the lake in which no vegetation grew. The second site (ECs) was an area invaded by an alien species of water hyacinth (*E. crassipes*), which covered 100% of the area. The third site (TNs) was covered by the floating plant *Trapa natans* (*T. natans*), with a total vegetation coverage of 95%. The fourth site (ZLs) consisted of the emergent aquatic plant *Zizania latifolia* (*Z. Latifolia*), with a total vegetation coverage of 90% and sparse areas containing *Nelumbo nucifera* and *T. natans*.

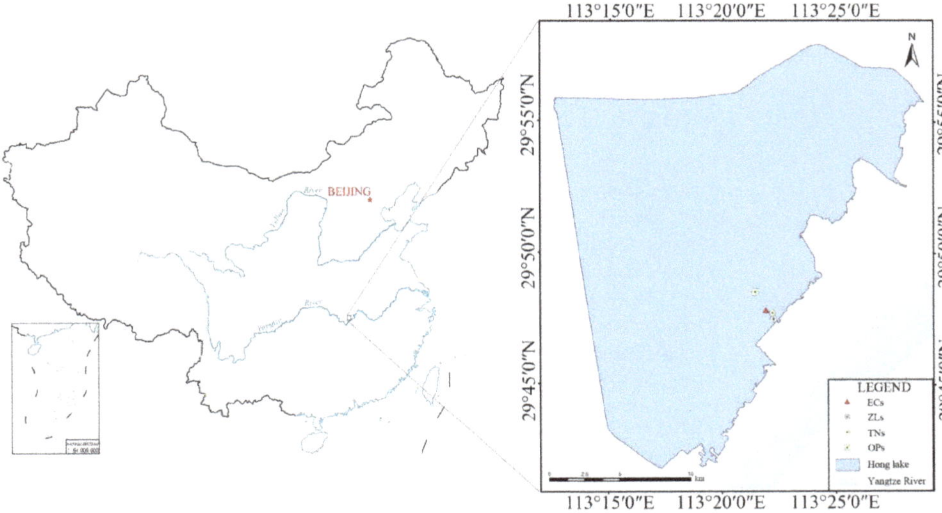

Figure 1. The study was conducted at Hong Lake in the middle reaches of the Yangtze River.

2.2. CH_4 Measurements

CH_4 flux measurements were taken at four sites in Hong Lake from April to December 2021. The measurements were carried out using floating chambers [29], which included three plastic opaque chambers (height above the water level 30 cm, volume 28.8 L) made of acrylic organic glass. The outside of the chamber contained a rubber plastic film to prevent an increase in the inner temperature of the chamber (Figure 2). In addition, the open-end of the chamber was fitted with a cystosepiment and tyre as floating equipment. The headspace of the chamber was equipped with a fan to mix the air and one sampling port and a temperature sensor. Before sampling, three chambers were placed upside down 50 to 100 cm apart on the water surface. Gas samples were drawn from each chamber every 5 min for 15 min with 60 mL polypropylene syringes equipped with three-way stopcocks and then transferred to a gas bag.

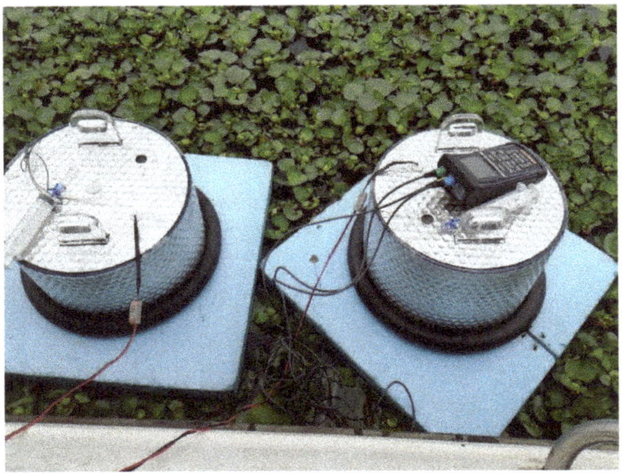

Figure 2. The measuring chambers at the invasive water hyacinth site (ECs) of Hong Lake.

Within one week, the CH_4 concentration of all samples was determined by a gas chromatography instrument (Agilent, 7890A, GC system, Agilent Co., Wilmington, DE, USA) equipped with a flame ionisation CH_4 detector from the Institute of Hydrobiology, Chinese Academy of Sciences. The fluxes were calculated using linear regression based on the concentration change as a function of time. 91% of all fluxes had a r-squared value of 0.70 or above (of which, 53% had a r-squared value of 0.9 or above). CH_4 flux at each site was calculated using the following equation:

$$F = \frac{d_c}{d_t} \times \frac{M}{V_0} \times \frac{T_0}{T} \times \frac{V}{A} \times 60 \quad (1)$$

F, flux at the time of chamber closure (mg m^{-2} h^{-1});
d_c/d_t, time derivative (slope) CH_4 concentration change over time (ppm min^{-1});
M, molecular mass of CH_4 (g mol^{-1});
V_0, ideal gas mole volume (0.0224 m^3 mol^{-1});
T_0, absolute temperature (273.15 K);
T, absolute temperature inside of chamber at sampling (K);
V, chamber volume (m^3) above the water surface;
A, chamber area (m^2).

2.3. Measurement of Environmental Factors

The water depth at each site was measured with a ruler and bamboo during sampling. The air temperature was measured using a digital thermometer (TM-902C, Factory of Lihuajin Instrument, Guangzhou, China). Conductivity, water temperature, pH and DO concentration at a water depth of 10 cm were measured using a portable multi-parameter water quality meter (Multi 3630 IDS, WTW Co., Munich, Germany).

At each site, the plants were sampled in September 2021 to measure biomass. Three 50 cm × 50 cm plots were randomly selected for these measurements. The plant samples were oven-dried at 70 °C for 48 h, and then weighed. In addition, three soil samples at each site were collected at a depth of 10 cm, and all soil samples were transferred to the laboratory, air-dried indoors and then dried at 70 °C for 48 h. The samples were milled and passed through a 0.125 mm sieve to determine the organic carbon concentration (SOC, g kg^{-1}) using the wet oxidation method with K_2CrO_7, and the soil pH was measured using the potentiometric method. In addition, total nitrogen (TN) concentration (g kg^{-1}) was measured using the Kjeldahl method with H_2SO_4 digestion. The total phosphorus (TP) concentration of the soil was determined by colorimetry by alkali fusion with NaOH.

2.4. Statistical Analysis

The significant differences in CH_4 fluxes at the four sites were analysed using SPSS software (18.0 version) based on Bonferroni's test as obtained by one-way analysis of variance. The relationship between CH_4 fluxes and environmental variables was determined using Pearson's rank correlation. $p < 0.05$ was considered statistically significant.

3. Results

3.1. Environmental Factors

The mean air temperature at the ECs, OPs and ZLs sites from April to December was not significantly different ($p > 0.05$), with mean values of 26.5 °C, 25.5 °C and 26.9 °C, respectively, whereas the mean air temperature at the ECs and ZLs sites was significantly higher ($p < 0.05$) than that of the TNs site (mean value of 24.7 °C, $p < 0.05$) (Figure 3a). The mean water temperature at the ECs, OPs and ZLs sites was not significantly different ($p > 0.05$), with mean values of 23.5 °C, 23.2 °C and 23.2 °C, respectively, but it was higher ($p < 0.05$) than that at the TNs site (23.0 °C) (Figure 3b). The mean water depth at the ECs, OPs, TNs and ZLs sites was 151, 192, 141 and 97.8 cm, respectively, and significant differences were observed ($p < 0.05$); however, there were no significant differences ($p > 0.05$) between the ECs and TNs sites (Figure 3c). The mean pH of the water at the ECs site

(7.9) was significantly lower ($p < 0.05$) than that at the OPs and TNs sites (8.5 and 8.2, respectively), and higher ($p < 0.05$) compared with that at the ZLs site (7.5) (Figure 3d). The mean DO at the ECs site (7.0 mg L^{-1}) was significantly lower ($p < 0.05$) than that at the Ops site (8.6 mg L^{-1}), but significantly higher ($p < 0.05$) than that of the ZLs site (3.2 mg·L^{-1}). There were no significant differences ($p > 0.05$) between the ECs and TNs sites (mean value of 7.6 mg L^{-1}) (Figure 3e). The mean conductivity of water at the ECs site (352.5 μS cm^{-1}) was significantly higher ($p < 0.05$) than that at the OPs (339.5 μS cm^{-1}) and TNs sites (347.6 μS cm^{-1}); however, there were no significant differences ($p > 0.05$) between the ECs and ZLs sites (355.5 μS cm^{-1}) (Figure 3f).

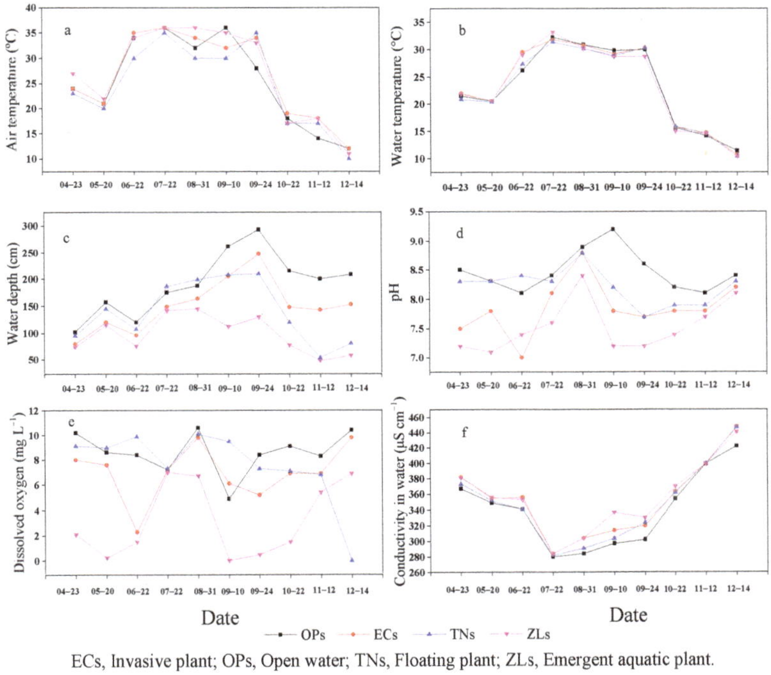

Figure 3. The spatial dynamics of environmental factors, including (**a**) air temperature, (**b**) water temperature, (**c**) water depth, (**d**) pH, (**e**) dissolved oxygen concentration and (**f**) water conductivity in the four community study zones.

The vegetation biomass at the ECs site was significantly higher ($p < 0.05$) than that at the Ops site, but lower than that at the ZLs site, and higher than that at the TNs site, whereas there were no significant differences among the ECs, TNs and ZLs sites ($p > 0.05$) (Table 1). The soil pH in the ZLs site was significantly lower than that of the Ops site ($p < 0.05$), and no significant differences were observed for the others ($p > 0.05$). The SOC and TP concentration in the top 10 cm of the soil at the ZLs were significantly higher than those at the OPs and ECs ($p < 0.05$); however, there were no significant differences between the ZLs and TNs sites ($p > 0.05$). Finally, the carbon-to-nitrogen ratio and the TP concentration were not significantly different between the sites ($p > 0.05$) (Table 1).

3.2. CH_4 Emission Fluxes

Temporal variations in CH_4 emission fluxes were recorded at the four sites, and the peak values occurred in the spring and/or summer, whereas the lowest values occurred in winter. The CH_4 emission fluxes at the ECs, OPs, TNs and ZLs sites ranged from 0.15 to 102.68, 0.12 to 59.75, −5.38 to 43.14 and 0.06 to 90.19 mg m^{-2} h^{-1}, respectively (Figure 4).

Table 1. The physicochemical characteristics at the four sites.

Sites	Vegetation		Soil				
	Types	Biomass (g m^{-2})	pH	SOC/g kg^{-1}	TN/g kg^{-1}	C:N Ratio	TP/g kg^{-1}
Open water (OPs)	—	No grown vegetations	8.12 ± 0.05 a	16.63 ± 1.54 a	1.33 ± 0.14 a	12.54 ± 0.21 a	0.64 ± 0.01 a
Invasive plant (ECs)	*E. crassipes*	270.02 ± 20.64 a	7.96 ± 0.04 ab	29.10 ± 1.71 a	2.42 ± 0.05 a	12.01 ± 0.65 a	0.63 ± 0.01 a
Floating plant (TNs)	*T. natans*	211.08 ± 17.63 a	7.57 ± 0.29 ab	46.47 ± 11.31 ab	3.44 ± 0.78 ab	13.37 ± 0.49 a	0.61 ± 0.01 a
Emergent aquatic plant (ZLs)	*Z. latifolia*, *N. nucifera*, *T. natans*	618.30 ± 187.50 a	7.03 ± 0.07 b	56.63 ± 2.84 b	4.49 ± 0.30 b	12.61 ± 0.17 a	0.67 ± 0.01 a

Note: different lowercase letters indicate a significant difference exists among the three sites. Significance level: 0.05.

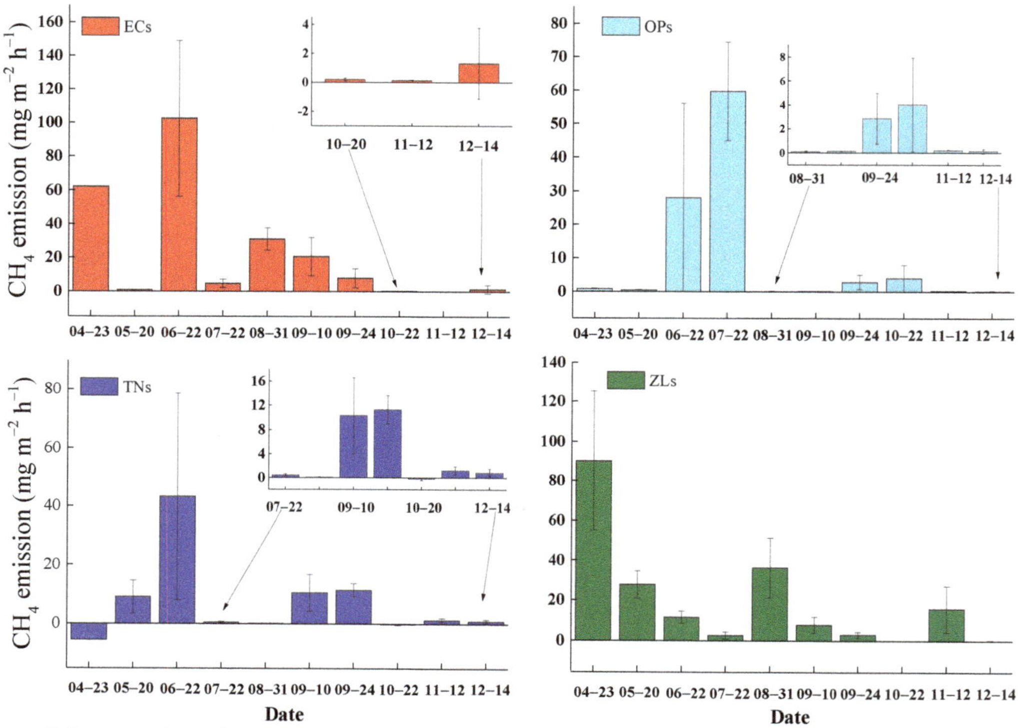

ECs, Invasive plant; OPs, Open water; TNs, Floating plant; ZLs, Emergent aquatic plant.

Figure 4. Seasonal variations in CH_4 emission fluxes at the four sites.

Mean CH_4 emissions at the ECs, OPs, TNs and ZLs sites from April to December were 23.16, 9.68, 7.08 and 19.48 mg m^{-2} h^{-1}, respectively (Figure 5). The highest CH_4 emissions were observed at the ECs site. These differences indicate that the invasive plant, *E. crassipes*, exerts an enormous influence on CH_4 emissions at Hong Lake. A one-way analysis of variance showed that there was a significant difference in CH_4 emissions between the ECs and TNs sites ($P = 0.03$), slight differences between the ECs and OPs sites ($p < 0.1$) and no significant differences between the ECs and ZLs sites ($p > 0.10$).

3.3. Dependence of CH_4 Fluxes on Environmental Factors

CH_4 emission fluxes at all four sites were significantly positively correlated with air and water temperature (Figure 6a, $p < 0.01$), and significantly negatively correlated with water depth (Figure 6b, $p < 0.01$). The CH_4 emission fluxes had no significant correlation with the DO concentration in the water (Figure 7, $p > 0.05$).

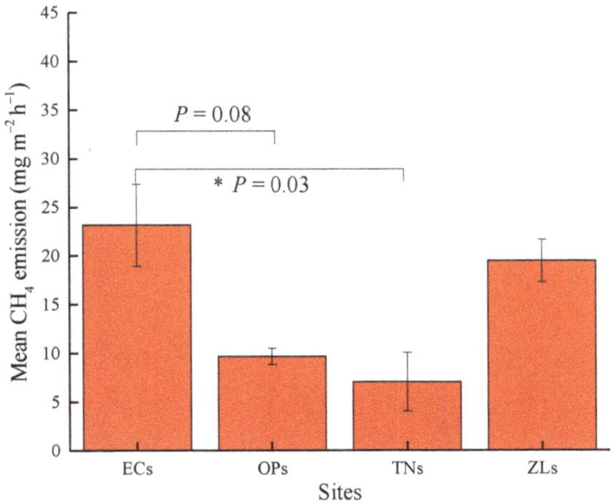

ECs, Invasive plant; OPs, Open water; TNs, Floating plant; ZLs, Emergent aquatic plant.

Figure 5. Mean CH_4 emission fluxes at the four sites (* indicated the significant levels at 0.05).

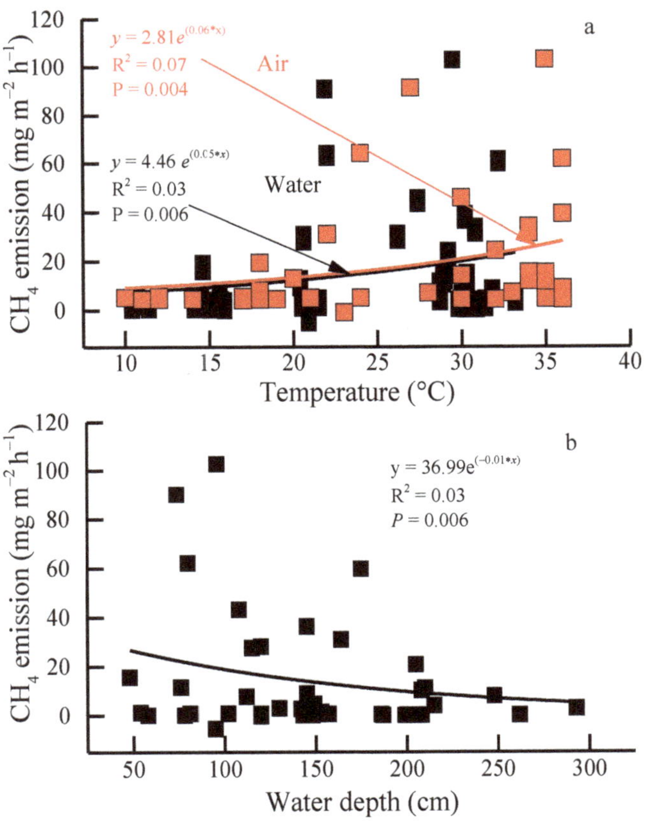

Figure 6. The correlation between CH_4 emissions, temperature (air and water) and water depth.

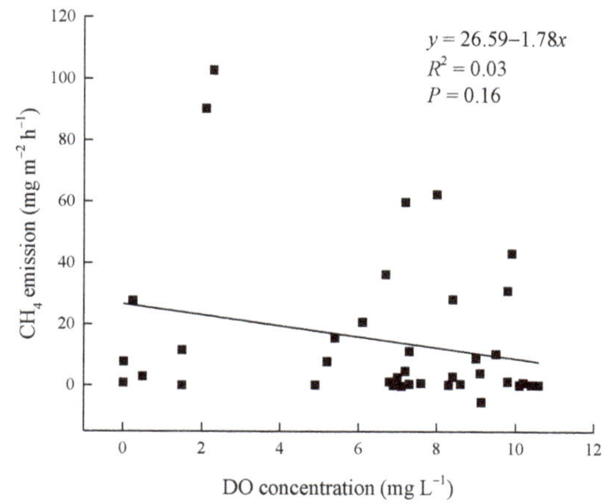

Figure 7. The correlation between CH_4 emissions and dissolved oxygen concentration.

4. Discussion

In this study, CH_4 emission fluxes at Hong Lake ranged from −5.38 to 102.68 mg m^{-2} h^{-1} (Figure 4). The results indicate that this shallow lake releases a large amount of CH_4 into the atmosphere, but the flux values matched the ranges recorded in other lakes in tropical and subtropical regions (ranging from −1.7 to 326 mg m^{-2} h^{-1}) [27,30]. Moreover, Gondwe and Masamba [31] reported that the maximum diffusive CH_4 emission rates in tropical wetlands varied between 0.24 and 293 mg m^{-2} h^{-1}, with a mean of 23.2 mg m^{-2} h^{-1}. The CH_4 flux at Hong Lake was one to two orders of magnitude higher than that at Dong Lake in the Yangtze River in China (ranging from 0.06 to 5.53 mg m^{-2} h^{-1}) [32], and higher than that at lakes in North America (ranging from 0.002 to 0.826 mg m^{-2} h^{-1}) [29,33] and a reservoir in the boreal region (ranging from 18 to 36 mg m^{-2} h^{-1}) [34]. A hydro-electric reservoir in French Guyana released CH_4 at flux intensities ranging from 8 to 35 mg m^{-2} h^{-1} [35].

Over the last five decades, Hong Lake has been extensively altered by flood regulation, irrigation, fish aquaculture and water supply demands [23]. This has resulted in a deterioration of water quality and caused the TP and TN to exceed their protective targets (TP $\leq$ 0.05 mg L^{-1} and TN $\leq$ 1.0 mg L^{-1}) [21]. According to a report in 2004, approximately 80% (250 km^2) of the lake has been used for large-scale aquaculture since the 1990s [36]. In 2017, the area used for aquaculture decreased to 40% of the lake area as a consequence of wetland protection and a restoration project at Hong Lake beginning in 2004 [21,23]. Based on our measurements, the TP at the lake was 0.088 mg L^{-1} in August and 0.239 mg L^{-1} in November 2021, which could have altered the CH_4 emissions of the lake [3]. Many studies have found that eutrophic lakes release more CH_4 emissions into the atmosphere [14,37,38]. Zhou et al. [37] reported that emission values ranged from 0.1 to 351.9 mg h^{-2} h^{-1} for shallow lakes in the Yangtze River Basin and were related to more enriched waters. Pickard et al. [12] also reported that severe pollution in urban lakes resulted in higher CH_4 emissions, including a large amount of untreated sewage input, with the highest flux recorded at 335 mg m^{-2} h^{-1}.

We consulted internal data from the Administration of the Hong Lake National Nature Reserve, beginning with the water hyacinth weed invasion in the 1990s. The coverage of the invading hyacinth increased from approximately 1300 ha in 2009 to 6000 ha in 2014, but decreased to 1200 ha in 2017 because of a project to remove water hyacinth from the lake. To date, it covers approximately 300 ha of the lake (Figure 8). The invasive water hyacinth weed exhibited high growth rates and a mean net CO_2 exchange (−3.4 to −5.4 g C-CO_2 m^{-2} d^{-1}, negative values indicate ecosystem CO_2 uptake) compared with open water (2.3 to 5.1 g C-CO_2 m^{-2} d^{-1}, positive values indicate ecosystem efflux) [39]. However, the extensive water hyacinth coverage of the lake surface resulted in the eutrophication of the lake [40]. Furthermore, the large water hyacinth mats prevented the transfer of oxygen from the atmosphere to the water and decreased oxygen production by other plants and algae [10,15]. This changes other ecological factors, such as the DO concentration in water and sediment temperature of the lake [19,20], which may be linked to CH_4 production and emissions from the lake [10]. In the present study, CH_4 emission fluxes at the ECs site were 139–227% higher on average than those at the OPs and TNs sites, and were approximately 20% higher than that at the ZLs site, although there was no significant difference between the ECs and ZLs sites (Figure 5). Our results are consistent with those of previous studies [27]. For the ECs site, higher CH_4 release rates during sampling may be explained by higher biomass. Wang et al. [30] found that the vegetation in the inundated area played an important role in CH_4 production and represented "hotspots" of CH_4 fluxes in water systems. Table 1 shows that the plant biomass in the ECs and the ZLs sites was higher than that in the OPs and TNs sites, and the former two sites had increased CH_4 release into the atmosphere. In general, numerous studies indicate that vegetation is a key factor of CH_4 release in wetlands and is attributed to primary production, which supplies organic matter to the sediment and induces the production of CH_4 by methanogenic bacteria [41,42]. For example, Furlanetto et al. [43] reported that higher CH_4 emissions in eutrophic lakes were attributed to higher organic matter concentration, resulting from higher primary production rates. Other studies indicated that CH_4 emissions were positively correlated with net primary production in two lakes [3,32]. Our results reveal that the invasive plants strongly enhance freshwater CH_4 emissions through an increase in plant productivity, thus contributing to global warming. Therefore, to reduce CH_4 emissions, we suggest the extensive removal of water hyacinth in lakes through wetland protection and restoration projects.

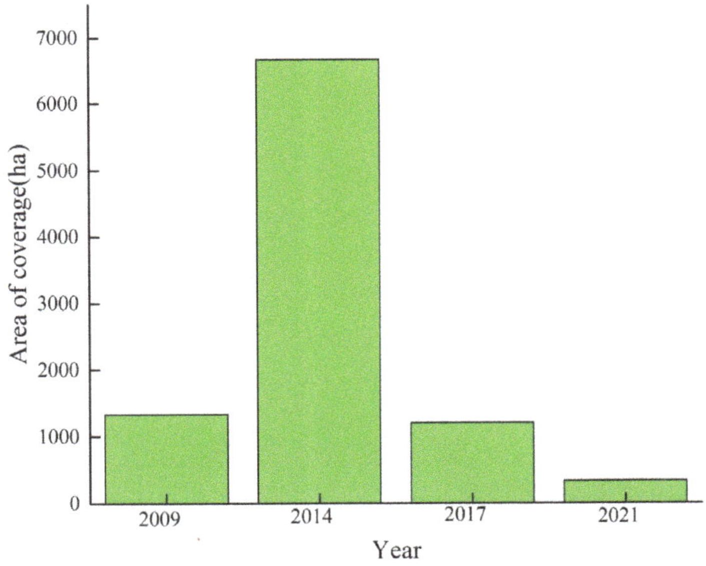

Figure 8. The area covered with water hyacinth in Hong Lake.

CH_4 is produced in sediment under anoxic conditions by methanogens and is released into the atmosphere through three pathways, including ebullition, diffusion and plant-mediated emissions [3,44,45]. Two studies showed that lower DO concentrations in water overlaying the sediment in the lake and marsh resulted in higher CH_4 production [46,47] and greater CH_4 emissions [10,48]. In the present study, the mean DO concentration at the four sites of the lake, from highest to lowest, was as follows: OPs > TNs > ECs > ZLs (Figure 3e). Thus, lower DO concentrations were observed in surface water with areas covered by macrophytes, which may have caused an increase in CH_4 emission fluxes at the ZLs and the ECs, compared with the TNs and Ops (Figures 4 and 5). The results indicate that CH_4 emissions had no significant correlation with DO concentration in the lake water, but the emissions decreased with increasing DO concentration in the lake water (Figure 7). For example, Bolpagni et al. [49] reported that the oxygen saturation in a stand of *T. natans* was lower than that in control areas that were devoid of plants, indicating vegetation could lead to a reduction in anoxic conditions and an increase in CH_4 emissions, although the aerenchyma in the plants may contribute to the transport of CH_4 from the water column into the atmosphere. In contrast, Attermeyer et al. [10] reported that the lower DO concentrations in the surface water of areas covered by *E. crassipes* led to a significantly lower flux of CH_4 emissions compared with that of open water, because the CH_4 in the surface water beneath the vegetation was oxidised by methanotrophic bacteria. Thus, the effect of DO concentrations in water on CH_4 emissions requires further study.

Air and water temperatures are important factors that limit the seasonal variation in CH_4 emissions [6,8,32,50]. It is well-established that methanogenesis in lake sediments increases exponentially with temperature because of the increased microbial activity at higher temperatures [10,51,52]. Figure 4 shows that the highest CH_4 emission fluxes occurred in June and July (Summer) at the three sites (ECs, OPs and TNs), whereas the lowest were seen in December (Winter). Figure 6a indicates that CH_4 emissions increase exponentially with air and water temperatures in Hong Lake (N = 40, $p < 0.01$), but R^2 was much lower. Although the highest CH_4 emissions occurred in April (Spring) for the ZLs site, the results suggest that other factors affect the CH_4 emissions at Hong Lake, such as water depth, plant growth and the availability of organic matter in sediments [6,45,53].

Water depth or the water table level in wetlands is usually a major factor affecting the spatial and temporal variation in the CH_4 emission flux [42,48,53]. We observed a negative correlation between CH_4 emission fluxes and water depth at all sites; however, R^2 was also much lower (Figure 6b). The results are consistent with the findings of several previous studies [8,53]. Because Hong Lake is a shallow lake with a mean water depth of 1.5 m [28], the variation in water depth in the lake from April to December ranged from 48 cm to 293 cm (Figure 3c). However, emergent plants, including *Z. latifolia* and *N. nucifera* plants, which were more dominant in the ZLs site compared with the other three sites, could extend their root systems into deeper and more anaerobic sediment and transport CH_4-rich gas into the atmosphere [53].

5. Conclusions

In this study, we determined how the invasive water hyacinth weed affects CH_4 emissions in subtropical Chinese lakes. The results indicate that CH_4 emissions in the zones invaded by water hyacinth were 20%–220% higher than those in the exotic-plant-free areas, because of the higher productivity of the invasive water hyacinth. The CH_4 emission fluxes at Hong Lake ranged from -5.38 mg m^{-2} h^{-1} (a sink of atmospheric CH_4) to 102.68 mg m^{-2} h^{-1} (a larger source of atmospheric CH_4). In addition, the results indicate that CH_4 emissions exhibited a weaker correlation with water temperature, water depth and dissolved oxygen concentration. Therefore, it is essential to further intensively study the CH_4 emissions of lakes in tropical and subtropical regions.

Author Contributions: Conceptualisation, W.Z., S.X., Y.S., X.X., H.L., W.O. and J.Y.; formal analysis, W.Z., Y.S., H.L. and J.Y.; funding acquisition, W.Z. and Y.S.; investigation, W.Z., S.X., X.X., W.O. and J.Y.; supervision, S.X., H.L. and W.O.; validation, W.Z.; writing—original draft, W.Z.; writing—review and editing, W.Z. and H.L. All authors have read and agreed to the published version of the manuscript.

Funding: This work was supported by the National Natural Foundation of China (NO. 31971474, 31800374).

Institutional Review Board Statement: Not applicable.

Data Availability Statement: Not applicable.

Acknowledgments: We thank Yanxia Zuo for their valuable field and laboratory analyses assistance from the Institute of Hydrobiology, Chinese Academy of Sciences. The authors would like to thank local fishermen and the staff from Administration of Hong Lake National Nature Reserve for their help in collecting data.

Conflicts of Interest: The authors declare no conflict of interest relevant to this study.

References

1. Downing, J.A.; Prairie, Y.T.; Cole, J.J.; Duarte, C.M.; Tranvik, L.J.; Striegl, R.G.; McDowell, W.H.; Kortelainen, P.; Caraco, N.F.; Melack, J.M.; et al. The global abundance and size distribution of lakes, ponds, and impoundments. *Limnol. Oceanogr.* **2006**, *51*, 2388–2397. [CrossRef]
2. Verpoorter, C.; Kutser, T.; Seekell, D.A.; Tranvik, L.J. A global inventory of lakes based on high-resolution satellite imagery. *Geophys. Res. Lett.* **2014**, *41*, 6396–6402. [CrossRef]
3. Bastviken, D.; Cole, J.J.; Pace, M.L.; Tranvik, L.J. Methane emissions from lakes: Dependence of lake characteristics, two regional assessments, and a global estimate. *Glob. Biogeochem. Cycl.* **2004**, *18*, 1–12. [CrossRef]
4. Bastviken, D.; Tranvik, L.J.; Downing, J.A.; Crill, P.M.; Enrich-Prast, A. Freshwater Methane Emissions Offset the Continental Carbon Sink. *Science* **2011**, *331*, 50. [CrossRef]
5. IPCC. Contribution of Working Group I to the Sixth Assessment Report of the Intergovernmental Panel on Climate Change. In *Climate Change 2021: The Physical Science Basis*; Masson-Delmotte, V., Zhai, P., Pirani, A., Connors, S.L., Péan, C., Berger, S., Caud, N., Chen, Y., Goldfarb, L., Gomis, M.I., Eds.; Cambridge University Press: Cambridge, UK; New York, NY, USA, 2021; pp. 1–2391.
6. Aben, R.C.H.; Barros, N.; van Donk, E.; Frenken, T.; Hilt, S.; Kazanjian, G.; Lamers, L.P.M.; Peeters, E.T.H.M.; Roelofs, J.G.M.; Domis, L.N.D.S.; et al. Cross continental increase in methane ebullition under climate change. *Nat. Commun.* **2017**, *8*, 1682. [CrossRef] [PubMed]
7. Yang, H.; Xie, P.; Ni, L.; Flower, R.J. Underestimation of CH_4 Emission from Freshwater Lakes in China. *Environ. Sci. Technol.* **2011**, *45*, 4203–4204. [CrossRef] [PubMed]
8. Natchimuthu, S.; Sundgren, I.; Gålfalk, M.; Klemedtsson, L.; Crill, P.; Danielsson, Å.; Bastviken, D. Spatio-temporal variability of lake CH_4 fluxes and its influence on annual whole lake emission estimates. *Limnol. Oceanogr.* **2015**, *61*, S13–S26. [CrossRef]
9. Johnson, M.S.; Matthews, E.; Du, J.; Genovese, V.; Bastviken, D. Methane Emission from Global Lakes: New Spatiotemporal Data and Observation-Driven Modeling of Methane Dynamics Indicates Lower Emissions. *J. Geophys. Res. Biogeosci.* **2022**, *127*, e2022JG006793. [CrossRef] [PubMed]
10. Attermeyer, K.; Flury, S.; Jayakumar, R.; Fiener, P.; Steger, K.; Arya, V.; Wilken, F.; van Geldern, R.; Premke, K. Invasive floating macrophytes reduce greenhouse gas emissions from a small tropical lake. *Sci. Rep.* **2016**, *6*, 20424. [CrossRef] [PubMed]
11. Marotta, H.; Pinho, L.; Gudasz, C.; Bastviken, D.; Tranvik, L.J.; Prast, A.E. Greenhouse gas production in low-latitude lake sediments responds strongly to warming. *Nat. Clim. Change* **2014**, *4*, 467–470. [CrossRef]
12. Pickard, A.; White, S.; Bhattacharyya, S.; Carvalho, L.; Dobel, A.; Drewer, J.; Jamwal, P.; Helfter, C. Greenhouse gas budgets of severely polluted urban lakes in India. *Sci. Total Environ.* **2021**, *798*, 149019. [CrossRef] [PubMed]
13. Pu, Y.; Zhang, M.; Jia, L.; Zhang, Z.; Xiao, W.; Liu, S.; Zhao, J.; Xie, Y.; Lee, X. Methane emission of a lake aquaculture farm and its response to ecological restoration. *Agric. Ecosyst. Environ.* **2022**, *330*, 107883. [CrossRef]
14. Palma-Silva, C.; Marinho, C.C.; Albertoni, E.F.; Giacomini, I.B.; Barros, M.P.F.; Furlanetto, L.M.; Trindade, C.R.T.; Esteves, F.D.A. Methane emissions in two small shallow neotropical lakes: The role of temperature and trophic level. *Atmos. Environ.* **2013**, *81*, 373–379. [CrossRef]
15. Villamagna, A.M.; Murphy, B.R. Ecological and socio-economic impacts of invasive water hyacinth (*Eichhornia crassipes*): A review. *Freshw. Biol.* **2010**, *55*, 282–298. [CrossRef]
16. Ding, J.; Ren, W.; Fu, W.; Zhang, G. Water hyacinth in China: Its Distribution, Problems and Control Status. In *Biological and Integrated Control of Water Hyacinth, Eichhornia Crassipes*; Julien, M.H., Hill, M.P., Center, T.D., Jianqing, D., Eds.; Australian Centre for International Agricultural Research: Canberra, ACT, Australia, 2001; pp. 29–32.
17. Owens, C.S.; Madsen, J.D. Low temperature limits of water hyacinth. *J. Aquat. Plant Manag.* **1995**, *33*, 63–68.
18. Hu, W.; Salomonsen, J.; Xu, F.-L.; Pu, P. A model for the effects of water hyacinths on water quality in an experiment of physico-biological engineering in Lake Taihu, China. *Ecol. Model* **1998**, *107*, 171–188. [CrossRef]
19. Rai, D.N.; Munshi, J.D. The influence of thick floating vegetation (Water hyacinth: *Eichhornia crassipes*) on the physic-chemical environment of a fresh water wetland. *Hydrobiologia* **1979**, *62*, 65–69. [CrossRef]
20. Wilson, J.R.; Holst, N.; Rees, M. Determinants and patterns of population growth in water hyacinth. *Aquat. Bot.* **2005**, *81*, 51–67. [CrossRef]
21. Zhang, T.; Ban, X.; Wang, X.; Cai, X.; Li, E.; Wang, Z.; Yang, C.; Lu, X. Analysis of nutrient transport and ecological response in Honghu Lake, China by using a mathematical model. *Sci. Total Environ.* **2017**, *575*, 418–428. [CrossRef] [PubMed]
22. Zhou, W.C.; Shi, Y.H.; Pan, L. Current status and controlling strategies of water pollution in Honghu Lake wetland in Jianghan plain of Middle Reaches of Yangtze River. *Wetl. Sci. Manag.* **2019**, *15*, 31–34. (In Chinese)
23. Han, M.; Dsouza, M.; Zhou, C.; Li, H.; Zhang, J.; Chen, C.; Yao, Q.; Zhong, C.; Zhou, H.; A Gilbert, J.; et al. Agricultural Risk Factors Influence Microbial Ecology in Honghu Lake. *Genom. Proteom. Bioinform.* **2019**, *17*, 76–90. [CrossRef] [PubMed]

24. Tong, C.; Morris, J.T.; Huang, J.; Xu, H.; Wan, S. Changes in pore-water chemistry and methane emission following the invasion of *Spartina alterniflora* into an oliogohaline marsh. *Limnol. Oceanogr.* **2017**, *63*, 384–396. [CrossRef]
25. Yin, S.; An, S.; Deng, Q.; Zhang, J.; Ji, H.; Cheng, X. *Spartina alterniflora* invasions impact CH_4 and N_2O fluxes from a salt marsh in eastern China. *Ecol. Eng.* **2015**, *81*, 192–199. [CrossRef]
26. Zhang, Y.; Ding, W.; Cai, Z.; Valerie, P.; Han, F. Response of methane emission to invasion of *spartina alterniflora* and exogenous n deposition in the coastal salt marsh. *Atmos. Environ.* **2010**, *44*, 4588–4594. [CrossRef]
27. Banik, A.; Sen, M.; Sen, S.P. Methane emissions from water hyacinth-infested freshwater ecosystems. *Chemosphere* **1993**, *27*, 1539–1552. [CrossRef]
28. Wang, Z.; Du, Y.; Yang, C.; Liu, X.; Zhang, J.; Li, E.; Zhang, Q.; Wang, X. Occurrence and ecological hazard assessment of selected antibiotics in the surface waters in and around Lake Honghu, China. *Sci. Total Environ.* **2017**, *609*, 1423–1432. [CrossRef] [PubMed]
29. Bellido, J.L.; Tulonen, T.; Kankaala, P.; Ojala, A. CO_2 and CH_4 fluxes during spring and autumn mixing periods in a boreal lake (Pääjärvi, southern Finland). *J. Geophy. Res.* **2009**, *114*, G04007.
30. Wang, H.; Lu, J.; Wang, W.; Yang, L.; Yin, C. Methane fluxes from the littoral zone of hypereutrophic Taihu Lake, China. *J. Geophys. Res. Earth Surf.* **2006**, *111*, D17. [CrossRef]
31. Gondwe, M.J.; Masamba, W.R.L. Spatial and temporal dynamics of diffusive methane emissions in the Okavango Delta, northern Botswana, Africa. *Wetl. Eco. Manag.* **2014**, *22*, 63–78. [CrossRef]
32. Xing, Y.; Xie, P.; Yang, H.; Ni, L.; Wang, Y.; Rong, K. Methane and carbon dioxide fluxes from a shallow hypereutrophic subtropical Lake in China. *Atmos. Environ.* **2005**, *39*, 5532–5540. [CrossRef]
33. Fernandez, J.M.; Townsend-Small, A.; Zastepa, A.; Watson, S.B.; Brandes, J.A. Methane and nitrous oxide measured throughout Lake Erie over all seasons indicate highest emissions from the eutrophic Western Basin. *J. Great Lakes Res.* **2020**, *46*, 1604–1614. [CrossRef]
34. Eugster, W.; DelSontro, T.; Sobek, S. Eddy covariance flux measurements confirm extreme CH_4 emission from a Swiss hydropower reservoir and resolve their short-term variability. *Biogeosciences* **2011**, *8*, 2815–2831. [CrossRef]
35. Abril, G.; Guérin, F.; Richard, S.; Delmas, R.; Galy-Lacaux, C.; Gosse, P.; Tremblay, A.; Varfalvy, L.; Dos Santos, M.A.; Matvienko, B. Carbon dioxide and methane emissions and the carbon budget of a 10-year old tropical reservoir (Petit Saut, French Guiana). *Glob. Biogeochem. Cycl.* **2005**, *19*, GB4007. [CrossRef]
36. Lu, S.; Wang, S.H.; Yuan, W.B.; Wen, Y.L. Lake environment evolution of Honghu and consideration on development of wetland ecological industry. *Ecol. Econ.* **2009**, *218*, 157–159. (In Chinese)
37. Zhou, Y.; Song, K.; Han, R.; Riya, S.; Xu, X.; Yeerken, S.; Geng, S.; Ma, Y.; Terada, A. Nonlinear response of methane release to increased trophic state levels coupled with microbial processes in shallow lakes. *Environ. Pollut.* **2020**, *265*, 114919. [CrossRef] [PubMed]
38. Sun, H.; Yu, R.; Liu, X.; Cao, Z.; Li, X.; Zhang, Z.; Wang, J.; Zhuang, S.; Ge, Z.; Zhang, L.; et al. Drivers of spatial and seasonal variations of CO_2 and CH_4 fluxes at the sediment water interface in a shallow eutrophic lake. *Water Res.* **2022**, *222*, 118916. (In Chinese) [CrossRef] [PubMed]
39. Peixoto, R.B.; Marotta, H.; Bastviken, D.; Enrich-Prast, A. Floating Aquatic Macrophytes Can Substantially Offset Open Water CO_2 Emissions from Tropical Floodplain Lake Ecosystems. *Ecosystems* **2016**, *19*, 724–736. [CrossRef]
40. Mwamburi, J. Spatial variations in sedimentary organic matter in surficial lake sediments of Nyanza Gulf (Lake Victoria, Kenya) after invasion of water hyacinth. *Lakes Reserv. Sci. Policy Manag. Sustain. Use* **2016**, *21*, 94–113. [CrossRef]
41. Whiting, G.J.; Chanton, J.P. Primary production control of methane emission from wetlands. *Nature* **1993**, *364*, 794–795. [CrossRef]
42. Juutinen, S.; Alm, J.; Larmola, T.; Huttunen, J.T.; Morero, M.; Martikainen, P.J.; Silvola, J. Major implication of the littoral zone for methane release from boreal lakes. *Global Biogeochem. Cycl.* **2003**, *28*, 1–11. [CrossRef]
43. Furlanetto, L.M.; Marinho, C.C.; Palma-Silva, C.; Albertoni, E.F.; Figueiredo-Barros, M.P.; de Assis Estevesb, F. Methane levels in shallow subtropical lake sediments: Dependence on the trophic status of the lake and allochthonous input. *Limnologica* **2012**, *42*, 151–155. [CrossRef]
44. Capone, D.G.; Kiene, R.P. Comparison of microbial dynamics in marine and freshwater sediments: Contrasts in anaerobic carbon catabolism. *Limnol. Oceanogr.* **1998**, *33*, 725–749. [CrossRef]
45. Ding, W.; Cai, Z.; Tsuruta, H.; Li, X. Key factors affecting spatial variation of methane emissions from freshwater marshes. *Chemosphere* **2003**, *51*, 167–173. [CrossRef]
46. Flury, S.; McGinnis, D.F.; Gessner, M.O. Methane emissions from a freshwater marsh in response to experimentally simulated global warming and nitrogen enrichment. *J. Geophys. Res. Earth Surf.* **2010**, *115*, G1. [CrossRef]
47. Maruya, Y.; Nakayama, K.; Sasaki, M.; Komai, K. Effect of dissolved oxygen on methane production from bottom sediment in a eutrophic stratified lake. *J. Environ. Sci.* **2023**, *125*, 61–72. [CrossRef] [PubMed]
48. Zhou, W.; Cui, L.; Wang, Y.; Li, W. Methane emissions from natural and drained peatlands in the Zoigê, eastern Qinghai-Tibet Plateau. *J. Forestry Res.* **2017**, *28*, 539–547. [CrossRef]
49. Bolpagni, R.; Pierobon, E.; Longhi, D.; Nizzoli, D.; Bartoli, M.; Tomasellli, M.; Viaroli, P. Diurnal exchanges of CO_2 and CH_4 across the water–atmosphere interface in a water chestnut meadow (*Trapa natans* L.). *Aquat. Bot.* **2007**, *87*, 43–48. [CrossRef]
50. Xiao, Q.; Zhang, M.; Hu, Z.; Gao, Y.; Hu, C.; Liu, C.; Liu, S.; Zhang, Z.; Zhao, J.; Xiao, W.; et al. Spatial variations of methane emission in a large shallow eutrophic lake in subtropical climate. *J. Geophys. Res. Biogeosci.* **2017**, *122*, 1597–1614. [CrossRef]

51. Zeikus, J.G.; Winfrey, M.R. Temperature limitation of methanogenesis in aquatic sediments. *Appl. Environ. Microb.* **1976**, *31*, 99–107. [CrossRef] [PubMed]
52. Duc, N.T.; Crill, P.; Bastviken, D. Implications of temperature and sediment characteristics on methane formation and oxidation in lake sediments. *Biogeochemistry* **2010**, *100*, 185–196. [CrossRef]
53. Hirota, M.; Tang, Y.; Hu, Q.; Hirata, S.; Kato, T.; Mo, W.; Cao, G.; Mariko, S. Methane emissions from different vegetation zones in a Qinghai-Tibetan Plateau wetland. *Soil Biol. Biochem.* **2004**, *36*, 737–748. [CrossRef]

diversity

MDPI

Article

Microbial Community Abundance Affects the Methane Ebullition Flux in Dahejia Reservoir of the Yellow River in the Warm Season

Yi Wu [1,2], Xufeng Mao [1,2,*], Liang Xia [1,2], Hongyan Yu [3], Yao Yu [3], Wenjia Tang [4], Feng Xiao [5] and Haichuan Ji [5]

1 MOE Key Laboratory of Tibetan Plateau Land Surface Processes and Ecological Conservation, Qinghai Normal University, Xining 810008, China
2 School of Geographical Science, Academy of Plateau Science and Sustainability, Qinghai Normal University, Xining 810008, China
3 Management and Service Center of Qilian Mountain National Wetland Park, Xining 810008, China
4 State Key Laboratory for Environmental Protection Monitoring and Assessment of the Qinghai-Xining Plateau, Xining 810007, China
5 Qinghai Forestry and Grass Bureau, Xining 810007, China
* Correspondence: maoxufeng@yeah.net; Tel.: +86-139-0978-7689

Abstract: Reservoirs are an integral part of the global carbon cycle and generally considered to be methane (CH_4) emission hot spots. Although remarkable research achievements have been made concerning CH_4 ebullition from inland waters, such as rivers, lakes, and ponds, few have been devoted to CH_4 ebullition from plateau reservoirs. The present study focused on CH_4 ebullition from the Dahejia Reservoir located in the upper reaches of the Yellow River. We analyzed the spatial and temporal characteristics of CH_4 ebullition flux across the water-atmosphere interface between July and August 2021. We also evaluated the influence of microbes on CH_4 ebullition flux. The results showed that (1) CH_4 ebullition was the dominant mode of CH_4 emissions in the study site, which contributed to 78.85 ± 20% of total CH_4 flux. (2) The mean CH_4 ebullition flux in the nighttime (0.34 ± 0.21 mg m^{-2} h^{-1}) was significantly higher than that in the daytime (0.19 ± 0.21 mg m^{-2} h^{-1}). The mean CH_4 ebullition flux first decreased and then increased from the upstream (0.52 ± 0.57 mg m^{-2} h^{-1}) to the downstream (0.43 ± 0.3 mg m^{-2} h^{-1}) of the Yellow River. (3) Sediment microbes affected the CH_4 ebullition flux primarily by changing the microbial community abundance. The regression analysis showed that CH_4 ebullition flux had a significantly linear negative correlation with microbial abundance in sediments. The redundancy analysis further showed CH_4 ebullition flux was significantly positively correlated with the abundances of Firmicutes and Actinobacteria, and negatively with that of Proteobacteria and Chloroflexi. Among abiotic variables, CH_4 ebullition flux was closely related to total phosphorus, total organic carbon, pH and nitrate nitrogen.

Keywords: CH_4; ebullition flux; plateau reservoir; microbes

Citation: Wu, Y.; Mao, X.; Xia, L.; Yu, H.; Yu, Y.; Tang, W.; Xiao, F.; Ji, H. Microbial Community Abundance Affects the Methane Ebullition Flux in Dahejia Reservoir of the Yellow River in the Warm Season. *Diversity* **2023**, *15*, 154. https://doi.org/10.3390/d15020154

Academic Editors: Peng Hou, Weiguo Jiang, Wei Li and Li Zhang

Received: 24 October 2022
Revised: 24 December 2022
Accepted: 18 January 2023
Published: 21 January 2023

1. Introduction

Global warming has been deteriorating since the age of industrial revolution. From 1750 to 2021, the atmospheric methane (CH_4) concentration increased from 722 ppb to 1889 ppb, by 162% [1]. CH_4 is one of the three primary greenhouse gases. On the centennial scale, the warming potential of CH_4 is maximally 28 times that of carbon dioxide [2]. In the past century, CH_4 has accounted for roughly 20% of the global temperature increase. CH_4 is not only an important food and energy source in the freshwater food web, but also occupies an irreplaceable place in the carbon cycle of aquatic ecosystems [3]. The increase in atmospheric CH_4 concentration has undergone three stages on the temporal scale. The first stage is the growth stage (8.4 ± 0.6 ppb $year^{-1}$) and the second is the stable stage (0.4 ± 0.5 ppb $year^{-1}$). The third stage, which is believed to have started in 2007, is another rapid growth stage (7.9 $\pm$ 0.6 ppb year^{-1}) [4]. The increasing atmospheric CH_4 concentration has caused a series of environmental safety problems, such as aggravating climate change, glacier melt, and permafrost thaw. However, the natural or human factors that are dominant in the above processes, remains unknown and attracts growing academic interest.

Currently, most estimates of CH_4 balance are usually carried out by replacing the total CH_4 emissions with CH_4 diffusion, while ignoring CH_4 ebullition [5]. The global freshwater ecosystem emitted 93.1 Tg CH_4 into the atmosphere every year, with CH_4 ebullition accounting for about 59% (55.3 Tg CH_4) [6]. Under extreme conditions, CH_4 ebullition may be several dozen times that of emissions by diffusion [7]. The most intensively studied influence factors of CH_4 emission fluxes from inland water bodies include meteorological factors (wind speed, atmospheric pressure), water quality factor (pH, dissolved oxygen, nutrient salts), hydrological factors (water depth, water velocity, water temperature), and ecosystem productivity [8–10]. In addition to environmental factors, CH_4 ebullition flux is also influenced by microbes. The global aquatic ecosystem emitted 576 Tg CH_4 every year, most of which was produced by methanogens, but more than half of which was oxidized to carbon dioxide by methanotrophs. In marine ecosystems, more than 90% of CH_4 was consumed by methanotrophs [11–13]. Therefore, mitigation of CH_4 emissions from aquatic ecosystem requires more attention to the relevant microbes [14].

The Qinghai-Tibet Plateau has been substantially influenced by global climate change, and the influence pattern is considered more complex than in many other places in the world. Global climate change has brought significant changes to the ecosystem composition, including hydrology, soil, climate, and biology [15,16]. Although the Qinghai-Tibet Plateau shows a lower temperature than that of many tropical zones, there are abundant psychrotolerant methanogens, indicating sustained metabolic activity even during the cold season [17]. In addition, low atmospheric pressure due to the high altitude facilitates significantly CH_4 ebullition emissions from the lakes and rivers of the Qinghai-Tibet Plateau [18,19]. Unlike natural aquatic ecosystems such as lakes and rivers, the water level of plateau reservoirs fluctuates more greatly and frequently due to human control [20]. The corresponding changes in hydrostatic pressure and alteration of redox conditions in the reservoir ecosystem have prompted CH_4 production and emissions [21]. These things considered, reservoir fluids are in a static state and associated with longer hydraulic retention time and lower dissolved oxygen concentration in water bodies and sediments compared with natural aquatic ecosystem [22]. Therefore, the CH_4 supersaturation and ebullition are common occurrences in reservoirs. A growing body of evidence has shown that water bodies in the Qinghai-Tibet Plateau are CH_4 emissions hotspots [19,23]. A more accurate estimate of global CH_4 emissions is hardly possible without the basic data from the Qinghai-Tibet Plateau. In the present study, we first analyzed the spatial and temporal characteristics of CH_4 ebullition flux across the water—atmosphere interface between July and August 2021. We then evaluated the influence of biotic and abiotic factors on CH_4 ebullition flux, in the hope of laying the foundation for in-depth investigation of CH_4 ebullition flux in plateau reservoirs.

2. Study Sites

The Dahejia Reservoir (102°45′8.80″ E, 35°50′10.44″ N) of the Yellow River is located in the hinterland area inhabited by Chinese ethnic minorities (Figure 1). Sitting along the trunk stream of the Yellow River at the junction between Guanting Town of Huzhu Tu Autonomous County in the Qinghai Province and Dahejia Town of Jishishan Bonan, Dongxiang and Salar Autonomous County, the Dahejia Reservoir is the twelfth cascade hydropower station under the hydropower development plan for the Longyang Gorge-Qingtong Gorge subregion of the Yellow River. The geomorphology around the reservoir is featured by the alternation between gorges and plains. The terrain is flat with conspicuous gullies on the two banks. The reservoir is 75 to 120 m wide during the flat water period. The water depth is mainly controlled by the upstream hydropower station and fluctuates widely, ranging from 1.8 to 5.6 m. The study site has a continental climate. Located deep in the northwest inland and far from the sea, the Dahejia Reservoir has large diel and annual temperature ranges, with little rainfall throughout the year and large evaporation. Due to high frequency of harsh weather events in winter, including Siberian high and cold waves, the average minimum temperature over the years is as low as −11.6 °C. The precipitation in this region increases as temperature increases in spring. The local precipitation further increases in summer and autumn as the Pacific subtropical high extends towards the west. Precipitation in these two seasons makes a greater contribution to annual precipitation compared with spring and winter, accounting for about 70%. The subregion surrounding the reservoir is dominated by cultivated plants, which are occasionally interspersed with natural vegetation, including coniferous forests, brush, grassland, and meadow. Light gray calcareous soil is the predominant soil type in the study site. Despite the high land use diversity, the land utilization level remains low [24].

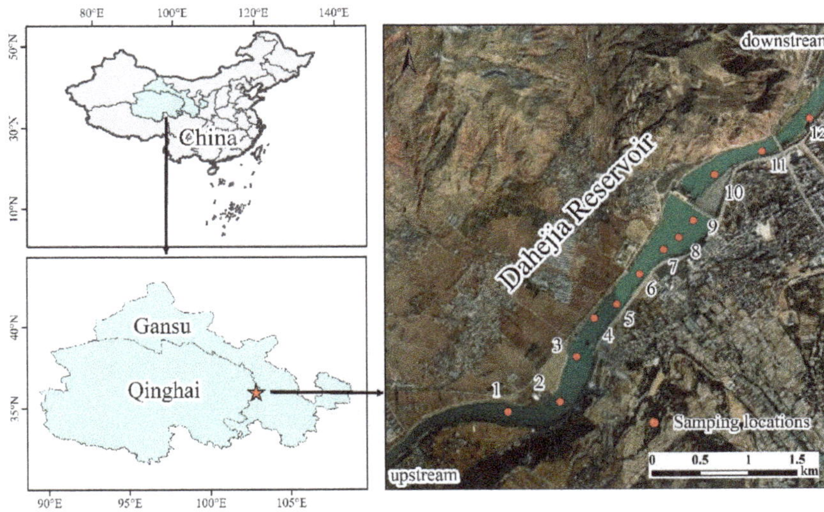

Figure 1. Location map of the study site. Twelve sampling points were subdivided into four subregions: River (1–3), Middle (4–6), Lake (7–9), and Down (10–12).

3. Materials and Methods

3.1. CH_4 Fluxes Monitoring

The study site extended from 2 km upstream to 1 km downstream of the Dahejia Reservoir. From upstream to downstream of the reservoir, the study site was subdivided into four subregions: River, Middle, Lake, and Down, as shown in Figure 1. In each subregion three sampling points were set up, and there were 12 sampling points in total. From July 29 to 31, 2021, CH_4 fluxes across the water-atmosphere interface were monitored on the diel scale for River, Lake, and Down subregions at a time interval of six hours. From August 1 to 18, 2021, 12 sampling points were monitored during the daytime. CH_4 fluxes across the water-atmosphere interface were measured using the static chamber-gas chromatography-based method [17]. Using the headspace equilibrium technique [25], the concentration of CH_4 in water samples was measured on a gas chromatography (Agilent 7890B, Santa Clara, CA, USA) equipped with a flame ionization detector.

Total CH_4 flux (F_t, μmol m^{-2} d^{-1}) was monitored using a static chamber and calculated according to the equation below [17]:

$$F_t = \frac{n_t - n_0}{At} \tag{1}$$

n_t and n_0 are CH_4 gas concentrations in the static chamber at time t and the initial moment 0 (mol), respectively. A is the surface area of the static chamber above water (m^2). t is the monitoring duration (min).

The CH_4 diffusion flux (F_d, μmol m^{-2} d^{-1}) is calculated using the thin-boundary layer method [26]:

$$F_d = k \times \left(C_{water} - C_{equilibrium}\right) \quad (2)$$

k is the gas diffusion rate (cm h^{-1}); $C_{water}$is the dissolved CH_4 concentration in the water body (μmol L^{-1}); $C_{equilibrium}$is the equilibrium concentration of CH_4 in the water body under actual conditions (μmol L^{-1}). CH_4 saturation in the surface water (%) is C_{water}divided by $C_{equilibrium}$. CH_4 ebullition flux (F_e, μmol m^{-2} d^{-1}) is defined as the total CH_4 flux minus the diffusion flux.

3.2. High-Throughput Sequencing

High-throughput sequencing was conducted by Beijing Biomarker Technologies using the Illumina NovaSeq 6000 system [27]. After microbial total DNA extraction from the sediments with a FastDNA SPIN Kit for Soil (MP Biomedicals, Santa Ana, CA, USA), we carried out real-time PCR, followed by product purification, quantification, and homogenization. A sequencing library was built and subjected to a quality check. If the library was eligible, it was then sequenced using the Illumina NovaSeq 6000 sequencing platform (Illumina, San Diego, CA, USA). The sequencing data analysis consisted of the following steps: (1) raw read processing: the raw reads were subjected to preliminary screening, with the low-quality reads filtered out and leaving only the high-quality ones. (2) operational taxonomic unit (OTU) clustering and species annotation: Usearch was employed for OTU clustering at 97% identity threshold, and the number of OTU was determined [28]. Thus, the high-quality reads were denoised and clustered into OTU. Then, based on the sequence compositions of OTU, we obtained the species abundance for different taxonomic ranks (phylum, class, order, family, genus, and species).

3.3. Collection and Measurement of Environmental Data

A Van Veen grab sampler with a mouth measuring $1/40$ m^2 was used to collect sediments at a depth of 0–15 cm at the specified sampling sites. The sediments were passed through a 2 mm sieve, placed into a 20 mL centrifuge tube, and transported back to the laboratory in a car refrigerator at 4 °C (PHILIPS TB5301, Amsterdam, The Netherlands). The colorimetric method was used to measure the total phosphorus, total nitrogen, nitrate nitrogen concentrations with Autoanalyser-3 (Seal Analytical, Norderstedt, Germany). The total dissolved carbon of the sediment was measured by a total organic analyzer (Shimadzu Corp, Kyoto, Japan). In the meantime, environmental parameters were acquired at about 0.5 m below the water surface at each sampling site using HQ40d portable water quality monitor (Hach, Loveland, CO, USA), including dissolved oxygen, water temperature, pH, salinity, and total dissolved solid. Water samples were collected using a 2 L stainless steel water sampler into a 1 L water sample bottle. They were used to determine the water quality indicators. The water depth and velocity were directly measured with doppler velocity meter (BOYIDA LSH10-1QC, Xiamen, China). The wind speed, air temperature, and air pressure were measured at 1 m above the water surface using a portable anemometer (Testo 480, Lenzkirch, Germany).

3.4. Statistical Analyses

Correlation analysis and one-way analysis of variance (ANOVA) were carried out using SPSS 24.0. A p-value smaller than 0.05 was statistically significant. Three duplicate samples were collected to determine the above parameters and indicators, and the results were expressed as mean ± standard deviation. R-3.6.3 was run to carry out redundancy analysis (RDA) of the correlation between the primary microbes and environmental variables, and statistical graph plotting.

4. Results

4.1. The CH_4 Ebullition Flux Was Higher in the Nighttime than in the Daytime

Table 1 shows the dissolved CH_4 concentration, saturation, and contribution of CH_4 ebullition flux in the surface water of the Dahejia Reservoir in the daytime and the nighttime. The diel variation range of CH_4 concentration in the surface water of the Dahejia Reservoir was 0.11–0.43 μmol L^{-1}. In the daytime, the CH_4 concentration fell within the range of 0.26–0.43 μmol L^{-1}; in the nighttime, it varied within the range of 0.11–0.37 μmol L^{-1}. An independent-samples T-test showed no significant differences in the mean CH_4 concentration in the daytime (0.34 ± 0.06 μmol L^{-1}) and the nighttime (0.28 ± 0.08 μmol L^{-1}) ($p > 0.05$). Both in daytime and nighttime, the dissolved CH_4 concentration was significantly higher than the equilibrium concentration in the water body. The CH_4 saturation in the surface water of the Dahejia Reservoir varied within the range of 81–6885%. In the daytime, the CH_4 saturation fell within the range of 157–6885%; in the nighttime, it fell within the range of 81–5189%. An independent-samples T-test showed that the mean CH_4 saturation in the daytime (1732 ± 2803%) and the nighttime (1433 ± 2378%) was not significantly different ($p > 0.05$). The contribution of CH_4 ebullition flux in the daytime to total CH_4 flux at the reservoir varied within the range of 16.47–92.92%; In the nighttime, it varied within the range of 80.35–100%. An independent-samples T-test showed that the mean contribution of CH_4 ebullition flux in the nighttime (95.65 ± 7.97%) was significantly higher than that in the daytime (68.01 ± 26.23%) ($p < 0.05$).

Table 1. Dissolved CH_4 concentration, saturation, and contribution of CH_4 ebullition flux in the Dahejia Reservoir at different time.

Time	CH_4 Concentration (μmol L^{-1})	CH_4 Saturation (%)	Contribution of CH_4 Ebullition Flux (%)
Daytime (06:00–18:00)	0.34 ± 0.06 [a]	1732 ± 2803 [a]	68.01 ± 26.23 [a]
Nighttime (18:00–06:00)	0.28 ± 0.08 [a]	1433 ± 2378 [a]	95.65 ± 7.97 [b]

[a,b] Values having different subscript letters in the same column are significantly different from each other ($p < 0.05$).

Figure 2 shows the temporal characteristics of CH_4 diffusion and ebullition flux across the water—atmosphere interface of the Dahejia Reservoir in the daytime and the nighttime. According to our measurements, the diel variation range of CH_4 diffusion flux across the water—atmosphere interface was 0–0.14 mg m^{-2} h^{-1}. In the daytime, the CH_4 diffusion flux fell within the range of 0.01–0.14 mg m^{-2} h^{-1}; in the nighttime, it varied within the range of 0–0.1 mg m^{-2} h^{-1}. An independent-samples T-test showed no significant differences in the mean CH_4 diffusion flux in the daytime (0.06 ± 0.05 mg m^{-2} h^{-1}) and the nighttime (0.02 ± 0.04 mg m^{-2} h^{-1}) ($p > 0.05$). The diel variation range of CH_4 ebullition flux across the water—atmosphere interface was 0.03–0.69 mg m^{-2} h^{-1}. In the daytime, the CH_4 ebullition flux fell within the range of 0.03–0.69 mg m^{-2} h^{-1}; in the nighttime, it varied within the range of 0.07–0.6 mg m^{-2} h^{-1}. An independent-samples T-test showed that the mean CH_4 ebullition flux in the nighttime (0.34 ± 0.21 mg m^{-2} h^{-1}) was significantly higher than that in the daytime (0.19 ± 0.21 mg m^{-2} h^{-1}) ($p < 0.05$).

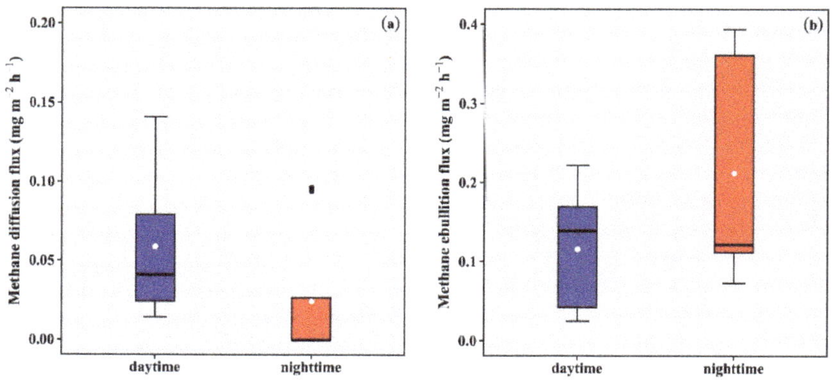

Figure 2. Diel differences in CH_4 diffusion flux (**a**) and ebullition flux (**b**). (The black lines represent the median and white dots indicate the mean; lower and upper edges represent 25th and 75th, respectively; whiskers represent confidence intervals of 1.5 times the interquartile range; black dots indicate outliers).

4.2. Spatial Variation of CH_4 Ebullition Flux from Upstream to Downstream

Table 2 shows the dissolved CH_4 concentration, saturation, and contribution of CH_4 ebullition flux in the surface water of the Dahejia Reservoir in different subregion. As shown by Table 2, the CH_4 concentration in the surface water varied spatially within the range of 0.12–0.47 μmol L^{-1}. Specifically, in the River subregion, Middle subregion, Lake subregion, and Down subregion, the CH_4 concentration in the surface water fell within the ranges of 0.39–0.42 μmol L^{-1}, 0.18–0.28 μmol L^{-1}, 0.3–0.47 μmol L^{-1}, and 0.12–0.24 μmol L^{-1}, respectively. ANOVA revealed that the mean CH_4 concentration in the River subregion (0.4 ± 0.03 μmol L^{-1}) and the Lake subregion (0.39 ± 0.09 μmol L^{-1}) was significantly higher than that of the Middle subregion (0.23 ± 0.05 μmol L^{-1}) and the Down subregion (0.19 ± 0.06 μmol L^{-1}) ($p < 0.05$). The CH_4 saturation varied spatially within the range of 2297–8951%. In the River, Middle, Lake, and Down subregions, the CH_4 saturation fell within the ranges of 6645–7246%, 3314–4730%, 5904–8951%, and 2297–4617%, respectively. ANOVA revealed that the mean CH_4 saturation in the River subregion (6695 ± 779%) and the Lake subregion (7720 ± 1565%) was significantly higher than that of the Middle subregion (3310 ± 808%) and Down subregion (3514 ± 1164%) ($p < 0.05$). The dissolved CH_4 concentrations in the surface water of all subregions of the Dahejia Reservoir were all significantly higher than the equilibrium concentration of CH_4 in the atmosphere. The contribution of CH_4 ebullition flux in the River subregion to total CH_4 flux at the reservoir varied within the range of 52.98–88.7%; for the Middle, Lake, and Down subregions, it varied within the range of 96.3–99.43%, 43.69–79.05%, and 63.67–94.61%, respectively. Thus, the Dahejia Reservoir was an emission source of CH_4.

Table 2. Dissolved CH_4 concentration, saturation, and contribution of CH_4 ebullition flux in the Dahejia Reservoir in different subregion.

Subregion	CH_4 Concentration (μmol L^{-1})	CH_4 Saturation (%)	Contribution of CH_4 Ebullition Flux (%)
River	0.40 ± 0.03 [a]	6695 ± 779 [a]	70.84 ± 25.25 [a]
Middle	0.23 ± 0.05 [b]	3310 ± 808 [b]	98.27 ± 1.73 [b]
Lake	0.39 ± 0.09 [a]	7720 ± 1565 [a]	61.58 ± 17.69 [a]
Down	0.19 ± 0.06 [b]	3514 ± 1164 [b]	75.39 ± 16.78 [a]

[a,b] Values having different subscript letters in the same column are significantly different from each other ($p < 0.05$).

Figure 3 shows the spatial characteristics of CH_4 diffusion flux and ebullition flux across the water—atmosphere interface in different subregions of the Dahejia Reservoir. The CH_4 diffusion flux across the water—atmosphere interface in the Dahejia Reservoir varied within the range of 0–0.21 mg m^{-2} h^{-1}. Specifically, the CH_4 diffusion flux in the River, Middle, Lake, and Down subregions varied within the range of 0.11–0.12 mg m^{-2} h^{-1}, 0–0.01 mg m^{-2} h^{-1}, 0.04–0.13 mg m^{-2} h^{-1}, and 0.04–0.21 mg m^{-2} h^{-1}, respectively. ANOVA showed that the mean CH_4 diffusion flux in the Down subregion (0.11 ± 0.09 mg m^{-2} h^{-1}) was significantly higher than that of the Middle subregion (0.01 ± 0.01 mg m^{-2} h^{-1}) ($p < 0.05$). However, the mean CH_4 diffusion flux of the Lake subregion (0.07 ± 0.05 mg m^{-2} h^{-1}) was not significantly different from that in the River subregion (0.11 ± 0.01 mg m^{-2} h^{-1}) ($p > 0.05$). As shown above, the mean CH_4 diffusion flux was higher in the upstream and downstream of the reservoir. The CH_4 ebullition flux varied within the range of 0.08–1.12 mg m^{-2} h^{-1}. Specifically, the CH_4 ebullition flux in the River, Middle, Lake, and Down subregions varied within the range of 0.12–0.92 mg m^{-2} h^{-1}, 0.08–1.12 mg m^{-2} h^{-1}, 0.08–0.15 mg m^{-2} h^{-1}, and 0.17–0.76 mg m^{-2} h^{-1}, respectively. ANOVA showed that the mean CH_4 ebullition flux in the River subregion (0.52 ± 0.57 mg m^{-2} h^{-1}), Middle subregion (0.51 ± 0.54 mg m^{-2} h^{-1}), Lake subregion (0.11 ± 0.04 mg m^{-2} h^{-1}), and Down subregion (0.43 ± 0.3 mg m^{-2} h^{-1}) were not significantly different ($p > 0.05$). From upstream to downstream, the mean CH_4 ebullition flux first decreased and then increased, with the minimum found in the Lake subregion. Taken together, ebullition was the primary mode of CH_4 emission in the Dahejia Reservoir.

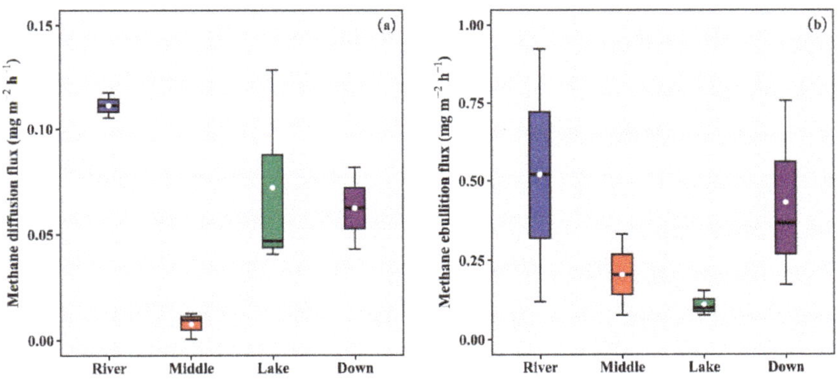

Figure 3. CH_4 diffusion flux (**a**) and ebullition flux (**b**) in different subregions (Symbols are similar to Figure 2).

5. Discussion

5.1. Influence of Abiotic Factors on CH_4 Ebullition Flux

We carried out the Spearman's rank correlation test for all data (Figure 4). The results showed that CH_4 ebullition flux was significantly positively correlated with total CH_4 flux (R = 0.94, $p < 0.05$). This finding demonstrated the considerable contribution made by CH_4 ebullition flux to total CH_4 flux. In other words, ebullition was a primary mode of CH_4 emissions from the reservoir. In the present study, we found that CH_4 ebullition flux was significantly negatively correlated with total organic carbon in the water (R = −0.66, $p < 0.05$) and total phosphorus in the sediment (R = −0.72, $p < 0.05$). Total organic carbon and total phosphorus have been identified as two limiting factors for primary productivity of the ecosystem [29,30]. The changes in ecosystem productivity may promote carbon dioxide generation and inhibit methane production in the Dahejia Reservoir. CH_4 ebullition flux was significantly positively correlated with sediment pH (R = 0.79, $p < 0.05$). It has been reported that pH variation resulted in changes in carbon source and sink of the aquatic ecosystem [31]. The higher the pH, the smaller the carbon dioxide emissions will be, which reduces the environmental pH. Within the suitable range of pH for methanogens, the lower the pH, the higher the activity of the methanogens and the greater the amount of CH_4 produced [32]. CH_4 ebullition flux was significantly negatively correlated with nitrate in sediment (R = −0.76, $p < 0.05$). Relevant studies have shown that nitrate nitrogen promoted nitrous oxide generation and emissions, but had a negative correlation with CH_4 emissions [33]. Nitrous oxide is involved in CH_4 oxidation as electron receptors and can be used for methane oxidation coupled to denitrification. [34,35]. Therefore, an excessively high level of nitrate nitrogen inhibits CH_4 production. We found no significant correlation between the other environmental variables and CH_4 ebullition flux. This was possibly because we only studied the warm season.

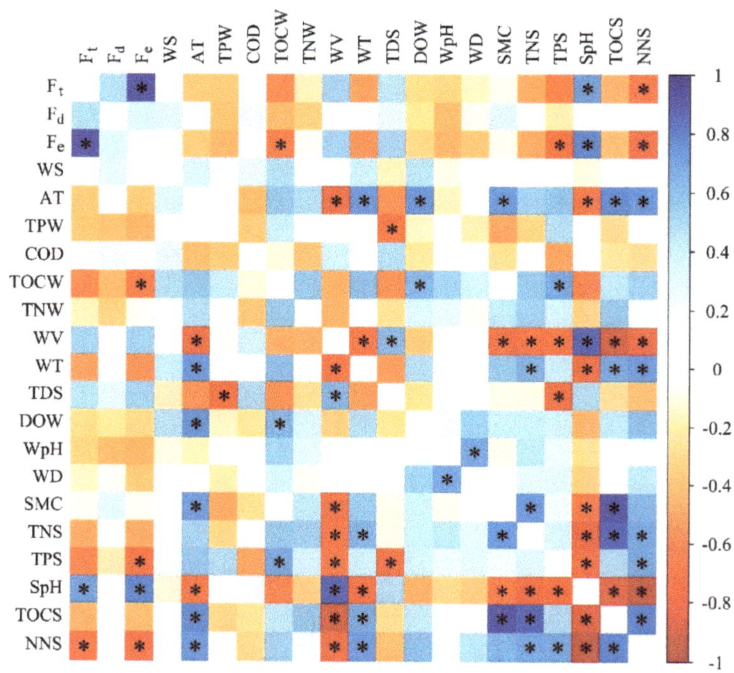

Figure 4. The correlations between CH_4 fluxes and physicochemical variables. Spearman's rank correlation coefficients are presented in the boxes in different colors. Blue represents positive correlation, red indicates negative correlation, and white means no correlation. The asterisks (*) mean the correlations are statistically significant ($p < 0.05$). (Abbreviation: F_t, total CH_4 flux; F_d, CH_4 diffusion flux; F_e, CH_4 ebullition flux; WS, wind speed; AT, Air temperature; TPW, total phosphorus in water; COD, chemical oxygen demand; TOCW, total organic carbon in water; TNW, total nitrogen in water; WV, water velocity; WT, water temperature; TDS, total dissolved solid; DOW, dissolved oxygen in water; WpH, water pH; WD, water depth; SMC, sediment moisture content; TNS, total nitrogen in sediment; TPS, total phosphorus in sediment; SpH, sediment pH; TOCS, total organic carbon in sediment; NNS, nitrate nitrogen in sediment).

5.2. Sediment Microbes Affected the CH_4 Ebullition Flux

CH_4 emissions from reservoirs implicate complex interactions between various factors, among which physicochemical variables only account for a certain proportion of CH_4 ebullition. CH_4 production and emissions from inland waters are largely associated with microbial activities [36]. To clarify the influence of microbial community composition and structure in sediments on CH_4 ebullition flux, we first performed a regression analysis between total microbial abundance in sediments and CH_4 ebullition flux. The results of the analysis are shown in Figure 5. CH_4 ebullition flux had a significantly negative correlation with the total microbial abundance in sediments (R^2 = 0.499, p = 0.009).

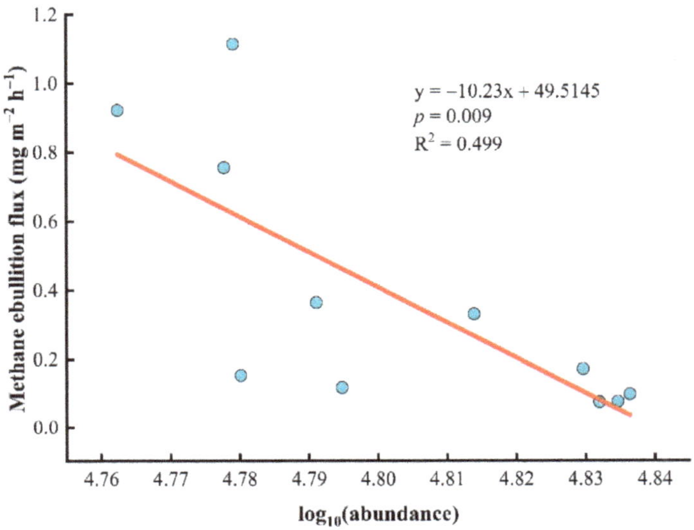

Figure 5. Regression analysis of CH_4 ebullition flux and the logarithm of microbial community abundance in sediments. The red line represents the fit of a linear regression through the observed data.

We further analyzed the influence of the abundance of different species on CH_4 ebullition flux. The microbial communities in sediments in different subregions were subjected to detrended correspondence analysis (DCA). The gradient lengths along the four axes were 0.1564, 0.0092, 0.0024, and 0, respectively, all of which were below 3. Therefore, we conducted a redundancy analysis (RDA) for the data. The results are shown in Figure 6. The explanation degrees of RDA1 and RDA2 for CH_4 ebullition flux were 96.68% and 0.56%, respectively. The two collectively explained 97.24% of CH_4 ebullition flux. Firmicutes and Verrucomicrobia were the main bacteria contributing to structural differentiation of microbial communities in sediments at the sampling points 1 and 3. Acidobacteria, Gemmatimonadetes, Epsilonbacteraeota, Chloroflexi, and Nitrospirae were the main bacteria contributing to structural differentiation of microbial communities in sediments at the sampling points 7, 8, and 9. Actinobacteria was the main contributor to structural differentiation of microbial communities in sediments at the sampling point 6. Bacteroidetes and Proteobacteria were the main contributors to structural differentiation of microbial communities in sediments at the sampling points 4 and 5 in the Middle subregion and at the sampling points 10, 11 and 12 in the Down subregion.

Total CH_4 flux and ebullition flux were positively correlated with the abundance of Firmicutes, Actinobacteria, and Verrucomicrobia, and negatively with Proteobacteria, Chloroflexi, Gemmatimonadetes, and Nitrospirae. The dissolved CH_4 concentration and CH_4 diffusion flux were positively correlated with the abundance of Firmicutes and Verrucomicrobia, and negatively with that of Bacteroidetes and Proteobacteria. Proteobacteria, Actinobacteria, Bacteroidetes, Gemmatimonadetes, and Nitrospirae actively participate in organic matter decomposition. These bacteria play crucial roles in the Earth's biochemical cycle of carbon, or even in sulfur cycle [37–39]. As shown in Figure 6, CH_4 ebullition flux and diffusion flux were negatively correlated with the abundance of Nitrospirae. Most species belonging to the Nitrospirae are nitrifying bacteria and are widely present in freshwater ecosystems. These bacteria can convert nitrites into nitrates and compete with methanogens for substrates [40]. It has been reported that nitrous oxide and CH_4 emissions are negatively correlated with each other [33]. Due to the joint action of methanotrophs (typically *Methylocaldum* species) and denitrifying bacteria (typically *Thauera* species), nitrous oxide reduced CH_4 production while accelerating its oxidation. Besides, as the temperature rose, there would be a transition from the predominance of reduction of nitrous oxide to CH_4 oxidation [41]. Furthermore, the CH_4 ebullition flux and diffusion flux were also negatively correlated with the abundance of Proteobacteria. This is because Proteobacteria competes with methanogens for substrates, which promotes carbon dioxide and nitrous oxide emissions simultaneously. A large number of studies have shown that the relative abundance of Proteobacteria had a significantly positive correlation with carbon dioxide and nitrous oxide emissions [35,42]. The higher the abundance of Proteobacteria and Bacteroidetes in soil, the higher the carbon dioxide emissions would be [43]. It has been found that as primary producers, Epsilonbacteraeota is not only backbones of the ecosystem, but also participates extensively in the carbon and nitrogen cycles [44]. These bacteria use hydrogen or reductive sulfides as energy sources and produce nitrogen via the denitrification pathway, where nitrates act as electron receptors. Besides, Epsilonbacteraeota fixes carbon dioxide via the reductive tricarboxylic acid cycle (rTCA) to achieve autotrophic growth [45]. As analyzed above, composition changes of microbial communities in sediments are important biotic factors influencing CH_4 ebullition flux.

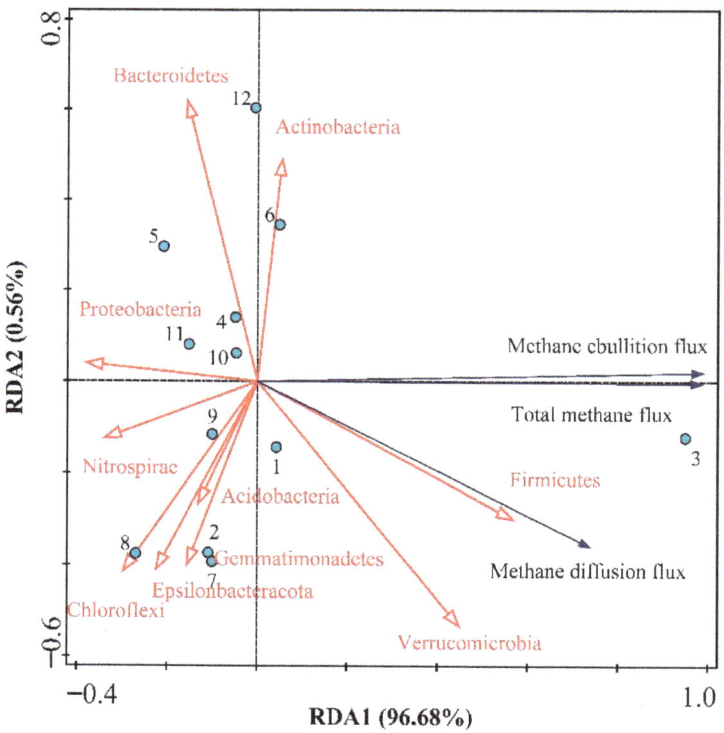

Figure 6. Redundancy analysis of the correlation between CH_4 fluxes and abundance of sediment microbes. Red arrows represent sediment microbial communities, black arrows indicate CH_4 fluxes, and blue dots represent sampling points (1–12).

5.3. Comparison of CH_4 Ebullition Flux at Reservoirs in Different Climate Zones

Table 3 shows the results of in-situ monitoring of CH_4 ebullition flux at several reservoirs in tropical, subtropical, and frigid zones. The CH_4 ebullition fluxes at reservoirs in frigid zones are generally lower than those in tropical and subtropical zones. This is because the lower the latitude and the higher the temperature, the higher the microbial activity and the faster the CH_4 production will be [46]. The CH_4 ebullition flux varies significantly across the reservoirs, especially for tropical and subtropical reservoirs. The reasons for the variability include reservoir age, drainage subregion, land use pattern, eutrophication, and primary productivity [47,48]. Recent studies have shown extensive CH_4 ebullition flux in cold region of the north, a phenomenon that tends to be ignored and has been rarely discussed [49]. CH_4 is very likely to accumulate in reservoirs located in cold regions. An explosive growth of CH_4 flux is common in melt season [50]. As global warming gets worse, the melting of permafrost rich in organic matter leads to an increase in CH_4 emissions from the local water bodies [17]. For reservoirs in frigid zones, the maximum CH_4 ebullition flux can reach up to 115.59 mg m^{-2} h^{-1}. Dahejia Reservoir is also located in the subtropical zone, like Samuel Reservoir in Brazil and Pengxi River/Reservoir in the Three Gorges Reservoir Area, China. The CH_4 ebullition flux is of the same order of magnitude at the three reservoirs. However, the Dahejia Reservoir is located in the Qinghai-Tibet Plateau, where the climate more resembles that in the frigid zone. Therefore, the CH_4 ebullition flux in the Dahejia Reservoir is closer to that in Porttipahta Reservoir in Finland (0.4 mg m^{-2} h^{-1}) and Miaowei Reservoir in tropical China (0.33 ± 0.56 mg m^{-2} h^{-1}). Compared with the existing reports at home and abroad, the CH_4 ebullition flux across the water—atmosphere interface in the Dahejia Reservoir is at a moderate level. The contribution made by CH_4 ebullition flux to total CH_4 flux is comparable to that at other reservoirs at home and abroad.

As for the temporal and spatial characteristics on the reservoir scale, Grinham et al. observed a significant difference in CH_4 ebullition flux in the daytime and nighttime [51]. However, we did not observe a similar pattern of variation for the Dahejia Reservoir. From the upstream to the downstream of the reservoir, Yang et al. found that ebullition was the dominant mode of CH_4 emissions in the upstream of the Xin'anjiang Reservoir, while diffusion was dominant in the downstream [52]. Mcclure et al. showed that the CH_4 ebullition flux decreased from the upstream to the downstream of the Falling Creek Reservoir [53]. We observed a similar variation for the Dahejia Reservoir, except that the CH_4 ebullition flux in the Down subregion was significantly increased. Bai et al. reported similar findings for the Three Gorges Reservoir during the low water period. That is, the CH_4 ebullition flux was higher in the downstream (167.173 mg m^{-2} h^{-1}) than in the upstream (12.23–123.05 mg m^{-2} h^{-1}) [54]. As analyzed above, CH_4 ebullition flux has displayed significant temporal and spatial heterogeneity, either on the global or the reservoir scale. In the context of global climate change, the temperature rise in the Qinghai-Tibet Plateau is twice as much as the global average during the same period [55]. Moreover, the precipitation and air temperature vary consistently in the plateau region. Both have been increasing over the years. The increase in precipitation will accelerate organic matter decomposition, while that in air temperature can lead to eutrophication. They work synergistically to promote the geochemical cycling of carbon in water bodies [56]. The uniqueness of the plateau environment plus the action of several other factors has aggravated the spatial and temporal heterogeneity in CH_4 ebullition. However, our study was geographically confined to the Dahejia Reservoir, and the findings may not be applicable to CH_4 ebullition from reservoirs over the entire Qinghai-Tibet Plateau. It is necessary to choose more representative reservoirs in the Qinghai-Tibet Plateau to clarify the spatial and temporal features of CH_4 ebullition and the associated influence factors.

Table 3. CH_4 ebullition flux from global reservoirs.

Reservoir	CH_4 Ebullition Flux ($mg\ m^{-2}\ h^{-1}$)	Contribution of CH_4 Ebullition Flux (%)	Note
Thirparappu [57]	114.47	95.07	Tropical
Miaowei [58]	0.33 ± 0.56	62.26	Tropical
Gatun Lake [59]	525.56	97.7	Tropical
Falling Greek [53]	0.67 ± 0.31	72.17	Subtropical
Xin'anjiang [52]	2.73 ± 2.02	92.86	Subtropical
Itaipu [60]	0.025	7	Subtropical
Samuel [60]	0.57	55.88	Subtropical
Pengxi River [7]	0.84	70	Subtropical
Saar River [61]	5.31 ± 7.46	97	Temperate zone
Eguzon [62]	0.24 ± 0.56	9.8	Temperate zone
Northern Québec [63]	0.1	83	Frigid zone
Porttipahta [6]	0.4	17.09	Frigid zone
Lokka [63]	115.59	83.63	Frigid zone
Dahejia	0.31 ± 0.29	86.11	This study

6. Conclusions

It should be noted that since global warming has been deteriorating, the uncertainty of CH_4 ebullition from reservoirs located in the Qinghai-Tibet Plateau will increase considerably [17,19]. We studied CH_4 ebullition flux from the Dahejia Reservoir located in the upper reaches of the Yellow River and analyzed the spatial and temporal characteristics of CH_4 ebullition flux, the contribution made by CH_4 ebullition flux to total CH_4 flux, and the influence factors during the warm season (from July–August 2021). We arrived at the following conclusions:

1. CH_4 ebullition was the dominant mode of CH_4 emissions at the study site and contributed to 78.85 ± 20% of total CH_4 flux.
2. The CH_4 ebullition flux in the nighttime (0.34 ± 0.21 $mg\ m^{-2}\ h^{-1}$) was significantly higher than that in the daytime (0.19 ± 0.21 $mg\ m^{-2}\ h^{-1}$).
3. The CH_4 ebullition flux first decreased and then increased from upstream to downstream. In the River, Middle, Lake, and Down subregions, the CH_4 ebullition flux was 0.52 ± 0.57 $mg\ m^{-2}\ h^{-1}$, 0.51 ± 0.54 $mg\ m^{-2}\ h^{-1}$, 0.11 ± 0.04 $mg\ m^{-2}\ h^{-1}$, and 0.43 ± 0.3 $mg\ m^{-2}\ h^{-1}$, respectively.
4. Among abiotic variables, the CH_4 ebullition flux was closely related to total phosphorus, total organic carbon, pH and nitrate nitrogen. Among biotic factors, CH_4 ebullition flux had a significant negative linear correlation with microbial abundance. The redundancy analysis showed that the CH_4 ebullition flux was significantly positively correlated with the abundances of Firmicutes and Actinobacteria and negatively with that of Proteobacteria and Chloroflexi.

Although we had studied the CH_4 ebullition flux of the Dahejia Reservoir on the diel scale, our study still had the following limitations due to limited time and manpower. We had discussed the diel and spatial variations of CH_4 ebullition flux across the water-atmosphere interface in the Dahejia Reservoir of the Yellow River. In the future, we need to further investigate the monthly, seasonal, and interannual variations of CH_4 ebullition flux in the study site, especially the variations in the cold season. Specifically, we will include all microbes in the water bodies of the reservoir into our study. Microbial genome data will be collected in a more comprehensive manner by metagenomic high-throughput sequencing, so as to precisely quantify CH_4 ebullition.

Author Contributions: Methodology, W.T., F.X. and H.J.; data curation, X.M. and L.X.; writing—review and editing, Y.W., Y.Y. and H.Y. All authors have read and agreed to the published version of the manuscript.

Funding: The research work was supported by the National Natural Science Foundation of China (Project No. 52070108), the Basic Research Program of Qinghai Province (2022-ZJ-718) and the Qinghai Province innovation platform construction project (2020-ZJ-Y06).

Institutional Review Board Statement: Not applicable.

Informed Consent Statement: Not applicable.

Data Availability Statement: The data presented in this study are available in the article.

Acknowledgments: We thank reviewers for their constructive comments and suggestions to improve the early version of this paper.

Conflicts of Interest: The authors declare no conflict of interest.

References

1. WMO. The State of Greenhouse Gases in the Atmosphere Based on Global Observations through 2020. *WMO Greenh. Gas Bull.* **2021**, *17*, 1–3.
2. IPCC. Climate Change 2013: The Physical Science Basis. In *Contribution of Working Group I to the Fifth Assessment Report of the Intergovernmental Panel on Climate Change*; Stocker, T.F., Qin, D., Plattner, G.-K., Tignor, M., Allen, S., Boschung, J., Nauels, A., Xia, Y., Bex, V., Midgley, P., Eds.; Cambridge University Press: Cambridge, UK; New York, NY, USA, 2013; p. 1535.
3. Agasild, H.; Zingel, P.; Tuvikene, L.; Tuvikene, A.; Timm, H.; Feldmann, T.; Salujõe, J.; Toming, K.; Jones, R.; Nõges, T. Biogenic methane contributes to the food web of a large, shallow lake. *Freshw. Biol.* **2014**, *59*, 272–285. [CrossRef]
4. Dlugokencky, E.J. Trends in Atmospheric Methane (NOAA/GML, 2021). Available online: https://www.esrl.noaa.gov/gmd/ccgg/trends_ch4 (accessed on 18 September 2021).
5. Zhang, P.; Wang, X.F.; Yuan, X.Z. General characteristics and research progress of methane emissions from freshwater ecosystems in China. *China Environ. Sci.* **2020**, *40*, 3567–3579. [CrossRef]
6. Bastviken, D.; Tranvik, L.J.; Downing, J.A.; Crill, P.M.; Enrich-Prast, A. Freshwater Methane Emissions Offset the Continental Carbon Sink. *Science* **2011**, *331*, 50. [CrossRef]
7. Li, Z.; Zhang, C.; Liu, L. Ebullition fluxes of CO_2 and CH_4 in Pengxi River, Three Gorges Reservoir. *J. Lake Sci.* **2014**, *26*, 789–798.
8. Linkhorst, A.; Paranaíba, J.R.; Mendonça, R.; Rudberg, D.; DelSontro, T.; Barros, N.; Sobek, S. Spatially Resolved Measurements in Tropical Reservoirs Reveal Elevated Methane Ebullition at River Inflows and at High Productivity. *Glob. Biogeochem. Cycles* **2021**, *35*, e2020GB006717. [CrossRef]
9. DelSontro, T.; Boutet, L.; St-Pierre, A.; del Giorgio, P.A.; Prairie, Y.T. Methane ebullition and diffusion from northern ponds and lakes regulated by the interaction between temperature and system productivity. *Limnol. Oceanogr.* **2016**, *61*, S62–S77. [CrossRef]
10. West, W.E.; Creamer, K.P.; Jones, S.E. Productivity and depth regulate lake contributions to atmospheric methane. *Limnol. Oceanogr.* **2016**, *61*, S51–S61. [CrossRef]
11. Saunois, M.; Stavert, A.R.; Poulter, B.; Bousquet, P.; Canadell, J.G.; Jackson, R.B.; Raymond, P.A.; Dlugokencky, E.J.; Houweling, S.; Patra, P.K.; et al. The Global Methane Budget 2000-2017. *Earth Syst. Sci. Data* **2020**, *12*, 1561–1623. [CrossRef]
12. Houghton, K.M.; Carere, C.R.; Stott, M.B.; McDonald, I.R. Thermophilic methanotrophs: In hot pursuit. *Fems Microbiol. Ecol.* **2019**, *95*, fiz125. [CrossRef]
13. Ding, J.; Lu, Y.-Z.; Fu, L.; Ding, Z.-W.; Mu, Y.; Cheng, S.H.; Zeng, R.J. Decoupling of DAMO archaea from DAMO bacteria in a methane-driven microbial fuel cell. *Water Res.* **2017**, *110*, 112–119. [CrossRef] [PubMed]
14. Chowdhury, T.R.; Dick, R.P. Ecology of aerobic methanotrophs in controlling methane fluxes from wetlands. *Appl. Soil Ecol.* **2013**, *65*, 8–22. [CrossRef]
15. Ren, M.F.; Li, L.Y.; Chen, L.; Xing, T.T.; Liu, Y.Q.; Dong, X.Z. Methanogen communities and predominant methanogenic pathways in three saline-alkaline lakes on the Tibetan Plateau. *Acta Microbiol. Sin.* **2020**, *60*, 161–171. [CrossRef]
16. Ma, X.F.; Chen, S.Y.; Deng, J.; Feng, Q.S.; Huang, X.D. Vegetation phenology dynamics and its response to climate change on the Tibetan Plateau. *Acta Prataculturae Sin.* **2016**, *25*, 13–21. [CrossRef]
17. Zhang, L.; Xia, X.; Liu, S.; Zhang, S.; Li, S.; Wang, J.; Wang, G.; Gao, H.; Zhang, Z.; Wang, Q.; et al. Significant methane ebullition from alpine permafrost rivers on the East Qinghai-Tibet Plateau. *Nat. Geosci.* **2020**, *13*, 349–354. [CrossRef]
18. Natchimuthu, S.; Sundgren, I.; Gålfalk, M.; Klemedtsson, L.; Crill, P.; Danielsson, Å.; Bastviken, D. Spatio-temporal variability of lake CH_4 fluxes and its influence on annual whole lake emission estimates. *Limnol. Oceanogr.* **2016**, *61*, S13–S26. [CrossRef]
19. Wang, L.; Du, Z.; Wei, Z.; Xu, Q.; Feng, Y.; Lin, P.; Lin, J.; Chen, S.; Qiao, Y.; Shi, J.; et al. High methane emissions from thermokarst lakes on the Tibetan Plateau are largely attributed to ebullition fluxes. *Sci. Total Environ.* **2021**, *801*, 149692. [CrossRef]
20. Harrison, J.A.; Deemer, B.R.; Birchfield, M.K.; O'Malley, M.T. Reservoir Water-Level Drawdowns Accelerate and Amplify Methane Emission. *Environ. Sci. Technol.* **2017**, *51*, 1267–1277. [CrossRef]
21. DelSontro, T.; Perez, K.K.; Sollberger, S.; Wehrli, B. Methane dynamics downstream of a temperate run-of-the-river reservoir. *Limnol. Oceanogr.* **2016**, *61*, S188–S203. [CrossRef]
22. Conrad, R.; Ji, Y.; Noll, M.; Klose, M.; Claus, P.; Enrich-Prast, A. Response of the methanogenic microbial communities in Amazonian oxbow lake sediments to desiccation stress. *Environ. Microbiol.* **2014**, *16*, 1682–1694. [CrossRef]

23. Zhu, D.; Wu, Y.; Chen, H.; He, Y.; Wu, N. Intense methane ebullition from open water area of a shallow peatland lake on the eastern Tibetan Plateau. *Sci. Total Environ.* **2016**, *542*, 57–64. [CrossRef] [PubMed]
24. Li, Q.Q. Aquatic Ecological Environmental Impact Assessment of Dahejia Hydropower Station Project. Master's Thesis, Xi'an University of Technology, Xi'an, China, 2017.
25. Johnson, K.M.; Hughes, J.E.; Donaghay, P.L.; Sieburth, J.M. Bottle-calibration static head space method for the determination of methane dissolved in seawater. *Anal. Chem.* **1990**, *62*, 2408–2412. [CrossRef]
26. Wanninkhof, R. Relationship between wind speed and gas exchange over the ocean revisited. *Limnol. Oceanogr. Methods* **2014**, *12*, 351–362. [CrossRef]
27. Modi, A.; Vai, S.; Caramelli, D.; Lari, M. The Illumina Sequencing Protocol and the NovaSeq 6000 System. *Methods Mol. Biol.* **2021**, *2242*, 15–42. [CrossRef]
28. Edgar, R.C. Search and clustering orders of magnitude faster than BLAST. *Bioinformatics* **2010**, *26*, 2460–2461. [CrossRef] [PubMed]
29. Beaulieu, J.J.; DelSontro, T.; Downing, J.A. Eutrophication will increase methane emissions from lakes and impoundments during the 21st century. *Nat. Commun.* **2019**, *10*, 1375. [CrossRef]
30. Xiao, Q.; Zhang, M.; Hu, Z.; Gao, Y.; Hu, C.; Liu, C.; Liu, S.; Zhang, Z.; Zhao, J.; Xiao, W.; et al. Spatial variations of methane emission in a large shallow eutrophic lake in subtropical climate. *J. Geophys. Res. Biogeosciences* **2017**, *122*, 1597–1614. [CrossRef]
31. He, B.; He, J.; Wang, J.; Li, J.; Wang, F. Characteristics of GHG flux from water-air interface along a reclaimed water intake area of the Chaobai River in Shunyi, Beijing. *Atmos. Environ.* **2018**, *172*, 102–108. [CrossRef]
32. Garcia, J.-L.; Patel, B.K.C.; Ollivier, B. Taxonomic, Phylogenetic, and Ecological Diversity of Methanogenic Archaea. *Anaerobe* **2000**, *6*, 205–226. [CrossRef]
33. Smith, R.L.; Bohlke, J.K. Methane and nitrous oxide temporal and spatial variability in two midwestern USA streams containing high nitrate concentrations. *Sci. Total Environ.* **2019**, *685*, 574–588. [CrossRef]
34. Ettwig, K.F.; Butler, M.K.; Le Paslier, D.; Pelletier, E.; Mangenot, S.; Kuypers, M.M.; Schreiber, F.; Dutilh, B.E.; Zedelius, J.; de Beer, D.; et al. Nitrite-driven anaerobic methane oxidation by oxygenic bacteria. *Nature* **2010**, *464*, 543–548. [CrossRef] [PubMed]
35. Yang, B. N_2O and CH_4 Fluxes and Their Influencing Factors in Spartina Alterniflora Salt Marsh in Nanhui Shore of Yangtze Estuary under High Nitrogen Background. Ph.D. Thesis, East China Normal University, Shanghai, China, 2021.
36. Yvon-Durocher, G.; Allen, A.P.; Bastviken, D.; Conrad, R.; Gudasz, C.; St-Pierre, A.; Thanh-Duc, N.; Del Giorgio, P.A. Methane fluxes show consistent temperature dependence across microbial to ecosystem scales. *Nature* **2014**, *507*, 488–491. [CrossRef] [PubMed]
37. Ma, S.; Fang, C.; Sun, X.; Han, L.; He, X.; Huang, G. Bacterial community succession during pig manure and wheat straw aerobic composting covered with a semi-permeable membrane under slight positive pressure. *Bioresour. Technol.* **2018**, *259*, 221–227. [CrossRef] [PubMed]
38. Emerson, J.B.; Varner, R.K.; Wik, M.; Parks, D.H.; Neumann, R.B.; Johnson, J.E.; Singleton, C.M.; Woodcroft, B.J.; Tollerson, R.; Owusu-Dommey, A.; et al. Diverse sediment microbiota shape methane emission temperature sensitivity in Arctic lakes. *Nat. Commun.* **2021**, *12*, 5815. [CrossRef]
39. Bolhuis, H.; Cretoiu, M.S.; Stal, L.J. Molecular ecology of microbial mats. *Fems Microbiol. Ecol.* **2014**, *90*, 335–350. [CrossRef]
40. Kuypers, M.M.M.; Marchant, H.K.; Kartal, B. The microbial nitrogen-cycling network. *Nat. Rev. Microbiol.* **2018**, *16*, 263–276. [CrossRef]
41. Cheng, C. Research on Simultaneous Reduction and Mechanisms of Methane and Nitrous Oxide. Ph.D. Thesis, Shandong University, Jinan, China, 2019.
42. Li, J. Key Driving Factors Of Greenhouse Gas Fluxes Under Diffrent Microhabitats in Ulansuhai Wetland. Master's Thesis, Inner Mongolia University, Huhehaote, China, 2021.
43. Wei, H. Soil Greenhouse Gas Emission And Its Influencing Factors from Temperate Deciduous Broad-Leaved Forest And Tropical Mountain Rain Forest. Ph.D. Thesis, Northwest A & F University, Xianyang, China, 2018.
44. Hou, J.L. Metagenomic Study of the Structure, Evolution and Metabolic Potential of Microbial Communities Inhabiting Deep Sea Hydrothermal Chimneys. Master Thesis, Shanghai Jiao Tong University, Shanghai, China, 2018.
45. Campbell, B.J.; Engel, A.; Porter, M.L.; Takai, K. The versatile ε-proteobacteria: Key players in sulphidic habitats. *Nat. Rev. Microbiol.* **2006**, *4*, 458–468. [CrossRef]
46. Gao, Y.H. Study on CH_4 and N_2O Emission Characteristics and Influencing Factors of Reservoirs in Cold Regions. Master's Thesis, Dalian University of Technology, Dalian, China, 2019.
47. Isidorova, A.; Grasset, C.; Mendonça, R.; Sobek, S. Methane formation in tropical reservoirs predicted from sediment age and nitrogen. *Sci. Rep.* **2019**, *9*, 11017. [CrossRef]
48. Borges, A.V.; Darchambeau, F.; Teodoru, C.R.; Marwick, T.R.; Tamooh, F.; Geeraert, N.; Omengo, F.O.; Guérin, F.; Lambert, T.; Morana, C.; et al. Globally significant greenhouse-gas emissions from African inland waters. *Nat. Geosci.* **2015**, *8*, 637–642. [CrossRef]
49. Deemer, B.R.; Harrison, J.A.; Li, S.; Beaulieu, J.J.; DelSontro, T.; Barros, N.; Bezerra-Neto, J.F.; Powers, S.M.; dos Santos, M.A.; Vonk, J.A. Greenhouse Gas Emissions from Reservoir Water Surfaces: A New Global Synthesis. *BioScience* **2016**, *66*, 949–964. [CrossRef]

50. Neumann, R.B.; Moorberg, C.J.; Lundquist, J.D.; Turner, J.C.; Waldrop, M.P.; McFarland, J.W.; Euskirchen, E.S.; Edgar, C.W.; Turetsky, M.R. Warming Effects of Spring Rainfall Increase Methane Emissions From Thawing Permafrost. *Geophys. Res. Lett.* **2019**, *46*, 1393–1401. [CrossRef]
51. Grinham, A.; Dunbabin, M.; Albert, S. Importance of sediment organic matter to methane ebullition in a sub-tropical freshwater reservoir. *Sci. Total Environ.* **2018**, *621*, 1199–1207. [CrossRef] [PubMed]
52. Yang, L. Contrasting methane emissions from upstream and downstream rivers and their associated subtropical reservoir in eastern China. *Sci. Rep.* **2019**, *9*, 8072. [CrossRef]
53. McClure, R.P.; Lofton, M.E.; Chen, S.; Krueger, K.M.; Little, J.C.; Carey, C.C. The Magnitude and Drivers of Methane Ebullition and Diffusion Vary on a Longitudinal Gradient in a Small Freshwater Reservoir. *J. Geophys. Res. Biogeosciences* **2020**, *125*, 18. [CrossRef]
54. Bai, X.; Xu, Q.; Li, H.; Cheng, C.; He, Q. Lack of methane hotspot in the upstream dam: Case study in a tributary of the Three Gorges Reservoir, China. *Sci. Total Environ.* **2020**, *754*, 142151. [CrossRef]
55. Yao, T.; Dong. A comprehensive study of Water-Ecosystem-Human activities reveals unbalancing Asian Water Tower and accompanying potential risks. *Chin. Sci. Bull.* **2019**, *64*, 2761–2762. [CrossRef]
56. Sinha, E.; Michalak, A.M.; Balaji, V. Eutrophication will increase during the 21st century as a result of precipitation changes. *Science* **2017**, *357*, 405–408. [CrossRef] [PubMed]
57. Selvam, B.P.; Natchimuthu, S.; Arunachalam, L.; Bastviken, D. Methane and carbon dioxide emissions from inland waters in India–implications for large scale greenhouse gas balances. *Glob. Chang. Biol.* **2014**, *20*, 3397–3407. [CrossRef]
58. Liu, L.; Yang, Z.; Delwiche, K.; Long, L.; Liu, J.; Liu, D.; Wang, C.; Bodmer, P.; Lorke, A. Spatial and temporal variability of methane emissions from cascading reservoirs in the Upper Mekong River. *Water Res.* **2020**, *186*, 116319. [CrossRef]
59. Keller, M.; Stallard, R.F. Methane emission by bubbling from Gatun Lake, Panama. *J. Geophys. Res. Atmos.* **1994**, *99*, 8307–8319. [CrossRef]
60. dos Santos, M.A.; Rosa, L.P.; Sikar, B.; Sikar, E.; dos Santos, E.O. Gross greenhouse gas fluxes from hydro-power reservoir compared to thermo-power plants. *Energy Policy* **2006**, *34*, 481–488. [CrossRef]
61. Maeck, A.; DelSontro, T.; McGinnis, D.; Fischer, H.; Flury, S.; Schmidt, M.; Fietzek, P.; Lorke, A. Sediment Trapping by Dams Creates Methane Emission Hot Spots. *Environ. Sci. Technol.* **2013**, *47*, 8130–8137. [CrossRef] [PubMed]
62. Descloux, S.; Chanudet, V.; Serça, D.; Guérin, F. Methane and nitrous oxide annual emissions from an old eutrophic temperate reservoir. *Sci. Total Environ.* **2017**, *598*, 959–972. [CrossRef] [PubMed]
63. Huttunen, J.T.; Väisänen, T.S.; Hellsten, S.K.; Heikkinen, M.; Nykänen, H.; Jungner, H.; Niskanen, A.; Virtanen, M.O.; Lindqvist, O.V.; Nenonen, O.S.; et al. Fluxes of CH_4, CO_2, and N_2O in hydroelectric reservoirs Lokka and Porttipahta in the northern boreal zone in Finland. *Glob. Biogeochem. Cycles* **2002**, *16*, 3-1–3-17. [CrossRef]

diversity

MDPI

Article

Assessment of Spatiotemporal Dynamics of Mangrove in Five Typical Mangrove Reserve Wetlands in Asia, Africa and Oceania

Cun Du [1], Shahbaz Khan [2,*], Yinghai Ke [1,*] and Demin Zhou [1]

[1] College of Resource Environment and Tourism, Capital Normal University, Beijing 100048, China
[2] UNESCO Beijing Cluster Office, Beijing 100600, China
* Correspondence: s.khan@unesco.org (S.K.); yke@cnu.edu.cn (Y.K.)

Abstract: Mangrove wetlands play a key role in global biodiversity conservation, though they have been damaged in recent decades. Therefore, mangrove habitats have been of great concern at the international level since the latter half of the 20th century. We focused on the key issue of the dynamics of mangrove habitats. A comprehensive review of their typicality and status from the global perspective was evaluated before the landscape dynamics of the mangrove habitats at the five sites were interpreted from Landsat satellite images covering 20 years, from 2000 to 2020. Ground-truthing was undertaken after comparing the results with the other published international mangrove datasets. We reached three conclusions: Firstly, within the period from 2000 to 2020, the mangrove area in Dongzhaigang increased by 414 ha, with an increase of 24.6%. In Sembilang NP, Sundarban, Kakadu NP, and RUMAKI, the mangrove area decreased by 1652 ha, 16,091 ha, 83 ha, and 2012 ha, with a decrease of 1.8%, 2.7%, 0.9%, and 3.9%, respectively. Secondly, other types of wetlands play a key role in degradating the mangrove wetlands in all of five protected areas. Thirdly, the rate of mangrove degradation has slowed dramatically based on the five sites over the past two decades, which are generally consistent with the findings of other researchers.

Keywords: mangrove conservation; wetland habitat; landscape changes; ecology & hydrology

Citation: Du, C.; Khan, S.; Ke, Y.; Zhou, D. Assessment of Spatiotemporal Dynamics of Mangrove in Five Typical Mangrove Reserve Wetlands in Asia, Africa and Oceania. *Diversity* **2023**, *15*, 148. https://doi.org/10.3390/d15020148

Academic Editor: Peng Hou

Received: 24 December 2022
Revised: 9 January 2023
Accepted: 12 January 2023
Published: 21 January 2023

1. Introduction

Mangroves are distributed in tropical, subtropical, and some temperate coasts and often in regions with high population density and intensive human activities. Mangroves provide essential ecosystem services, such as coastal protection, pollution control, and cultural values to hundreds of millions of people [1,2]. However, mangrove habitats have been drastically degraded due to various factors over the past few decades, such as climate change, urban expansion, aquaculture, mining, and logging [3,4]. Global mangrove area has decreased by 35–50% in the past half-century (equivalent to an annual loss of 1% to 2%) [5].

International organizations and governments have taken actions to protect mangrove wetlands. In 2015, the United Nations included the mangrove ecosystem as an important factor in achieving sustainable development [6]. As early as 1996, India implemented a project to restore mangrove wetlands along the east coast through a partnership between the Environment Foundation and the government [7]. In 2002, the State Forestry Administration of China began a series of mangrove conservation and restoration projects. In 2020, the Chinese government again launched Special Action Plan for Mangrove Protection and Restoration (2020–2025)" [2]. To prevent the continued degradation of mangrove ecosystems, the Indonesian government has also focused on implementing a community-based mangrove management program [8].

Many researchers have investigated mangrove wetlands' long-term change. The FAO published the global status of mangrove degradation in 2007 and reported that the global mangrove wetland area decreased by 20% from 1980 to 2005. Goldberg et al. [9] showed an average annual loss rate of 0.13% (2000–2016) of the global mangrove area. Global Mangrove Watching Version 3.0 [10] showed a global average annual mangrove loss rate of 0.21% (1996–2010) and 0.04% (2010–2020), respectively, and that the mangrove area in Australia has been declining. However, Lymburner et al. [11] indicated that the mangrove area in Australia maintained an increasing trend between 2000 and 2010. Most studies reported that the rate of mangrove decline has started to decrease, but inconsistencies still existed in some of the results. In addition, the existing studies have mainly focused on the area change of mangrove ecosystems, while the change in landscape characteristics has been under-investigated.

Mangroves are one of the world's most threatened and vulnerable ecosystems, with climatic factors and human activities as the primary degradation factors [12]. Climatic factors include precipitation, temperature, sea levels rise, natural disasters, etc. Human activities can be divided into agricultural encroachment, urban expansion, environmental pollution, coastal aquaculture, etc. [13–15]. Friess et al. [16] suggested that the intertidal mangroves are a dynamic ecosystem, where the range and habitat quality were undergoing rapid changes. Since their first appearance in the geological record 75 million years ago, climate and sea level change have entirely changed the spatial distribution of mangroves. However, climate change contributes less to mangrove degradation [17]. With economic development, human activities have become the most important factor for mangrove degradation. Unplanned shrimp farms and urban development mainly cause mangrove deforestation, but the fundamental problem is inappropriate systems and regulations in the past [18,19]. Water pollution caused by offshore oil extraction may also impact the degradation of mangroves [20]. These studies suggest that the factors driving mangrove change are complex and vary from region to region.

The rapid development of remote sensing technology has provided technical support to realize mangrove monitoring on a large scale. Hamilton and Casey [21] established new global data sets of the 21st-century continuous mangrove coverage by integrating the World Forest Change Database, the World Mangrove Database, and the World Terrestrial Ecosystem Database. Giri et al. [22] obtained the global mangrove distribution dataset in 2000 from the Global Land Survey (GLS) dataset and Landsat images, which were interpreted by using a hybrid supervised classification and unsupervised classification method. Goldberg et al. [9] used a random forest classification approach to map the extent of global mangroves from 2000–2016 based on Landsat images. Bunting et al. [10] used synthetic aperture radar data to generate GMW version 3.0, which is valuable for analyzing global mangrove changes.

Despite international conservation policies and ambitious global restoration goals, deforestation is still occurring at a large scale, especially in Asia and Africa. At the same time, the rapid development of remote sensing technology provides support for the dynamic monitoring of mangroves [23,24]. Our objective is to select mangrove reserve wetlands within five continents over 20 years, use remote sensing and Geographic Information System (GIS) techniques to analyze and evaluate the loss/gain and landscape characteristics of mangrove wetlands, and compare current mainstream mangrove data to obtain actual mangrove change trends.

2. Materials and Methods

2.1. Selection of Study Sites

2.1.1. Principles for Selecting Study Sites of Mangrove Importance

There are 298 Ramsar-listed sites worldwide, playing key roles as international mangrove habitats. Among them, some sites also play additional key roles in UNESCO Biosphere Plan or in the UNESCO Natural Heritage. These mangrove habitats have different functions and regional features. The following principles were considered to select representative sites in our study.

(1) Site importance. To select those sites which are listed in the list of Ramsar wetlands of international importance, the UNESCO marine heritage list, and the UN biosphere list as much as possible;
(2) Geographical representativeness. The selected sites present different regional characteristics and cover wide geographical regions of mangrove spatial distribution as much as possible, and the site network should cover Asia, Oceania, and Africa, linking the Indian Ocean and the Pacific Ocean;
(3) Difference in functional roles of mangrove habitats. The selected sites can present different functional roles of mangrove habitats, such as Storing carbon, biodiversity protection, tsunami risk reduction, coastal line protection, and tourism service;
(4) Different challenges or problems on the sites. The selected sites are facing different challenges or national or international issues from economic development, environmental change, urbanization, etc.

2.1.2. The Final Scheme of the Study Sites of Mangrove Importance

The selected five sites are shown in Figure 1 and Table 1. The serial number was then finally fixed as A to E. The Sundarban mangrove, including Sundarbans Reserved Forest (Bangladesh) and Sundarban Wetland (India), is the largest mangrove wetland, while Bangladesh takes the largest protection ratio of the national mangrove area [25]. Indonesia has around 20% global mangrove and is one of the countries with the largest mangrove wetlands. However, it has lost nearly half mangrove area due to aquaculture development according to a research report from Richards and Friess [12].

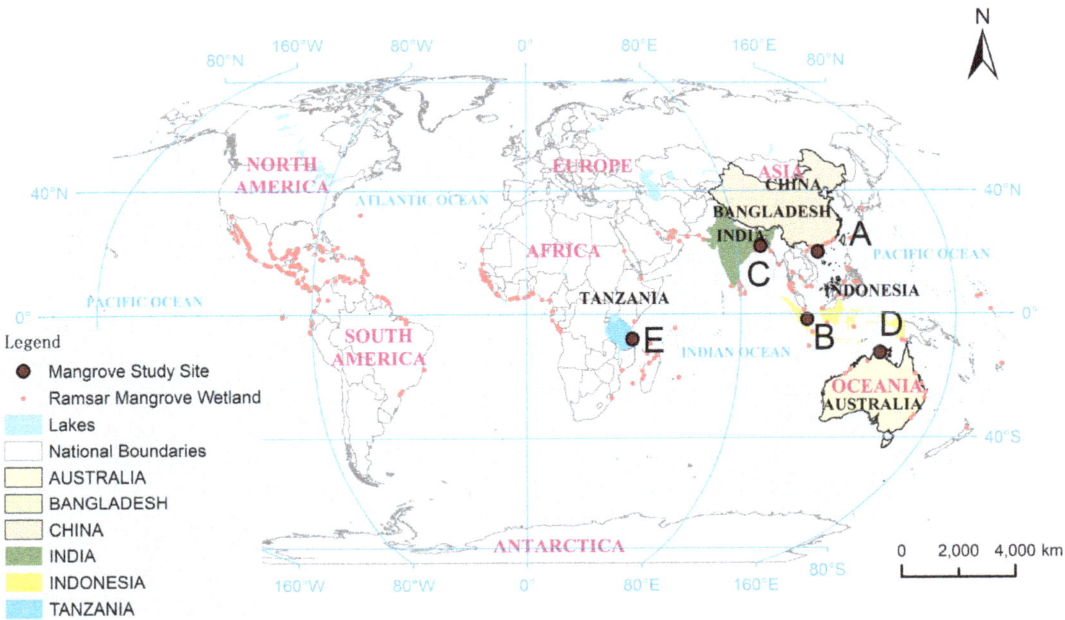

Figure 1. The map of the finally selected sites for the study. (A): Dongzhaigang; (B): Sembilang NP; (C): Sundarban; (D): Kakadu NP; (E): RUMAKI.

Table 1. The check list of five study sites to be finally selected.

Serial	Site Name	Geo-Location	Area (hm^2)	UNESCO Heritage	Ramsar List	Biosphere List	Country and Region	Major Threats
A	Dongzhaigang	19°58′ N 110°34′ E	5400	N	Y	N	China, Eastern Asia	Aquaculture; Urban expansion
B	Sembilang National Park	01°57′ S 104°36′ E	202,896	N	Y	Y	Indonesia, Southeast Asia	Aquaculture
C	Sundarban Wetland/ Sundarbans Reserved Forest	21°46′ N 88°42′ E	1,024,700	Y	Y	N	Bangladesh/India, South Asia	Cropland encroachment
D	Kakadu National Park	12°40′ S 132°45′ E	1,979,766	Y	Y	N	Australia, Oceania	Natural disasters
E	Rufiji-Mafia-Kilwa Marine Ramsar Site	08°07′ S 39°37′ E	596,908	Y	Y	N	Tanzania, Eastern Africa	Deforestation

2.2. Study Sites Introduction

As China's earliest and largest mangrove nature reserve, with an area of 5400 ha, Dongzhaigang Mangrove Nature Reserve is densely populated, surrounded by many villages and vast paddy fields. A mangrove restoration project has been underway since 2010 after this site's apparent loss of mangrove habitats. Dongzhaigang Mangrove Nature Reserve has been listed in the UNESCO World Heritage Sites Tentative List since 1996.

Sembilang Nation Park (Sembilang NP) has the largest mangrove habitats (88,046 ha) in the Indo-Malayan region [26]. Over 43% of mangrove species in Indonesia are found here. Sembilang NP has been acknowledged as a part of the UNESCO World Network of Biosphere Reserves since 2018.

The Sundarban mangrove wetland is the world's largest contiguous forest wetland system (579,446 ha). It locates in the deltas of the Ganga–Brahmaputra–Meghna rivers. It is the most diverse mangrove forest in the world and about 78 species of mangroves have been found at this site [27]. More than 12 million people live in and around the Sundarbans, of which 2.5 million depend almost entirely upon the mangroves for their livelihoods [28].

Kakadu National Park(Kakadu NP) is renowned for its exceptional beauty and unique biodiversity, with various landforms, habitats and wildlife. The mangrove habitat is a major ecosystem within Kakadu NP [29]. Kakadu NP was inscribed on the World Heritage List as a dual cultural and natural heritage in 1981.

Rufiji-mafia-kilwa marine ramsar site (RUMAKI) is located in the three districts of Rufiji, Mafia, and Kilwa, covering the most extensive mangrove habitat (48,991 ha) on the East African coast. According to the List of Wetlands of International Importance, RUMAKI is a good representative wetland for East Africa, for containing multiple and ecologically interconnected wetlands [30].

Five wetlands face different threats. Since the 1990s, the governments of China and Indonesia have encouraged the conversion of mangroves into aquaculture ponds to increase food security. Thus, aquaculture is a major driving factor in the degradation of Dongzhaigang and Sembilang NP. Because Dongzhaigang is close to the urban area and the tourism industry is relatively developed, urban expansion is another primary driver [31,32]. Development started late in Sundarban mangroves, and local governments and residents need to be aware of the conservation of mangrove wetland ecosystems. In addition, the presence of petroleum resources in the reserve and oil pollution from their exploitation are significant factors in mangrove degradation [14,33]. Australia is a developed country, and mangrove wetlands in protected areas are less affected. However, natural disasters (e.g., droughts, tsunamis) occur from time to time along the Australian coast, and this is a major factor in mangrove loss [29,34] Tanzania is developing. The inhabitants of the reserve use mangroves as economic forests to sustain their livelihoods, and indiscriminate deforestation is a major factor in the loss of mangroves in this reserve [35,36].

2.3. Data Preparing and Pre-Processing

2.3.1. Images Collection and Band Synthesis

Google earth engine (GEE) enables users to call a large number of remote sensing images online and perform calculations [37]. In this study, we used GEE to select Landsat and Sentinel images that meet cloud cover requirements (<15%) and eliminate images acquired at high tide level through visual interpretation. Too much cloud cover is not conducive to subsequent interpretation, and the high tide level will cover many mangroves. If there is no image meeting the requirements, we use all remote sensing images in the year to synthesize the maximum value of the NDVI index and the median value of the bands in this case (Figure 2). The detailed information of the used remote sensing images can be checked from Table 2.

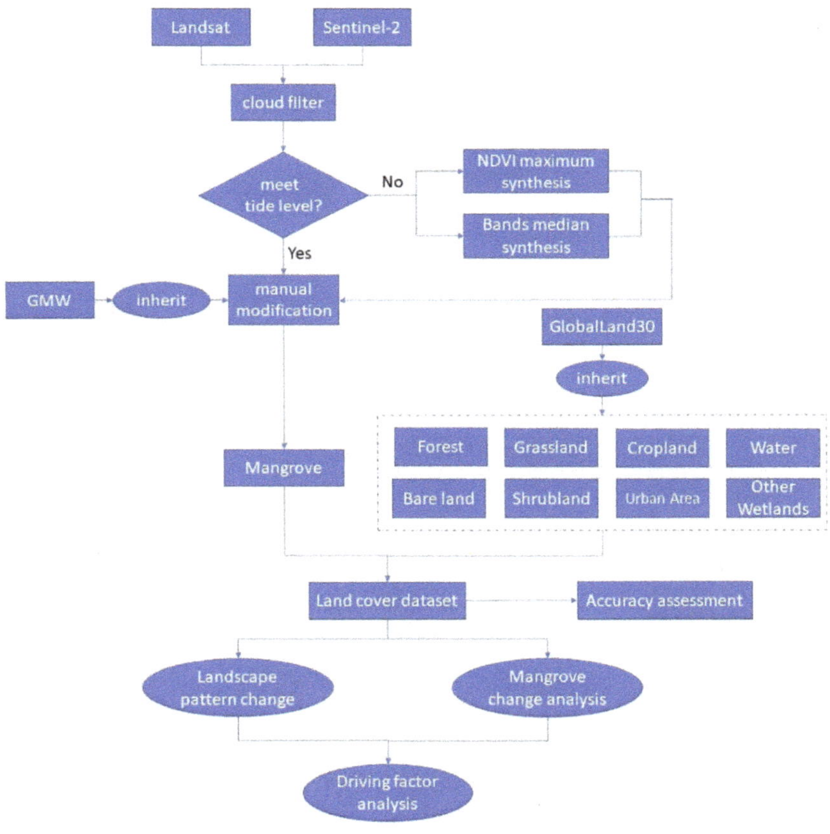

Figure 2. Flow chart of data collection and analysis procedures.

Table 2. The data source information of the remote sensing images used in five sites.

Serial	Year	Sensor	Number
A	1990	Landsat 5	6
	2000	Landsat 5/7	6
	2010	Landsat 5/7	6
	2020	Sentinel-2	40
B	2000	Landsat 7	42
	2010	Landsat 7	16
	2020	Sentinel2	146
C	2000	Landsat 7	2
	2010	Landsat 5	2
	2020	Landsat 8	2
D	2000	Landsat 7	2
	2010	Landsat 5	2
	2020	Landsat 8	2
E	2000	Landsat 7	4
	2010	Landsat 5	4
	2020	Landsat 8	74

2.3.2. Mangrove Dataset Collections

Some published mangrove data sets were evaluated carefully for selecting study sites and referring to our monitoring of the dynamics of mangrove habitats in the past 20 years. Among them, the most valuable data sets include the Global Mangrove Forests Distribution, v1 (2000, GMFD), the Global Mangrove Watch (GMW), and the 10 m GLOBAL MANGROVE CLASSIFICATION PRODUCTS OF 2018–2020 BASED ON BIG DATA (GMCP). The technical indicators are listed in Table 3.

Table 3. The basic information of free download data sources.

Data Set	Pixels	The Dates of Production	Data Format	Precision	Citation
GMFD	30 m	2000	Grid	90.75%	Giri et al. [22]
GMW	24 m	1996, 2007, 2008, 2009, 2010, 2015, 2016	Shape	95.25%	Bunting et al. [10]
GMCP	10 m, 100 m	2018–2020	Shape, Grid	91.62%	Xiao et al. [38]

2.4. *Manual Modification for Mangrove Boundary and Landcover Dataset Synthesis*

The mangrove extent provided by GMW is slightly inaccurate at five specific sites and does not provide the mangrove extent for 2000 and 2020. Then, we manually modified the mangrove boundary by referring to the selected or synthesized remote sensing images and the collected mangrove dataset with a high reference value in this study (Figure 2). Then, we mosaicked the Globalland30 and our manually modified mangrove datasets to obtain a landcover dataset, which contains nine land use types (Mangrove, Forest, Grassland, Cropland, Water, Bareland, Shrubland, Urban Area, and Other Wetlands) to subsequently analyze the conversion between mangroves and other land-use types. The Globalland30 dataset is available at http://www.globallandcover.com/home.html?type=data (accessed on 1 January 2014) [39]

2.5. Accuracy Assessment

In order to verify the developed mangrove dataset, we merged the nine land cover types into three types: mangrove, water, and others. Balanced number of validation samples were generated for each type. For instance, the area proportion of the water body and mangrove is very low in Kakadu NP, and we appropriately increased the validation samples in the area where the water body and mangrove forest are located in the typical area, rather than simply generating random test points according to the uniform distribution. Otherwise, most test points were distributed in other land types, such as forests and grassland. Fewer samples of test points in water and mangrove will result in accuracy errors. We determined the actual land use types represented by the sample points with the help of Google Earth, other mangrove datasets, and Sentinel data [22,38].

2.6. Landscape Pattern Change Analysis and Mangrove Change Analysis

We used ArcGIS software to statistically analyze the land cover dataset during different periods, including conversion between mangrove habitats and other land-use types, mangrove degradation direction, and changes in mangrove landscape patterns for different periods (Figure 2). We counted four landscape indexes using ArcMap10.2: Number of Patches (NP), Area of Maxsum Patch (MAXP); Mean Area of Patches (MPS); Total Area of Patches (CA). MPS reflects an average condition and demonstrates the degree of landscape fragmentation and heterogeneity in the landscape analysis. NP reflects the total number of patches in the landscape pattern, and generally, a larger NP indicates a higher degree of fragmentation. The four parameters can comprehensively reflect the degree of fragmentation of the landscape and other ecological information of the landscape. To show the degradation of mangroves more visually, we mapped the direction of mangrove degradation by mangrove to other land-use types.

3. Results

3.1. Accuracy Assessment

The accuracy assessment results are shown in Table 4. Both the mangrove accuracy and overall accuracy are above 95%, which meets the standards for analyzing landscape changes.

Table 4. Accuracy assessment table.

Serial	2000		2010		2020	
	Mangrove	Overall	Mangrove	Overall	Mangrove	Overall
A	98.3	96.6	96.7	95.4	97.5	96.7
B	97.2	95.1	97.5	95.4	97.1	95.2
C	98.0	97.0	98.1	97.2	98.4	97.4
D	97.1	96.1	97.2	95.9	97.5	96.3
E	97.4	95.2	97.5	95.4	97.8	95.8

3.2. Landscape Pattern Change of the Mangrove Habitats from 2000 to 2020

The landscape matrix of the changed mangrove habitats in the five sites was generated using the landcover dataset. The detailed dynamic information of the mangrove habitats from 2000 to 2020 can be checked in Table 5. From 1990 to 2000, the NP of the mangrove habitats in Dongzhaigang tripled from 27 to 87 ha; but the MPS decreased from 71 to 20 ha, nearly 70%. This indicates that the process of mangrove habitat fragmentation was very severe. Fortunately, the fragmentation of mangrove wetlands in the other four protected areas, except for Sundarbans, decreased or remained the same from 2000. The fragmentation process of Sundarbans mangrove only gradually stopped in 2010.

Table 5. Landscape matrix of the changed mangrove wetlands in five sites (ha).

Seria	Index	1990	2000	2010	2020
A	NP	27	87	41	64
	MAXP	742	247	683	505
	MPS	71	20	42	33
	CA	1930	1683	1733	2097
B	NP	-	1279	881	913
	MAXP	-	32,746	20,435	20,255
	MPS	-	70	100	96
	CA	-	89,698	87,965	88,046
C	NP	-	4859	6666	6788
	MAXP	-	26,154	23,251	23,102
	MPS	-	123	86	85
	CA	-	595,537	575,700	579,446
D	NP	-	1003	918	979
	MAXP	-	1140	1140	1137
	MPS	-	9	10	9
	CA	-	8965	8942	8882
E	NP	-	1668	1312	1185
	MAXP	-	6909	6801	6793
	MPS	-	31	38	41
	CA	-	51,003	49,748	48,991

"-": no data; The Dongzhaigang data are partially cited from the Ph.D. thesis of Jia (2014) with corrections and slightly adjusted classification.

3.3. Mangrove Habitats Change Analysis from 2000–2020 at Five Sites

3.3.1. Analyzing the Dynamics of Mangrove Habitats in Dongzhaigang

The mangrove habitats area decreased by 12.8% from 1990 to 2000 before increasing from 2000 to 2020. The mangrove area increased by 24.6% from 2000 to 2020 according to the statistical data (Figure 3). The degradation of mangroves has been presented from 1990 to 2020. However, the degradation area had been decreasing (Figures 4 and 5). The highest percentage of mangroves was converted to other wetlands (mainly aquaculture) from 1990 to 2000, but the highest percentage of mangroves was converted to forest from 2000 to 2010. In addition, there were larger areas of mangrove wetlands that had been encroached on by cropland (Figure 6). The degraded area of mangroves was already much lower than the new area of mangroves from 2010–2020, which should be attributed to mangrove planting and conservation policy of the local government. Mangrove degradation mainly occurred in the landward direction from 1990 to 2010, and mangrove degradation mainly occurred in the seaward direction from 2010–2020 (Figure 6). Comparing the landscape matrices between 1990 and 2000, it was obvious that the lost mangrove habitat was replaced by the pond, which might be used for fishery purposes. The area of mangrove habitat increased obviously after 2000, attributed to mangrove planting supported by the local government.

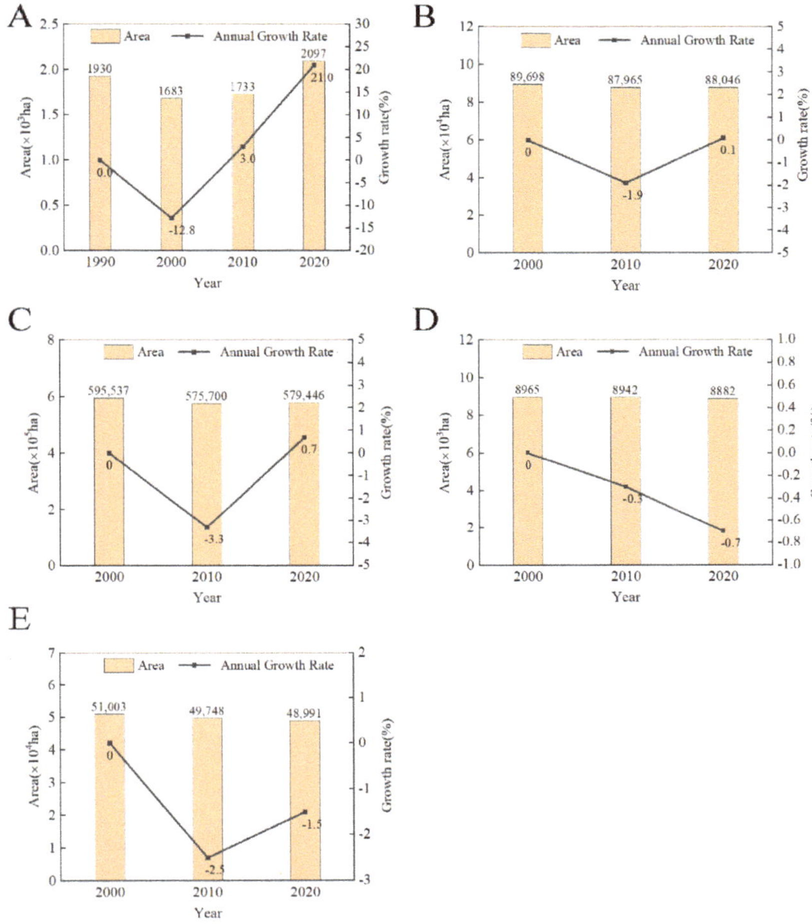

Figure 3. The dynamics of the mangrove habitats in five sites. (**A**): Dongzhaigang(1990–2020); (**B**): Sembilang NP; (**C**): Sundarban; (**D**): Kakadu NP; (**E**): RUMAKI (2000–2020).

3.3.2. Analyzing the Dynamics of Mangrove Habitats in Sembilang NP

Mangrove habitats are mainly dispersed in the central part of Sembilang NP (Figure 7). Mangrove area decreased by a total of 1733 ha from 2000 to 2010 with a 10-year decrease ratio of 1.9%, though it increased a total of 81 ha from 2010 to 2020. It indicated that the protection of mangroves in the latter 10 years was significantly more effective than that in the former, which not only curbed the previous trend of decreasing mangrove wetland area but also contributed to a slight increase of the mangrove habitats (Figure 3). Degradation occurred primarily at the landward edge of the northern mangroves, mainly degraded to forest and other wetlands, however, the increase in mangrove area also came mainly from forest and other wetlands (Figures 8 and 9). In the southeast coastal region of the reserve, there are other farming ponds that have not yet been retired.

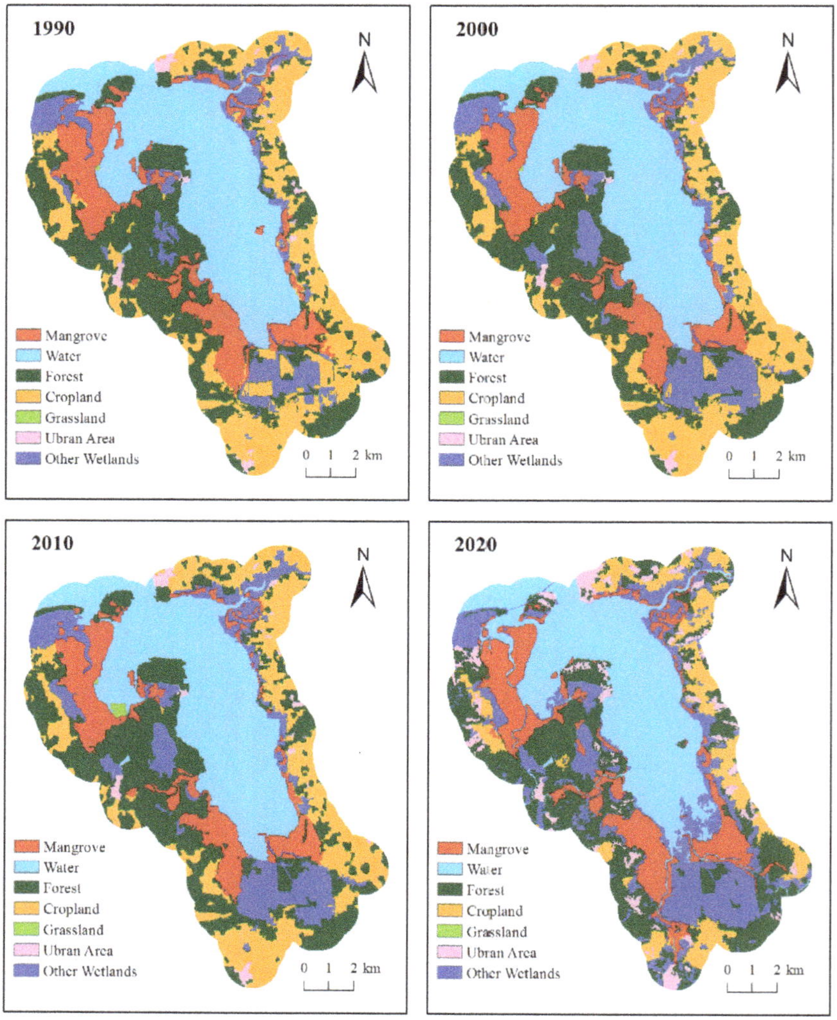

Figure 4. Landscape thematic maps in Dongzhaigang from 1990 to 2020.

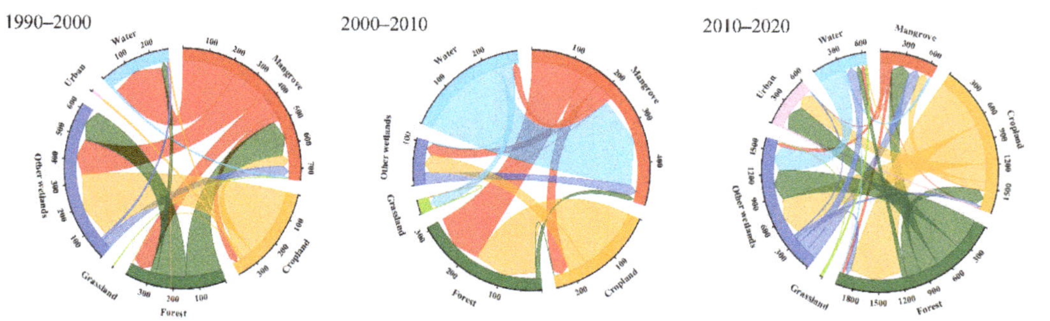

Figure 5. Transfer direction between the different land-use types in Dongzhaigang ((**left**): 1990−2000, (**middle**): 2000−2010, (**right**): 2010−2020).

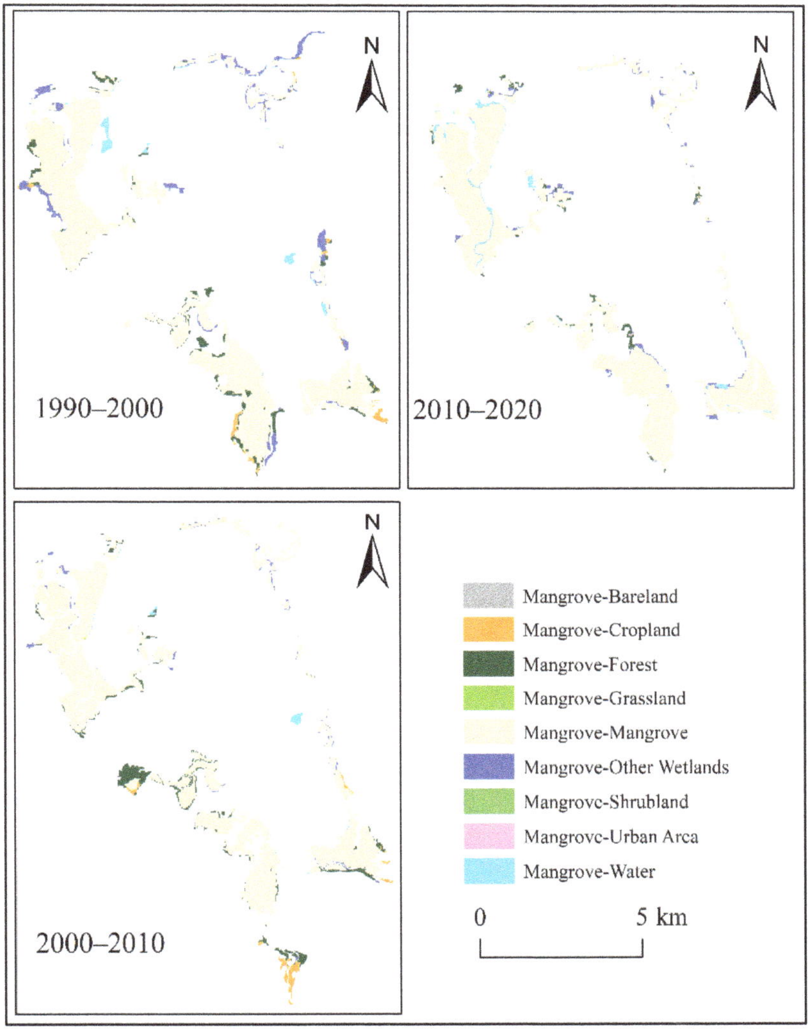

Figure 6. Maps of spatial–temporal distribution of mangrove degradation in Dongzhaigang from 1990 to 2020.

3.3.3. Analyzing the Dynamics of Mangrove Habitats in the Sundarban

The degradation of mangroves in Sundarban was concentrated in the northern (near the boundary) and the eastern part of the reserve. Mangroves were mainly degraded to cropland and other wetlands in the first ten years, and to wetlands during the second ten years. The degraded area was drastically reduced and more dispersed (Figures 10 and 11). Mangrove habitat area decreased by a total of 19,837 ha from 2000 to 2010, with a 10-year decreased rate of 3.3%. The decrease rate is higher than the other four sites in the same period. The degraded area in 2000–2010 was significantly larger than that in 2010–2020 (Figures 3 and 11). A total of 3746 ha increased with an increased rate of 0.7% from 2010 to 2020, indicating that the downward trend at the beginning of the 21st century was significantly contained, and the protection and restoration of mangrove habitats in the past ten years have achieved remarkable achievements (Figure 3).

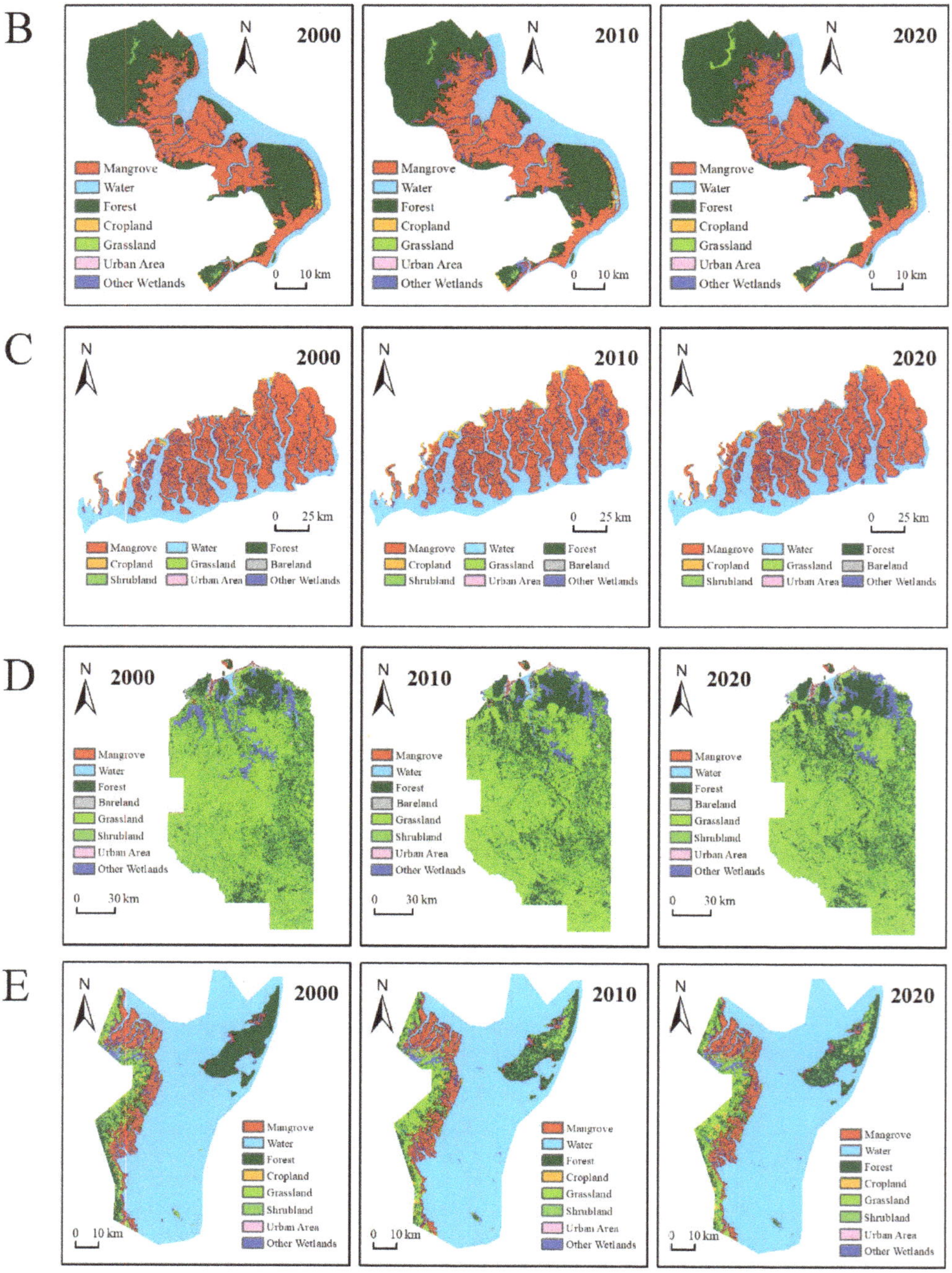

Figure 7. Landscape thematic maps in four sites from 2000 to 2020 (**B**): Sembilang NP, (**C**): Sundarban, (**D**): Kakadu NP, (**E**): RUMAKI.

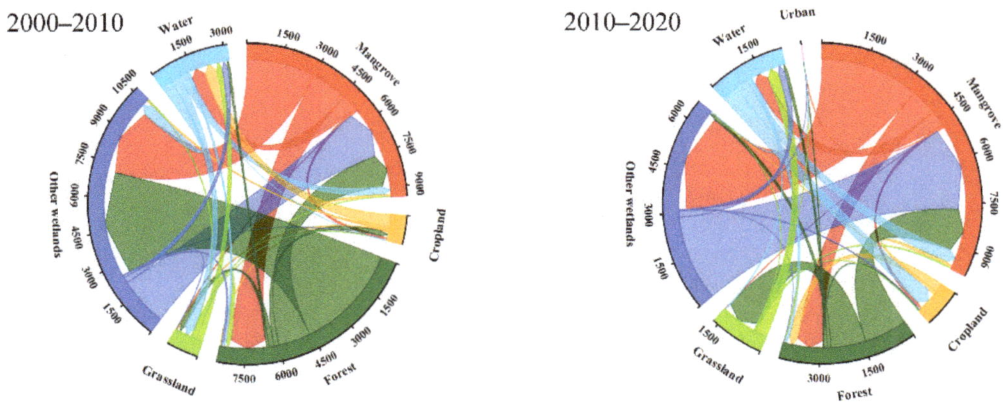

Figure 8. Transfer direction between the different land-use types in Sembilang NP ((**left**): 2000–2010, (**right**): 2010–2020).

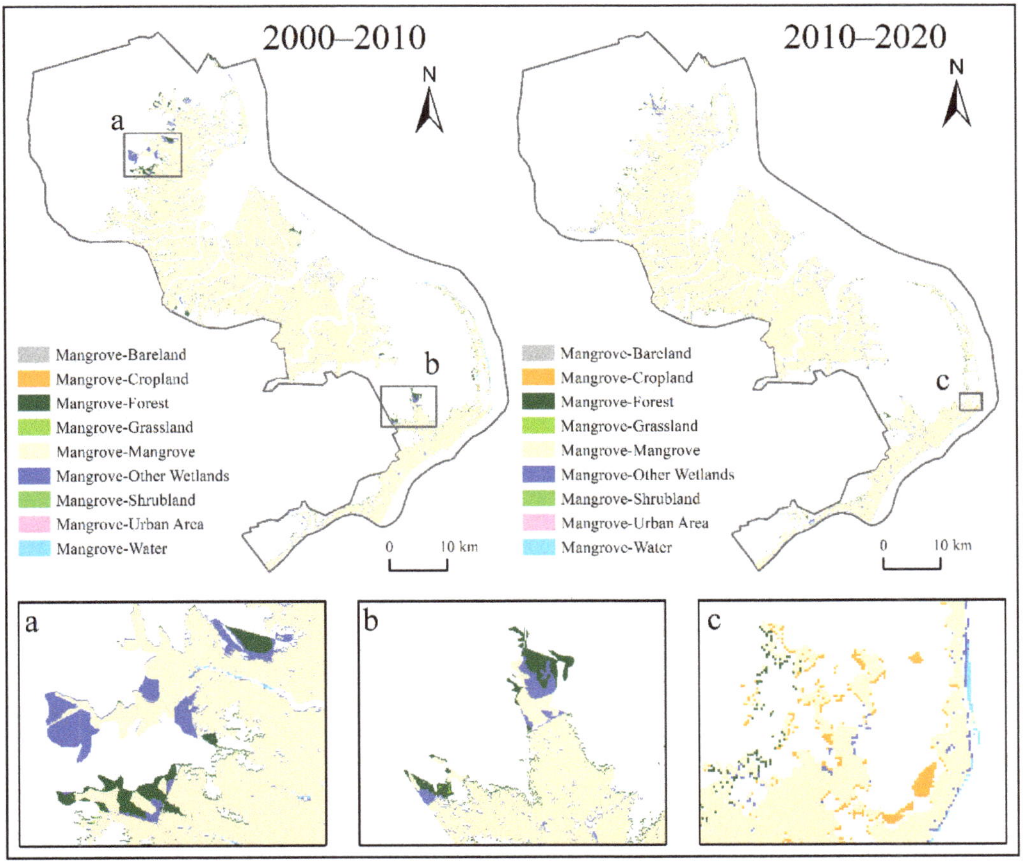

Figure 9. Maps of spatial–temporal distribution of mangrove degradation in Sembilang from 2000 to 2020. (**a**) first degradation detail picture; (**b**) second degradation detail picture; (**c**) third degradation detail picture.

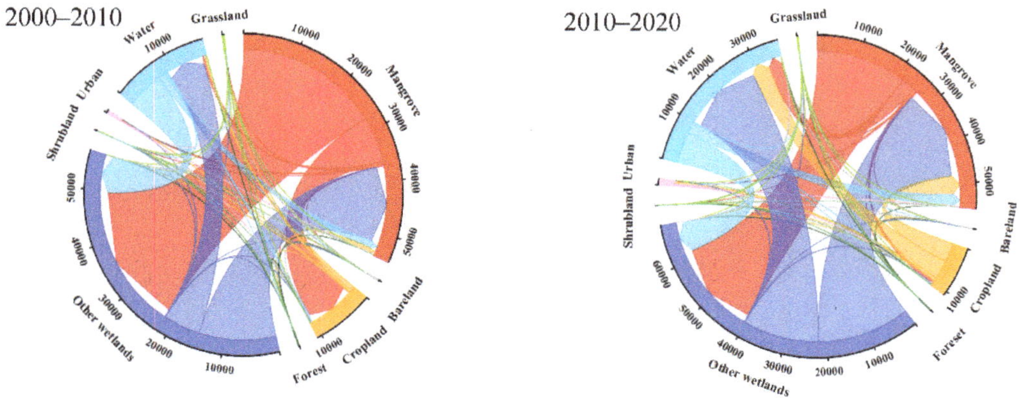

Figure 10. Transfer direction between the different land-use types in the Sundarban ((**left**): 2000–2010, (**right**): 2010–2020).

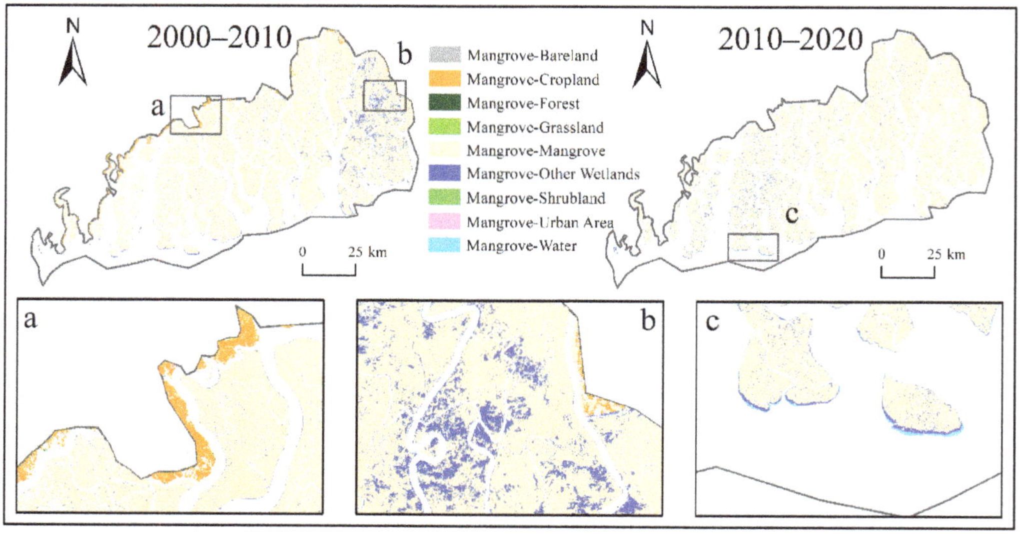

Figure 11. Maps of spatial–temporal distribution of mangrove degradation in the Sundarban from 2000 to 2020. (**a**) first degradation detail picture; (**b**) second degradation detail picture; (**c**) third degradation detail picture.

3.3.4. Analyzing the Dynamics of Mangrove Habitats in Kakadu NP

Mangrove habitats are mainly located in the northern part of Kakadu NP, along the Van Diemen's coastline. Forest and grassland are the most prominent landscape types in Kakadu NP, while mangrove habitats account for a relatively small area (Figure 7). Mangrove habitat area decreased by only 23 ha from 2000 to 2010, with a 10-year decrease rate of only 0.3% and decreased by 60 ha from 2010 to 2020, with a 10-year decrease rate of 0.7%, doubling the rate of reduction over the former 10-year period (Figure 3). As a whole, Kakadu NP was second only to Dongzhaigang in terms of the conservation effect from 2000 to 2010, and the total decrease of mangrove habitats area (83 ha) and decrease rate (0.9%) were lower than those of the other three sites. Unfortunately, mangrove habitats at other sites have shown an increasing trend over the last decade (except for sites in Tanzania), while mangroves within Kakadu NP have continuously decreased. Mangrove wetlands are mainly degraded to other wetlands, followed by forest and water (Figures 12 and 13).

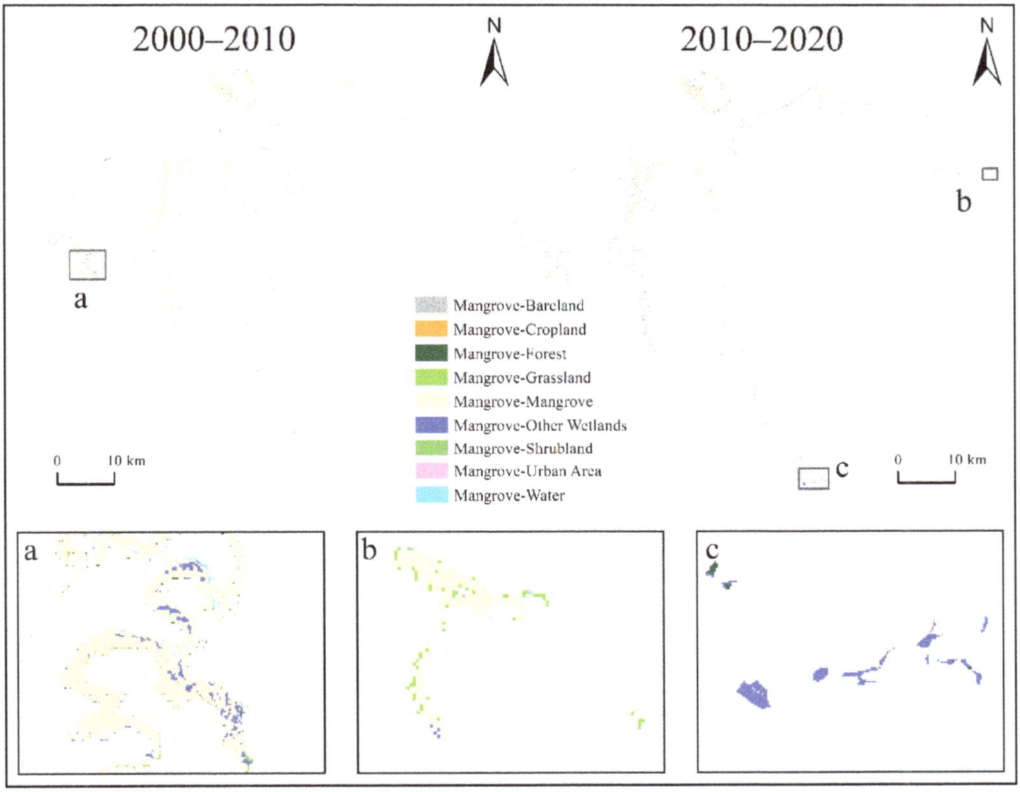

Figure 12. Maps of spatial–temporal distribution of mangrove degradation in Kakadu NP from 2000 to 2020. (**a**) first degradation detail picture; (**b**) second degradation detail picture; (**c**) third degradation detail picture.

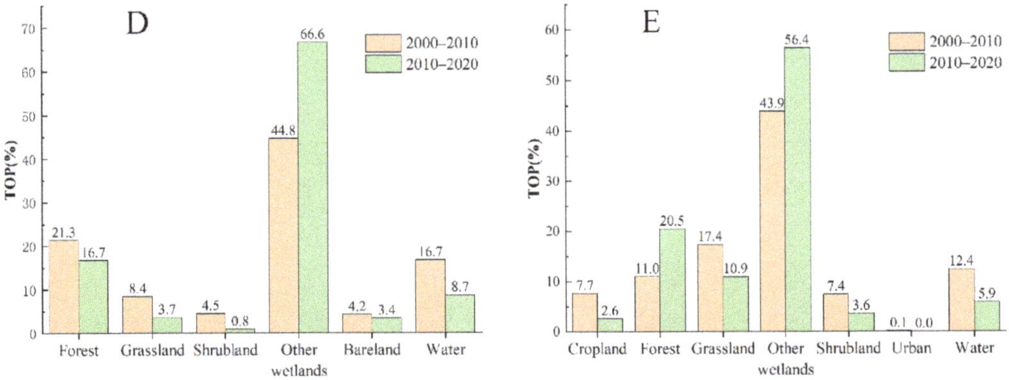

Figure 13. Maps of other land contributing to mangrove transfer out percent (TOP) in Kakadu NP (**D**) and RUMAKI (**E**) from 2000 to 2020 (e.g., a total of 100 ha of mangroves are transferred out, of which 50 ha are other wetlands, the corresponding TOP = 50%).

3.3.5. Analyzing the Dynamics of Mangrove Habitats in RUMAKI

Mangrove habitats within RUMAKI were dispersed north to south along the mafia channel, degraded mainly to cropland, other wetlands, and grassland (Figures 7 and 13). The extent of degradation was concentrated more in 2010–2020 compared to that in 2000–2010 (Figure 14). Mangrove habitat area decreased by 1255 ha with a 10-year decrease of 2.5% from 2000 to 2010, and by 757 ha with a 10-year decrease of 1.5% from 2010 to 2020 (Figure 3). It was still relatively higher compared to other regions except Kakadu NP, although the decrease proportion of mangrove habitats has slowed down in the past ten years.

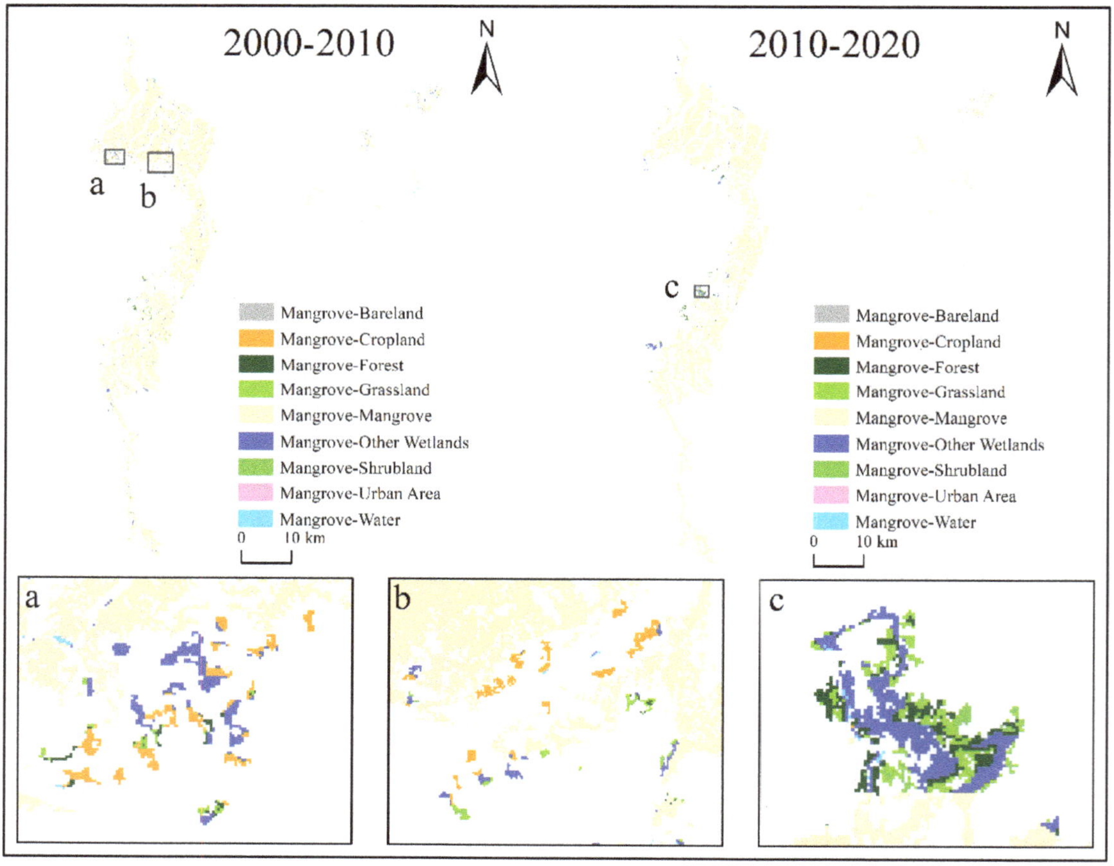

Figure 14. Maps of spatial–temporal distribution of mangrove degradation in RUMAKI from 2000 to 2020. (**a**) first degradation detail picture; (**b**) second degradation detail picture; (**c**) third degradation detail picture.

3.3.6. Comparison of Mangrove Habitats Changes from 2000 to 2020 in Five Sites

Dongzhaigang (China) is the only reserve where the mangrove area expanded between 2000 and 2020 at a rate of 24.6%. Several other reserves experienced different degrees of mangrove habitat loss. RUMAKI (Tanzania) has the most significant loss among the five reserves, but the loss proportion is at most 4.0%. The mangrove habitats in Sundarban (India) have experienced a shift from rapid degradation to positive growth. The degradation rate of mangrove wetlands in Kakadu NP (Australia) has remained the lowest (Figure 15).

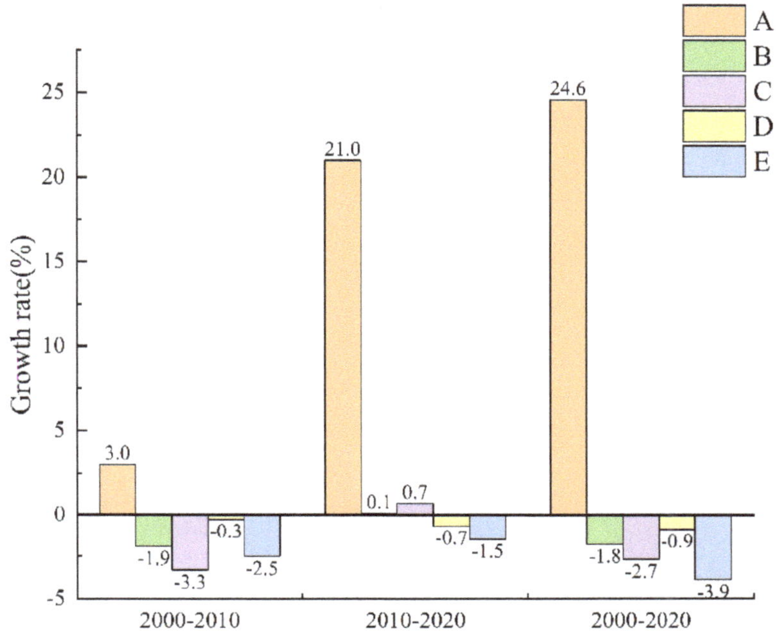

Figure 15. Comparison of the dynamics of mangrove habitats in five sites from 2000 to 2020. (A): Dongzhaigang; (B): Sembilang NP; (C): Sundarban; (D): Kakadu NP; (E): RUMAKI.

4. Discussion

4.1. Driving Forces of Mangrove Degradation in the Five Protection Areas

The driving forces of mangrove degradation are mainly divided into climatic factors (precipitation, temperature, sea level rise, natural disasters, etc.) and human activities (agricultural encroachment, urban expansion, aquaculture, etc.) [13–15]. Dongzhaigang is close to the urban area, and urban expansion and aquaculture were the main factors of mangrove wetland degradation. The Chinese government increased the protection of mangrove habitats and implemented mangrove restoration projects after 2000; thus, the mangrove wetlands were effectively restored [2]. The Indonesian government formulated and implemented a policy to vigorously develop the aquaculture industry in order to achieve economic development after the 1980s, which led to aquaculture being the most critical factor in the degradation of mangrove forests in Semmelang NP [40,41]. The Sundarban mangroves is the largest contiguous mangrove forest in the world, and the northern part of the reserve is adjacent to farmers' settlements. Farmers entering the reserve might develop the mangrove wetland into farmland for their livelihood, which becomes the most significant human activity factor for mangrove degradation [42,43]. Kakadu NP is the only mangrove reserve in a developed country with more stringent legislation and regulations for mangrove protection. However, the mangrove wetlands in reserve are mainly exposed to natural disasters because the northern part of Australia is highly influenced by climatic factors [11,43]. The laws on mangrove habitats protection made in Tanzania cannot be effectively implemented, and mangroves are used as a source of timber and woodland destroyed by local residents. On the one hand, the local population is considered by the government to be the destroyer of mangroves; on the other hand, the impoverished population has to rely on mangrove raiding for their livelihood. Poverty-induced agricultural encroachment is a major factor in mangrove degradation in this study area [44].

4.2. Comparison with Other Studies

The State of the World's Mangroves 2022 [45] notes a global mangrove loss of 5245 km^2 from 1996 to 2020, with an average annual decline rate of 0.34%; a global mangrove loss of 327 km^2 from 1996 to 2010, with an average annual decline rate of 0.21%; and a global mangrove loss of 66 km^2 from 2010 to 2020, with an average annual decline rate of 0.02%. The results are consistent with our findings, indicating that the rate of mangrove degradation has slowed down significantly after 2010. Dongzhaigang lost 62% of its mangrove area from 1973 to 2000, but achieved rapid growth after 2000, and the mangrove area recovered to the 1980 level by 2020 [2]. Correspondingly, our results indicate that the mangrove habitats of Dongzhai lost more than 10% from 1990 to 2000 and achieved a 24% increase from 2000 to 2020, a rate that is much higher than that of the entire East Asia region (Table 6). The State of the World's Mangroves 2022 [45] indicates that aquaculture has made Southeast Asia the fastest-degrading region for mangrove wetlands (1996–2020: 4.8% degradation proportion). However, our results show that RUMAKI in East Africa has the highest rate of mangrove loss among the five typical reserves, which is a side indication that mangroves in the Sembilang NP is better protected. The State of the World's Mangroves 2022 shows an overall degradation rate of 3.0% (1996–2020) in South Asia, and our results show a degradation rate of 2.7% (2000–2020) for mangrove habitats in the Sundarbans mangroves in South Asia, which is consistent with each other and the changes in degradation rates are uniform: the area of mangrove habitats started to increase after 2010. Unfortunately, the results are contrary to those of Giri et al. [42]. This may be an error in the mangrove extent interpretation process, and the authors consider mangrove degradation in protected areas to be insignificant and a case of dense population–nature symbiosis. The results of Bunting et al. [10] show that the average annual rate of loss of mangrove habitats in Australia is 0.23% (1996–2010) and 0.1% (2010–2020), while our results showed that the rate of mangrove loss within Kakadu NPark was 0.03% (2000–2010) with an increasing trend after 2010, suggesting that mangrove habitats within Kakadu NP is better protected.

Table 6. Changes in mangrove area by region (Leal & Spalding, [45]).

Region	Mangrove Area (ha)			Proportion of Change (%)		
	1996	2010	2020	1996–2010	2010–2020	1996–2020
Eastern Asia	257.2	223.6	227.7	−13.1	1.8	−11.5
Southern Asia	9960.7	9710.4	9661.1	−2.5	−0.5	−3.0
Southeastern Asia	50,678.8	48,440.9	48,222.3	−4.4	−0.5	−4.8
Australia & New Zealand	10,945.0	10,562.5	10,466.9	−3.5	−0.9	−4.4
Eastern Africa	7883.3	7688.6	7610.0	−2.5	−1.0	−3.5

4.3. Limitations

The tide levels could have an impact on the interpreted mangrove extent. Although we obtained synthetic NDVI and band information using the annual maximum and annual median methods, we were still unable to eliminate the inaccuracy caused by the high tide level conaealing the seaward mangroves. In addition, a comparison of conservation effectiveness cannot be made based solely on an analysis of the changing characteristics of mangrove ecosystems within protected areas. The conservation effect can be reflected better if comparing the change of mangrove wetlands inside and outside the protected area. Hence, it will be our next research direction.

5. Conclusions

(1) The dynamics of the mangrove habitats. From 2000 to 2020, the mangrove area in Dongzhaigang increased by 414 ha, with an increased percent of 24.6%. The mangrove area in Sembilang NP, Sundarban, Kakadu NP, and RUMAKI decreased by 1652 ha, 16,091 ha, 83 ha, and 2012 ha, with a decrease of 1.8%, 2.7%, 0.9%, and 3.9%, respectively.

(2) Direct driving factors of mangrove habitats. Other types of wetlands play a key role in degradating the mangrove wetlands in all five protected areas. The increase of mangroves in Dongzhaigang is a result of the local government-supported conservation scheme for returning ponds to forests and plantations. Agricultural encroachment is the other factor in reducing of mangrove areas in Sundarbans. Natural disaster is another factor to degrade the mangrove wetland in RUMAKI.

(3) The rate of mangrove degradation has slowed dramatically over the past two decades, based on five sites, which are generally consistent with those of other researchers. Among the five typical reserves, the mangrove area in Dongzhaigang is the only one to have expanded. The mangrove area in the other four locations has fallen to a lesser level, with the greatest decrease not surpassing 4%.

Recommendations

Our research did not cover mangroves in the Americas, and we hope to scale up to the global scale and encrypt more representative mangrove sites in the future.

Author Contributions: Conceptualization, D.Z., Y.K. and S.K.; methodology, C.D.; formal analysis, C.D.; data curation, C.D.; writing—original draft preparation, C.D.; writing—review and editing, D.Z. and Y.K. All authors have read and agreed to the published version of the manuscript.

Funding: This research was funded by UNESCO funding sources, grant number 4500447323.

Institutional Review Board Statement: Not applicable.

Data Availability Statement: Not applicable.

Acknowledgments: We would like to thank to associate researcher Mingming Jia (Northeast Institute of Geography and Agroecology, CAS), for providing some land-use data, and Yin Xiaolan, Jiang Yuxin and Ju Xiaoxiao for processing some data.

Conflicts of Interest: The authors declare no conflict of interest.

References

1. Barbier, E.B.; Hacker, S.D.; Kennedy, C.; Koch, E.W.; Stier, A.C.; Silliman, B.R. The value of estuarine and coastal ecosystem services. *Ecol. Monogr.* **2011**, *81*, 169–193. [CrossRef]
2. Jia, M.; Wang, Z.; Mao, D.; Huang, C.; Lu, C. Spatial-temporal changes of China's mangrove forests over the past 50 years: An analysis towards the Sustainable Development Goals (SDGs). *Chin. Sci. Bull.* **2021**, *66*, 3886–3901. [CrossRef]
3. Alongi, D.M. Present state and future of the world's mangrove forests. *Environ. Conserv.* **2002**, *29*, 331–349. [CrossRef]
4. Alongi, D.M. The Impact of Climate Change on Mangrove Forests. *Curr. Clim. Chang. Rep.* **2015**, *1*, 30–39. [CrossRef]
5. Bosire, J.O.; Kaino, J.J.; Olagoke, A.O.; Mwihaki, L.M.; Ogendi, G.M.; Kairo, J.G.; Berger, U.; Macharia, D. Mangroves in peril: Unprecedented degradation rates of peri-urban mangroves in Kenya. *Biogeosciences* **2014**, *11*, 2623–2634. [CrossRef]
6. Amezaga, J.; Bathurst, J.; Iroumé, A.; Jones, J.; Kotru, R.; Bhatta, L.D.; Springgay, E.; Katila, P.; Colfer, C.J.P.; de Jong, W.; et al. SDG 6: Clean Water and Sanitation—Forest-Related Targets and Their Impacts on Forests and People. In *Sustainable Development Goals: Their Impacts on Forests and People*; Cambridge University Press: Cambridge, UK, 2019; pp. 178–205. [CrossRef]
7. Shah, H.; Ramesh, R. Development-aligned mangrove conservation strategy for enhanced blue economy: A successful model from Gujarat, India. *Estuar. Coast. Shelf Sci.* **2022**, *274*, 107929. [CrossRef]
8. Damastuti, E.; de Groot, R.; Debrot, A.O.; Silvius, M.J. Effectiveness of community-based mangrove management for biodiversity conservation: A case study from Central Java, Indonesia. *Trees For. People* **2022**, *7*, 100202. [CrossRef]
9. Goldberg, L.; Lagomasino, D.; Thomas, N.; Fatoyinbo, T. Global declines in human-driven mangrove loss. *Glob. Chang. Biol.* **2020**, *26*, 5844–5855. [CrossRef]
10. Bunting, P.; Rosenqvist, A.; Hilarides, L.; Lucas, R.M.; Thomas, N.; Tadono, T.; Worthington, T.A.; Spalding, M.; Murray, N.J.; Rebelo, L.-M. Global Mangrove Extent Change 1996–2020: Global Mangrove Watch Version 3.0. *Remote Sens.* **2022**, *14*, 3657. [CrossRef]

11. Lymburner, L.; Bunting, P.; Lucas, R.; Scarth, P.; Alam, I.; Phillips, C.; Ticehurst, C.; Held, A. Mapping the multi-decadal mangrove dynamics of the Australian coastline. *Remote Sens. Environ.* **2019**, *238*, 111185. [CrossRef]
12. Richards, D.R.; Friess, D.A. Rates and drivers of mangrove deforestation in Southeast Asia, 2000–2012. *Proc. Natl. Acad. Sci. USA* **2016**, *113*, 344–349. [CrossRef]
13. Heumann, B.W. Satellite remote sensing of mangrove forests: Recent advances and future opportunities. *Prog. Phys. Geogr. Earth Environ.* **2011**, *35*, 87–108. [CrossRef]
14. Islam, M.M. Shamsuddoha Coastal and marine conservation strategy for Bangladesh in the context of achieving blue growth and sustainable development goals (SDGs). *Environ. Sci. Policy* **2018**, *87*, 45–54. [CrossRef]
15. Savari, M.; Damaneh, H.E. Factors involved in the degradation of mangrove forests in Iran: A mixed study for the management of this ecosystem. *J. Nat. Conserv.* **2022**, *66*, 126153. [CrossRef]
16. Friess, D.A.; Rogers, K.; Lovelock, C.E.; Krauss, K.W.; Hamilton, S.E.; Lee, S.Y.; Lucas, R.; Primavera, J.; Rajkaran, A.; Shi, S. The state of the world's mangrove forests: Past, present, and future. *Annu. Rev. Environ. Resour.* **2019**, *44*, 89–115. [CrossRef]
17. Omo-Irabor, O.O.; Olobaniyi, S.B.; Akunna, J.; Venus, V.; Maina, J.M.; Paradzayi, C. Mangrove vulnerability modelling in parts of Western Niger Delta, Nigeria using satellite images, GIS techniques and Spatial Multi-Criteria Analysis (SMCA). *Environ. Monit. Assess.* **2010**, *178*, 39–51. [CrossRef]
18. Mendoza-González, G.; Martínez, M.; Lithgow, D.; Pérez-Maqueo, O.; Simonin, P. Land use change and its effects on the value of ecosystem services along the coast of the Gulf of Mexico. *Ecol. Econ.* **2012**, *82*, 23–32. [CrossRef]
19. Nguyen, H.-H.; McAlpine, C.; Pullar, D.; Johansen, K.; Duke, N.C. The relationship of spatial–temporal changes in fringe mangrove extent and adjacent land-use: Case study of Kien Giang coast, Vietnam. *Ocean Coast. Manag.* **2013**, *76*, 12–22. [CrossRef]
20. Ishtiaque, A.; Myint, S.W.; Wang, C. Examining the ecosystem health and sustainability of the world's largest mangrove forest using multi-temporal MODIS products. *Sci. Total Environ.* **2016**, *569*, 1241–1254. [CrossRef]
21. Hamilton, S.E.; Casey, D. Creation of a high spatio-temporal resolution global database of continuous mangrove forest cover for the 21st century (CGMFC-21). *Glob. Ecol. Biogeogr.* **2016**, *25*, 729–738. [CrossRef]
22. Giri, C.; Ochieng, E.; Tieszen, L.L.; Zhu, Z.; Singh, A.; Loveland, T.; Masek, J.; Duke, N. Status and distribution of mangrove forests of the world using earth observation satellite data. *Glob. Ecol. Biogeogr.* **2011**, *20*, 154–159. [CrossRef]
23. Kovacs, J.M.; Wang, J.; Blanco-Correa, M. Mapping Disturbances in a Mangrove Forest Using Multi-Date Landsat TM Imagery. *Environ. Manag.* **2001**, *27*, 763–776. [CrossRef] [PubMed]
24. Li, W.; El-Askary, H.; Qurban, M.A.; Li, J.; ManiKandan, K.; Piechota, T. Using multi-indices approach to quantify mangrove changes over the Western Arabian Gulf along Saudi Arabia coast. *Ecol. Indic.* **2019**, *102*, 734–745. [CrossRef]
25. Spalding; Mark, D.; Leal; Maricé (Eds.) *The State of the World's Mangroves 2021*; Global Mangrove Alliance, 2021. Available online: https://www.mangrovealliance.org/wp-content/uploads/2021/07/The-State-of-the-Worlds-Mangroves-2021-FINAL.pdf (accessed on 14 January 2022).
26. Silvius, M.J.; Noor, Y.R.; Lubis, I.R.; Giesen, W.; Rais, D. *Sembilang National Park: Mangrove Reserves of Indonesia*; The Wetland Book; Springer: Dordrecht, The Netherlands, 2018. [CrossRef]
27. Rahman, M.R.; Asaduzzaman, M. Ecology of sundarban, bangladesh. *J. Sci. Found.* **2010**, *8*, 35–47. [CrossRef]
28. Gopal, B.; Chauhan, M. The transboundary Sundarbans mangroves (India and Bangladesh). In *The Wetland Book: II: Distribution, Description, and Conservation*; Springer: Dordrecht, The Netherlands, 2018; pp. 1733–1742.
29. Lucas, R.; Finlayson, C.M.; Bartolo, R.; Rogers, K.; Mitchell, A.; Woodroffe, C.D.; Asbridge, E.; Ens, E. Historical perspectives on the mangroves of Kakadu National Park. *Mar. Freshw. Res.* **2018**, *69*, 1047. [CrossRef]
30. East, D.M. *Information Sheet on Ramsar Wetlands (RIS)*; Joint Nature Conservation Committee: Peterborough, UK, 2004.
31. Herbeck, L.S.; Krumme, U.; Andersen, T.J.; Jennerjahn, T.C. Decadal trends in mangrove and pond aquaculture cover on Hainan (China) since 1966: Mangrove loss, fragmentation and associated biogeochemical changes. *Estuar. Coast. Shelf Sci.* **2019**, *233*, 106531. [CrossRef]
32. Zhang, R.; Jia, M.; Wang, Z.; Zhou, Y.; Mao, D.; Ren, C.; Zhao, C.; Liu, X. Tracking annual dynamics of mangrove forests in mangrove National Nature Reserves of China based on time series Sentinel-2 imagery during 2016–2020. *Int. J. Appl. Earth Obs. Geoinf.* **2022**, *112*, 102918. [CrossRef]
33. Ntiyakunze, M.S.; Stage, J. Forest dependence in Tanzania: Analysis of the determinants of perceived forest de-pendence. *Trees For. People* **2022**, *8*, 100277. [CrossRef]
34. Duke, N.C.; Kovacs, J.M.; Griffiths, A.D.; Preece, L.; Hill, D.J.; Van Oosterzee, P.; Mackenzie, J.; Morning, H.S.; Burrows, D. Large-scale dieback of mangroves in Australia's Gulf of Carpentaria: A severe ecosystem response, coincidental with an unusually extreme weather event. *Mar. Freshw. Res.* **2017**, *68*, 1816–1829. [CrossRef]
35. Ferreira, M.A.; Andrade, F.; Bandeira, S.O.; Cardoso, P.; Mendes, R.N.; Paula, J. Analysis of cover change (1995–2005) of Tanzania/Mozambique trans-boundary mangroves using Landsat imagery. *Aquat. Conserv. Mar. Freshw. Ecosyst.* **2009**, *19*, S38–S45. [CrossRef]
36. Gayo, L. Local community perception on the State Governance of mangroves in Western Indian coast of Kinondoni and Bagamoyo, Tanzania. *Glob. Ecol. Conserv.* **2022**, *39*, e02287. [CrossRef]
37. Hu, Y.; Dong, Y.; Batunacun. An automatic approach for land-change detection and land updates based on integrated NDVI timing analysis and the CVAPS method with GEE support. *ISPRS J. Photogramm. Remote Sens.* **2018**, *146*, 347–359. [CrossRef]

38. Xiao, H.; Su, F.; Fu, D.; Yu, H.; Ju, C.; Pan, T.; Kang, L. *10-M Global Mangrove Classification Products of 2018-2020 Based on Big Data (V1)*; Science Data Bank: Beijing, China, 2021. Available online: https://datapid.cn/31253.11.sciencedb.01019 (accessed on 16 January 2022).
39. Chen, J.; Ban, Y.; Li, S. China: Open access to Earth land-cover map. *Nature* **2014**, *514*, 434. [CrossRef]
40. Ilman, M.; Dargusch, P.; Dart, P. Onrizal A historical analysis of the drivers of loss and degradation of Indonesia's mangroves. *Land Use Policy* **2016**, *54*, 448–459. [CrossRef]
41. Hauser, L.T.; Vu, G.N.; Nguyen, B.A.; Dade, E.; Nguyen, H.M.; Nguyen, T.T.Q.; Le, T.Q.; Vu, L.H.; Tong, A.T.H.; Pham, H.V. Uncovering the spatio-temporal dynamics of land cover change and fragmentation of mangroves in the Ca Mau peninsula, Vietnam using multi-temporal SPOT satellite imagery (2004–2013). *Appl. Geogr.* **2017**, *86*, 197–207. [CrossRef]
42. Giri, C.; Pengra, B.; Zhu, Z.; Singh, A.; Tieszen, L.L. Monitoring mangrove forest dynamics of the Sundarbans in Bangladesh and India using multi-temporal satellite data from 1973 to 2000. *Estuar. Coast. Shelf Sci.* **2007**, *73*, 91–100. [CrossRef]
43. Jayanthi, M.; Thirumurthy, S.; Nagaraj, G.; Muralidhar, M.; Ravichandran, P. Spatial and temporal changes in mangrove cover across the protected and unprotected forests of India. *Estuar. Coast. Shelf Sci.* **2018**, *213*, 81–91. [CrossRef]
44. Monga, E.; Mangora, M.M.; Trettin, C.C. Impact of mangrove planting on forest biomass carbon and other structural attributes in the Rufiji Delta, Tanzania. *Glob. Ecol. Conserv.* **2022**, *35*, e02100. [CrossRef]
45. Leal; Maricé; Spalding; Mark, D. (Eds.) *The State of the World's Mangroves 2022*; Global Mangrove Alliance, 2022. Available online: https://www.mangrovealliance.org/wp-content/uploads/2022/09/The-State-of-the-Worlds-Mangroves-Report_2022.pdf (accessed on 30 October 2022).

diversity

Article

Assessment of Changes in Key Ecosystem Factors and Water Conservation with Remote Sensing in the Zoige

Peng Hou [1], Jun Zhai [1,*], Dian-Dian Jin [1,*], Yan Zhou [2], Yan Chen [1] and Hai-Feng Gao [1]

[1] Satellite Application Center for Ecology and Environment, Ministry of Ecology and Environment of People's Republic of China, Beijing 100094, China; houpcy@163.com (P.H.); chenyan30033@163.com (Y.C.); gaohf03@hotmail.com (H.-F.G.)

[2] Ministry of Education Key Laboratory of Geospatial Technology for the Middle and Lower Yellow River Regions, Henan University, Kaifeng 475004, China; yanzhou152@126.com

* Correspondence: zhaij@lreis.ac.cn (J.Z.); jin_diandian@163.com (D.-D.J.)

Abstract: As the largest alpine peat swamp wetland distribution area in the world, the Zoige has important ecological functions, including water conservation and biodiversity maintenance. In the past 20 years, the regional ecological protection and restoration measures continuously strengthened under the leadership of the local government have led to gradual improvements in the ecological environment of the region. In this study, multisource satellite remote-sensing image data were used to carry out quantitative monitoring and assessment of the main ecological elements (vegetation and water), as well as the regional leading ecosystem service function in the Zoige. Combined with local ecological protection management policies and measures, we analyzed the characteristics and effectiveness of ecological protection. We compared the ecosystem change trends of the Zoige reserve and the county, from 2001 to 2020, and found that the fractional vegetation cover (FVC) of Zoige county has increased at a rate of 0.25%/year. The growth rate was highest between 2015 and 2020, and the growth rate of FVC in the Zoige Wetland National Nature Reserve is approximately 1.89-fold that of the whole county. The water area also shows similar variation characteristics. On the whole, the water conservation capacity of the Zoige showed a significant increase from 2001 to 2020. We used high-resolution satellite remote-sensing images to capture the details of land use changes brought about by local ecological protection policies and measures, and together with macroecological indicators, we reflected on the effectiveness of regional ecological protection measures. We observed that the ecological effects of nature reserves are more direct and rapid, and the amount of water conservation within the nature reserve is about 1×10^4 m^3/km^2 higher than that of the surrounding grasslands. Satellite remote-sensing images can not only capture the multiscale change information of ecological indicators, such as vegetation and water, in a timely manner, but can also help us to identify the effectiveness of conservation measures by distinguishing and analyzing the causes of these changes.

Keywords: Zoige wetland; nature reserve; fractional vegetation cover; water conservation; ecological conservation

Citation: Hou, P.; Zhai, J.; Jin, D.-D.; Zhou, Y.; Chen, Y.; Gao, H.-F. Assessment of Changes in Key Ecosystem Factors and Water Conservation with Remote Sensing in the Zoige. *Diversity* **2022**, *14*, 552. https://doi.org/10.3390/d14070552

Academic Editor: Corrado Battisti

Received: 5 June 2022
Accepted: 6 July 2022
Published: 9 July 2022

Publisher's Note: MDPI stays neutral with regard to jurisdictional claims in published maps and institutional affiliations.

1. Introduction

Wetlands represent the transitional zone between terrestrial and aquatic ecosystems, and possess a unique ecosystem and functional characteristics [1,2]. At present, changes in vegetation and water bodies are basic factors that reflect the status of and changes in wetland ecosystems, and are key indicators of wetland shrinkage and restoration monitoring based on satellite remote sensing [3,4].

The Zoige wetland is located on the eastern edge of the Qinghai Tibet Plateau and is listed as an "International Important Wetland" by UNESCO [5,6]. It is an important water conservation area in the upper reaches of the Yellow and Yangtze Rivers, with the largest plateau peat swamp in the world [7]. It is a key area for biodiversity protection both in China and worldwide [8]; it is also an important habitat for wildlife species, such as the black-necked cranes and white-tailed sea eagles, and is highly sensitive to climate change [9,10]. The Zoige Wetland National Nature Reserve was established in 1998 and covers an area of 1665.70 km^2. The alpine swamp wetland ecosystems, black-necked cranes, and other rare animals are especially targeted for protective measures.

Owing to the joint impact of climate change and human activities, the Zoige wetland has been shrinking since the 1950s, resulting in the decline in the groundwater level, land degradation, reduction in biodiversity, and increased carbon emissions. Relevant research has mainly focused on regional climate change analysis [11–13], wetland area change monitoring [14], grassland degradation, land desertification [15–17], soil organic carbon change [18,19], and ecosystem services and health [20,21]. Most of these studies were published before 2016, and the conclusions were mainly based on the continuous degradation of the ecological environment. There is still a lack of analysis on the effectiveness of conservation measures.

Since the 1970s, China has successively launched a series of ecological protection and restoration projects, including the Natural Forest Protect Projection, the Grain for Green program, and the Returning Rangeland to Grassland, which have increased regional ecological functions [22,23]. Since 2003, ecological protection and restoration projects, such as grassland restoration and desertification land management, have significantly increased in the Zoige area [24–26], and a series of restoration and rectification works of nature reserves have been carried out in succession. In the past 30 years, the growth rate of regional desertification land has significantly slowed [17,25]. The project of returning grazing land to grassland in the county has had significant benefits: the vegetation height in the project area has elevated by 69.1% on average, compared with that in the nonproject area [26].

From the perspective of satellite remote sensing, we used the main elements of the ecosystem (vegetation, water bodies, etc.) and the service function of water conservation, as indicators to analyze the changing characteristics of the ecological environment in Zoige county and nature reserves [27]. Simultaneously, combined with high-resolution remote-sensing images and field investigations, we attempted to link the direct image evidence of ecological restoration with the above-mentioned remote-sensing indicators of ecological change, thus further elaborating the obvious ecological benefits brought about by ecological protection and supervision.

2. Materials and Methods

2.1. Study Area

Zoige county (102°08–103°39′ E, 32°56–34° N) is located in the northern part of the Northwest Sichuan Plateau, which is the core area of the Zoige wetland. The topography is low in the middle and high in the surrounding areas. It belongs to the Aba Tibetan and Qiang Autonomous Prefecture of Sichuan province, and has a total area of 1.04×10^4 km^2 and an average altitude of 3471 m (Figure 1). The regional altitude difference is largely attributed to the depth of the river. The study area has a plateau cold temperate humid monsoon climate, with an annual average temperature of 1.1 °C, and annual precipitation of 650 mm [28]. However, in recent years, there has been an obvious trend of warming and drying [29].

2.2. Vegetation Data Acquisition and Calculation Method

Fractional vegetation cover (FVC) reflects the vegetation coverage of a certain area, and is an important indicator for measuring the quality of vegetation and ecosystem change [30]. Based on the remote-sensing cloud platform of Google Earth Engine and the Moderate Resolution Imaging Spectroradiometer (MODIS) 13Q1 normalized difference vegetation index (NDVI) product (https://ladsweb.modaps.eosdis.nasa.gov, (accessed on 5 June 2021)), we used the pixel dichotomy model to calculate the vegetation coverage [31,32]. The MODIS

13Q1 NDVI product has a temporal resolution of 16 days with 23 cycles of data per year. In this study, FVC data for 23 periods were generated annually, and then aggregated into the average annual FVC. The specific formula of FVC is as follows:

$$FVC = \frac{NDVI - NDVI_s}{NDVI_v - NDVI_s} \tag{1}$$

where $NDVI_v$ and $NDVI_s$ are the NDVI pixel values of pure vegetation and bare land, respectively. With the NDVI value statistics, we determined the cumulative frequency of 1% $NDVI_s$ and cumulative frequency of 99% $NDVI_v$ [33].

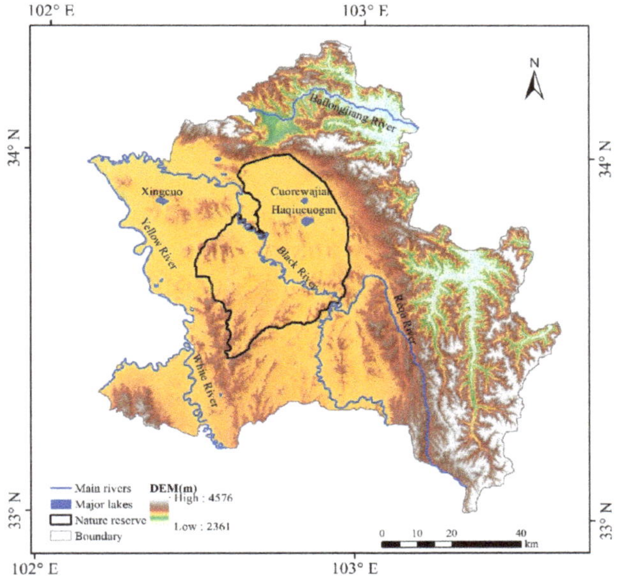

Figure 1. Topographical map of Zoige County.

2.3. Water Data and Acquisition Methods

Water information in this study mainly included open water surfaces, such as rivers and lakes. Using information based on the Google Earth Engine platform, long time-series Landsat remote-sensing image data (https://ladsweb.modaps.eosdis.nasa.gov, (accessed on 24 May 2021)) corresponding to the study area and study period were directly called through the database, from which the vegetation, water, and dry naked index were calculated [34]. Each index is calculated as follows:

$$NDVI = \frac{\rho_{NIR} - \rho_{red}}{\rho_{NIR} + \rho_{red}} \tag{2}$$

$$MNDWI = \frac{\rho_{Green} - \rho_{SWIR}}{\rho_{Green} + \rho_{SWIR}} \tag{3}$$

$$AWEI = \rho_{Blue} + 2.5 \times \rho_{Green} - 1.5 \times (\rho_{NIR} + \rho_{SWIR1}) - 0.25 \times \rho_{SWIR2} \tag{4}$$

$$DBSI = \frac{\rho_{SWIR1} - \rho_{Green}}{\rho_{SWIRI1} + \rho_{Green}} - NDVI \tag{5}$$

where *NDVI* is the normalized vegetation index, *MNDWI* is the improved normalized difference water body index, *AWEI* is the automatic water body extraction index, *DBSI* is the dry bare index, ρ_{Blue} is the blue band, ρ_{Green} is the green band, ρ_{NIR} is the near-infrared band, ρ_{red} is the red band, ρ_{SWIR1} is the short-wave infrared band 1, and ρ_{SWIR2} is the short-wave infrared band 2.

Secondly, the regional water body area was extracted by combining the remote-sensing water body index and the basin relative elevation setting, and the non-water-body information was filtered by combining the vegetation index and drought exposure index [34]. Among them, 30 m spatial resolution digital elevation model (DEM) products came from the geospatial data cloud platform (http://www.gsclod.cn, (accessed on 24 May 2021)).

Finally, the annual data product of flood frequency (the ratio of the number of times a place is submerged to the number of effective observation) was constructed using annual multiperiod water body data to determine the flood frequency [34,35]. Then, the human–computer interaction method was used to determine the data threshold of flooding frequency, which could maximize the avoidance of seasonal water and noise (the threshold in the study is 70%), and obtain the stable water distribution of the study area in the current year.

2.4. Ecosystem Function Assessment

Water conservation means that ecosystems (such as forests, grassland, etc.) intercept infiltrate, accumulate precipitation through their unique structures and water interactions, and control water flow and the water cycle through evapotranspiration. We used the water balance equation to calculate water conservation capacity [36]:

$$TQ = \sum_{i=1}^{j} (P_i - R_i - ET_i) \times A_i \times 10^3 \quad (6)$$

where TQ is water conservation (m^3), P_i is rainfall (mm), R_i is surface runoff (mm), ET_i is evapotranspiration (mm), A_i is the type i ecosystem area (km^2), i is the type i ecosystem in the study area, and j is the number of ecosystem types in the study area.

Rainfall data were obtained from a TerraClimate dataset with a spatial resolution of 1 km (https://climate.northwestknowledge.net/TERRACLIMATE/index, (accessed on 10 June 2021)). Evapotranspiration data were the result of the MODIS MOD16A3 product, with a spatial resolution of 1 km (https://ladsweb.modaps.eosdis.nasa.gov, (accessed on 24 June 2021))). The surface runoff was obtained by multiplying the rainfall by the surface runoff coefficient, and the surface runoff coefficient was obtained from the literature [36]. The ecosystem area was derived from data comprising remote-sensing surveys and assessments of the national ecological status, which mainly included forest, shrub, grassland, garden, and wetland types [37].

To avoid the influence of the interannual precipitation fluctuations, based on the calculation of annual water conservation, we took the average every five years to represent the regional water conservation capacity in different periods. Then, we comprehensively analyzed the change characteristics of water conservation function over the past 20 years.

2.5. Change Trend Analysis Method

For trend analysis, we used the Theil–Sen median method, a trend analysis method for nonparametric statistics. The calculation method of the Sen trend degree is as follows [38]:

$$\beta = \text{mean}\left(\frac{x_j - x_i}{j - i}\right), \forall j > i \quad (7)$$

where x_j and x_i are time series data, a β greater than 0 indicates an increasing trend in the time series, and a β of less than 0 indicates a decreasing trend in the time series.

The Mann–Kendall trend test is a nonparametric statistical test method for which measured values do not need to follow a normal distribution. The trend is not required to be linear, and is not affected by missing values and outliers [39,40]. The statistical test method used is as follows:

$$Z_c = \begin{cases} \frac{S-1}{\sqrt{Var(s)}} & S > 0 \\ 0 & S = 0 \\ \frac{S+1}{\sqrt{Var(s)}} & S < 0 \end{cases} \tag{8}$$

$$S = \sum_{i=1}^{n-1} \sum_{j=i+1}^{n} sgn(x_j - x_i) \tag{9}$$

$$sgn(x_j - x_i) = \begin{cases} 1 & x_j - x_i > 0 \\ 0 & x_j - x_i = 0 \\ -1 & x_j - x_i = 0 \end{cases} \tag{10}$$

$$Var(s) = \frac{n(n-1)(2n+5)}{8} \tag{11}$$

where *sgn* is a symbolic function, x_j and x_i are sequential data sets, and n is the length of the data samples. Z_c followed a standard normal distribution. If $Z_c > Z_{(1-\alpha)/2}$, there is a significant trend change. $Z_{(1-\alpha)/2}$ is the corresponding value of the standard normal function distribution table at confidence level α. The confidence level α was set at 0.05.

3. Results

3.1. Vegetation Change Characteristics

The spatial distribution of fractional vegetation cover in Zoige county is shown in Figure 2. The FVC of the county in 2020 was 57.56%. Areas with large FVC (>70%) were mainly distributed in the eastern Zoige forest ecosystem, while areas of low FVC (<20%) were distributed in the northwest of the county where sandy, urban, and rural land were concentrated.

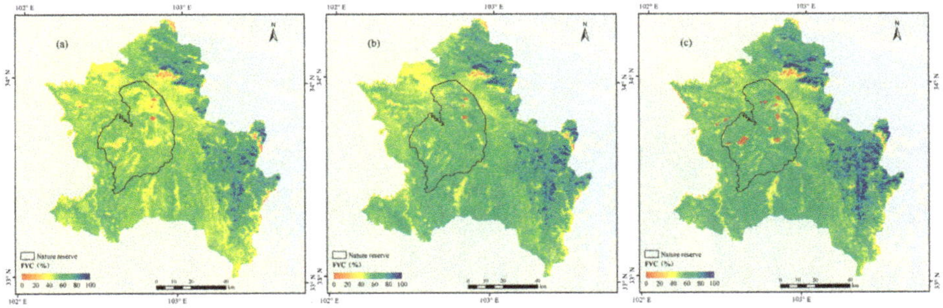

Figure 2. Spatial distribution of fractional vegetation cover (FVC) in Zoige in (**a**) 2001, (**b**) 2010, and (**c**) 2020.

In 2020, the average FVC of the nature reserve was 53.22%. Because the terrain was relatively flat, the grassland in the area was evenly distributed. Although the average vegetation coverage in the nature reserve was 4% lower than that of the whole county, the vegetation coverage of grassland ecosystems inside the reserve was 1% higher than that outside of the reserve.

Between 2001 and 2020, the FVC in Zoige demonstrated an overall upward trend, with an increase of 5.38% in 2020 compared with 2001. In particular, the rate of increase from 2015 to 2020 was significant compared with previous years (an increase of 1.35%). The average annual increase in FVC in this period was approximately 1.64-fold higher than that of the previous 15 years (2001–2015; Figure 3).

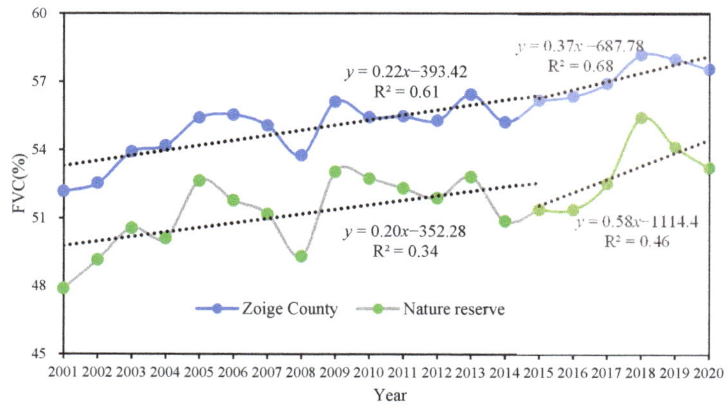

Figure 3. Interannual variation in fractional vegetation cover (FVC) in Zoige county and the nature reserve from 2001 to 2020.

It can be seen that the FVC of the Zoige Wetland National Nature Reserve also demonstrated an increasing trend, consistent with the changing trend of the whole county, and reflecting the overall improvement in vegetation growth. The mean FVC in the protected area was smaller than the average FVC of the whole county, mainly due to the existence of a forest ecosystem in the east of the county. However, from 2015 to 2020, the FVC in the reserve significantly improved, and was approximately 1.35-fold larger than the increase of the county value. This suggests stronger ecological resilience in the reserve.

From 2001 to 2020, the FVC in Zoige county increased at an annual rate of 0.25%; the FVC of 95.23% of the county increased, of which 63.6% of the area experienced significant increases (Figure 4). The rapidly increasing area was concentrated in the east, where forest and grassland met, and the nature reserve is located. Across the county, the area of FVC decrease accounted for 4.5%, mainly distributed in the northwest, where urban and rural land was concentrated. By comparing high-resolution remote-sensing images, some illegal facilities in the reserve were demolished (Figure 5).

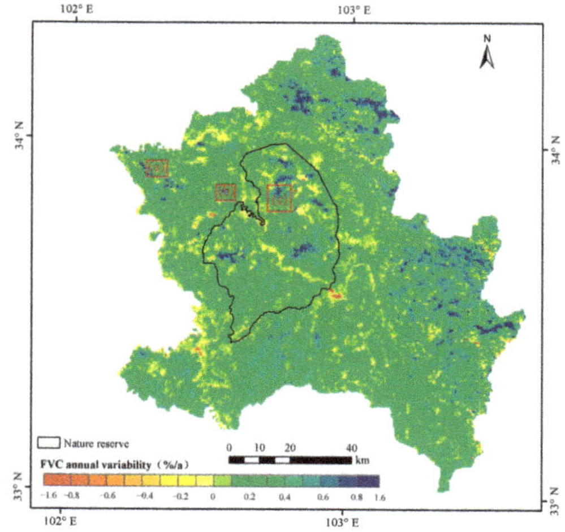

Figure 4. *Cont.*

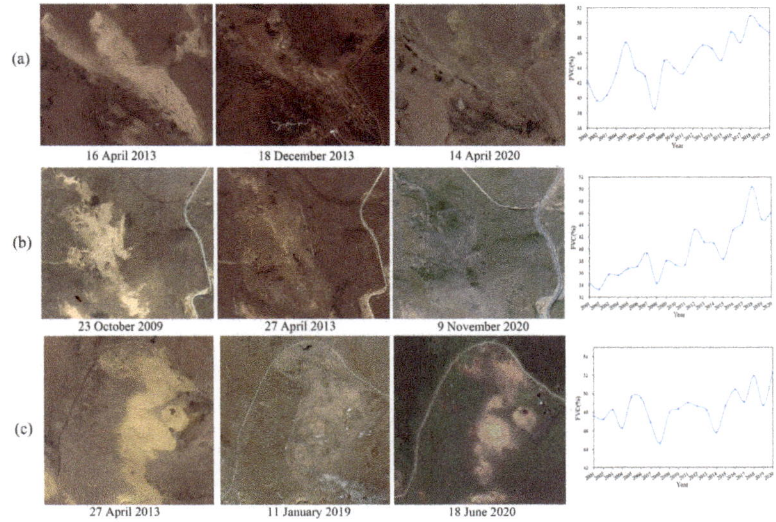

Figure 4. Changes in trend of fractional vegetation cover (FVC) in Zoige county from 2001 to 2020: (**a–c**) are the remote-sensing image comparison maps and vegetation coverage change time-series map of three typical ecological engineering implementation points. High-resolution remote-sensing images are from Google Earth and Chinese GF-1 satellites.

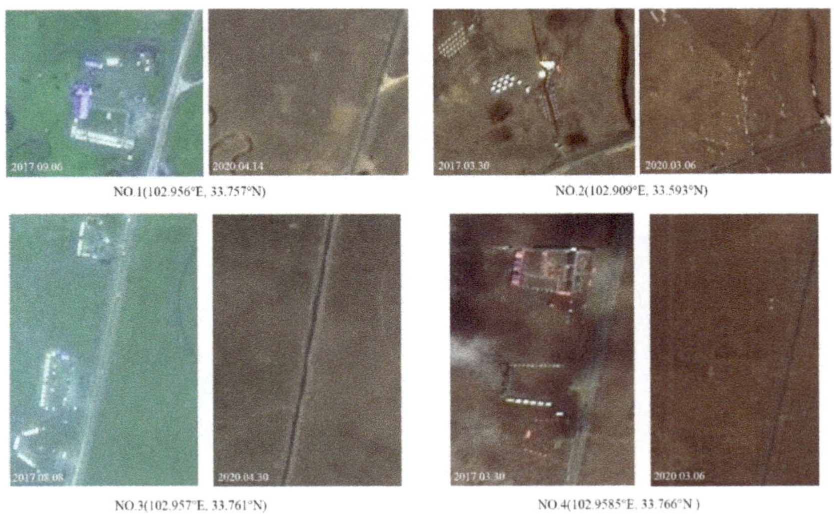

Figure 5. Comparison of remote-sensing images of demolition points of typical illegal facilities (high-resolution remote-sensing images from Chinese GF-1 and GF-2 satellites).

From 2015 to 2020, the rate of FVC increase escalated in both Zoige county and the reserve. The growth rate of the reserve (0.68% per year) was greater than that of the county as a whole (0.36%/a), reflecting the positive effects of targeted vegetation protection.

3.2. Water Body Change Characteristics

Remote-sensing data revealed that the water area of Zoige county in 2020 was 71.8 km^2, and mainly comprising the Yellow, Baihe, and Heihe Rivers in the west, along with the Huahu and Cuorewajian Lakes, and surrounding water bodies in the northwest.

The average water area of Zoige county from 2001 to 2020 was approximately 42.77 km^2. It exhibited an increasing trend. The water body area was lowest in 2002 (29.7 km^2). In 2020, the water body area was approximately 32.66 km^2 larger than that in 2001, an increase of 83.44% (Figure 6). By comparing high-resolution remote-sensing images, we noted that the area of some water bodies in Zoige county has surged (Figure 7).

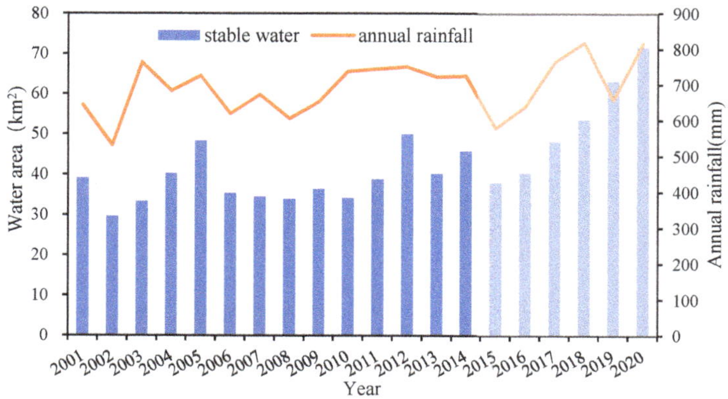

Figure 6. Changes in the stable water area in Zoige county from 2001 to 2020.

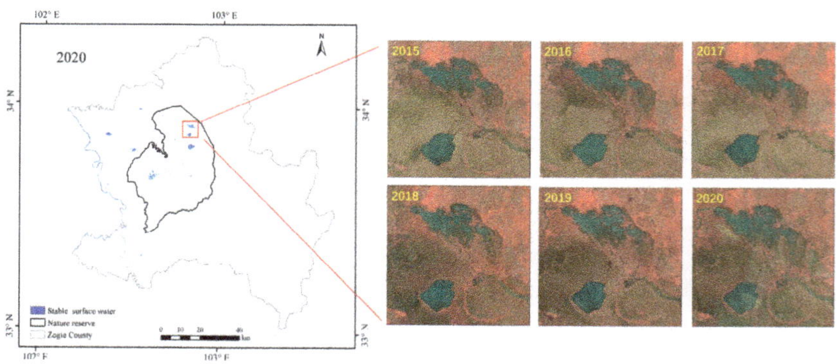

Figure 7. Spatial distributions of water area in Zoige county in 2020 and Landsat remote-sensing images of the Huahu Lake.

Viewed by stages, the water body area in Zoige county was basically stable from 2001 to 2015, and continued to increase after 2015. From 2015 to 2020, the county's water body area rapidly increased, reaching nearly 1.83 times the level of early 2001.

3.3. Characteristics of Water Conservation Function

From 2016 to 2020, the mean annual water conservation capacity of Zoige county was approximately 21.2×10^4 m^3/km^2, and high-value areas were mainly distributed in the southern herbaceous swamp and lake area and eastern mountainous forest area (Figure 8). The mean annual water conservation capacity in the Nature Reserve from 2016 to 2020 was 19×10^4 m^3/km^2.

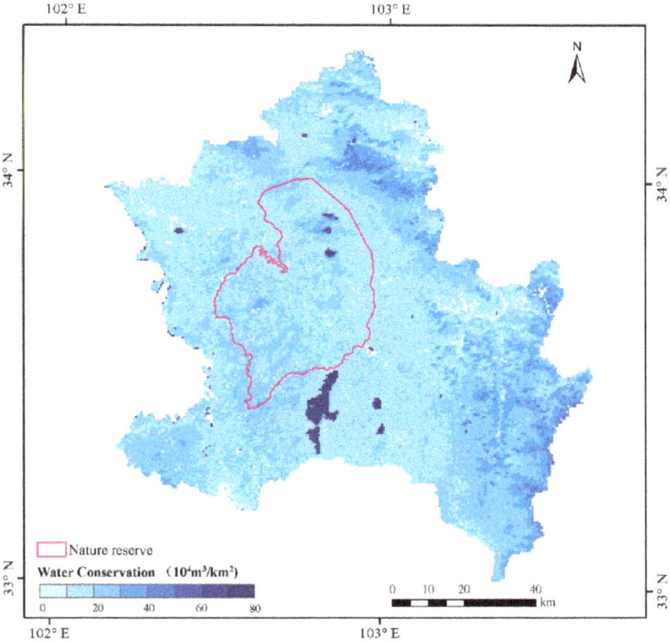

Figure 8. The mean value of water conservation in Zoige county (2016–2020).

From 2001 to 2020, the mean annual water conservation capacity of the county was 18.41×10^4 m^3/km^2, and showed a decreasing and then increasing trend (Figure 9). Water conservation capacity in the county increased from 19.8×10^4 m^3/km^2 in 2001–2005 to 21.2×10^4 m^3/km^2 in 2015–2020, an increase of 15.23%.

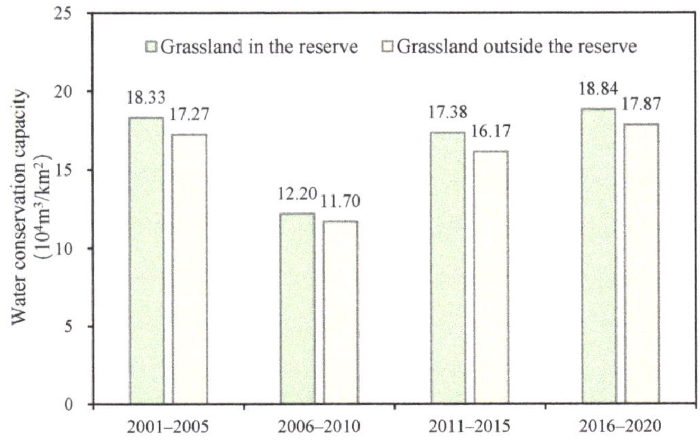

Figure 9. Annual change of water conservation in Zoige county from 2001 to 2020.

From 2001 to 2020, water conservation in Zoige county increased at a rate of 0.25×10^4 m^3/km^2/year, and 84.55% of the region exhibited an increasing trend. Areas with the fastest growth rates were mainly located in the northern part of the reserve and the forested areas in the eastern part of Zoige county. Overall, water conservation in 15.02% of the county decreased, and the areas with the fastest decrease were mainly distributed in the west and south of Zoige county (Figure 9).

Between 2001 and 2020, the annual average water conservation of the nature reserve was 16.89×10^4 m^3/km^2; water conservation of the reserve increased at a rate of approximately 0.2×10^4 $m^3/km^2/year$, with 80.9% of the area experiencing an increasing trend. The reserve consisted of mainly grassland ecosystems. Although the average water conservation in the entire reserve was slightly lower than that in the whole county, the average water conservation in the grasslands of the reserve was about 1×10^4 m^3/km^2 higher than that of the surrounding grasslands (Figure 10).

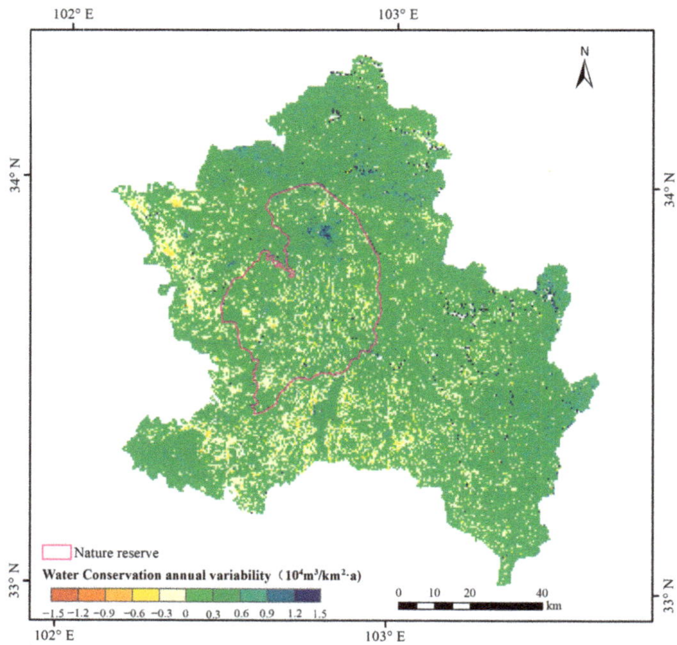

Figure 10. Variation trend in water conservation of Zoige county from 2001 to 2020.

4. Discussion

The purpose of this study was to make full use of satellite remote-sensing images with different resolutions to try to capture the changes in the ecosystem in Zoige from different scales, in order to corroborate changes in statistical analysis indicators and determine the intuitive characteristics of surface information. Furthermore, the objective of this study was also to analyze and evaluate the causes of ecosystem changes at the regional scale, and help us put forward specific and quantitative measures for regional ecological protection and restoration. Remote sensing has proven to be both feasible and effective. However, the existing indicators still require improvement.

The FVC describes key parameters of land surface vegetation, and reflects the structural characteristics of vegetation [30]. For grassland ecosystems, FVC is the basic index that reflects changes in ecological status. Here, we used FVC to monitor and evaluate the basic vegetation status of Zoige county. Furthermore, it was utilized to describe the regional ecological protection status by comparing high-resolution images of grassland and sand variations. However, FVC has obvious limitations in identifying the succession direction of the vegetation community. For example, FVC cannot directly reflect the characteristics of forage productivity, nor can it reflect the quantitative characteristics of dominant forage. Therefore, for future monitoring and evaluation of grassland ecosystem restoration, as well as for desertification control effectiveness analysis, it will be necessary to continuously carry out long-term tracking of ground communities in the Zoige area, especially in the grassland ecosystem restoration area. Additionally, it is necessary to comprehensively evaluate the effectiveness of regional ecological protection in terms of quantity, productivity, and grassland quality.

Due to the influence of climate warming and humidification, glacier melting and surface runoff have increased to a certain extent, which has contributed to the increase in the area of regional lakes and other water bodies, and consequently to the amount of water conservation [35,41]. We used Landsat satellite imagery to extract the water body area with a spatial resolution of 30 m. However, water bodies of subpixel size are widely distributed in the peat swamp area, which is not only an important indicator of regional water conservation function, but also very important for climate change research, including regional greenhouse gas change monitoring [41]. Therefore, to carry out a more refined regional ecological function assessment of the Zoige wetland, it will be necessary to perform higher-resolution identification of water information. In the future, we will explore the driving effects of human activities and climate change on ecological conditions.

Water conservation is the dominant ecosystem function in the Zoige area. As an internationally important wetland, the Zoige wetland is also a habitat for rare species such as black-necked cranes, and the function of maintaining biodiversity is also worthy of attention. Therefore, in the future, we will also evaluate ecosystem protection effectiveness and develop technical methods for researching regional biodiversity based on satellite remote-sensing images [41].

5. Conclusions

From 2001 to 2020, the changes in vegetation and water bodies showed that the ecological status of Zoige improved, especially within the last five years (2015–2020). The vegetation coverage of the county showed an increasing trend; about 95.23% of the county experienced increasing vegetation coverage. During 2015–2020, the growth rate of vegetation coverage was 1.64 times higher than that of the previous 15 years (2001–2015). In the last five years, the water area increased by 77.35%, which is the fastest period of increase in the past two decades.

Local land desertification control, wetland restoration, restrictions on illegal construction in nature reserves, facility demolition, and other protective measures not only reflect the local government's determination and willingness to protect the ecological environment, but also to improve the quality of the environment. Based on an interannual comparison of high-resolution satellite images, some land desertification control areas in the county have been restored to green. For example, the vegetation coverage of typical patches for desertification control had increased by about 5–10% since the implementation of the project in 2016. Some dried lakes and marshes have also been significantly restored. For example, multiphase remote-sensing images showed that the water body of Huahu Lake in the reserve has increased, and the surrounding land has become more swampy.

Nature reserves generally represent areas with good ecological background and fragile ecology. Driven by regional ecological protection policies, the protection effects of nature reserves will be more prominent and direct (for example, faster vegetation restoration, priority of water body restoration, greater force of human interference reduction measures, etc.) In addition, nature reserves can play a leading and exemplary role in the protection of the overall ecological function of the region. Taking the water conservation function of grassland ecosystem as an example, the amount of water conservation within the nature reserve was about 1×10^4 m^3/km^2 higher than that of the surrounding grasslands.

In the future, with the continuous enrichment in satellite remote-sensing data sources, especially with the increase in high-resolution images, it will be possible to extract high-spatial resolution and long time series of vegetation, water and other information, which will be very useful to improving the accuracy of regional ecological protection and assessment, and even replace most of the ground investigation and engineering implementation assessment tasks. Information pertaining to regional ecological function changes will be further refined with this new technology.

Author Contributions: Conceptualization, P.H. and J.Z.; methodology, Y.Z.; formal analysis, J.Z. and H.-F.G.; data curation, D.-D.J. and Y.C.; writing—original draft preparation, D.-D.J.; writing—review and editing, D.-D.J., J.Z. and P.H.; funding acquisition, P.H. All authors have read and agreed to the published version of the manuscript.

Funding: This research was funded by National Key R&D Program of China (grant number 2021YFF0703903) and Major Projects of High-Resolution Earth Observation Systems of National Science and Technology (grant number 05-Y30B01-9001-19/20-4).

Institutional Review Board Statement: Not applicable.

Data Availability Statement: Not applicable.

Acknowledgments: Thanks to the local environmental protection bureau for its help with the field investigation.

Conflicts of Interest: The authors declare no conflict of interest.

References

1. Zakharov, I.; Kapfer, M.; Hornung, J.; Kohlsmith, S.; Puestow, T.; Howell, M.; Henschel, M.D. Retrieval of Surface Soil Moisture from Sentinel-1 Time Series for Reclamation of Wetland Sites. *IEEE J. Sel. Top. Appl. Earth Obs. Remote Sens.* **2020**, *13*, 3569–3578. [CrossRef]
2. Mleczko, M.; Mroz, M.; Fitrzyk, M. Riparian wetland mapping and inundation monitoring using amplitude and bistatic coherence data from the tandem-x mission. *IEEE J. Sel. Top. Appl. Earth Obs. Remote Sens.* **2021**, *14*, 2432–2444. [CrossRef]
3. Guo, M.; Li, J.; Sheng, C.-L.; Xu, J.-W.; Wu, L. A review of wetland remote sensing. *Sensors* **2017**, *17*, 777. [CrossRef] [PubMed]
4. Hu, S.; Niu, Z.; Chen, Y. Global Wetland Datasets: A Review. *Wetlands* **2017**, *37*, 807–817. [CrossRef]
5. Jiang, N.; Wang, Y.-F.; Dong, X.-Z. Methanol as the Primary Methanogenic and Acetogenic Precursor in the Cold Zoige Wetland at Tibetan Plateau. *Microb. Ecol.* **2010**, *60*, 206–213. [CrossRef]
6. Li, M.; Xu, R.; Huang, W.; Sun, H.; Luo, L. A study on the effects of the surrounding faults on water loss in the Zoige Wetland, China. *J. Mt. Sci.* **2011**, *8*, 518–524. [CrossRef]
7. Tan, Y.-Y.; Wang, X.; Li, C.-H.; Cai, Y.-P.; Yang, Z.-F.; Wang, Y.-L. Estimation of ecological flow requirement in Zoige Alpine Wetland of Southwest China. *Environ. Earth Sci.* **2012**, *66*, 1525–1533. [CrossRef]
8. Guo, B.; Wang, S.; Wang, M.-T. Spatio-temporal variation of NPP from 1999 to 2015 in Zoige grassland wetland. *Chin. J. Appl. Ecol.* **2020**, *31*, 424–432.
9. Yu, H.; Luedeling, E.; Xu, J. Winter and spring warming result in delayed spring phenology on the Tibetan Plateau. *Proc. Natl. Acad. Sci. USA* **2010**, *107*, 22151–22156. [CrossRef]
10. Fei, S.-M.; Cui, L.-J.; He, Y.-P.; Chen, X.-M.; Jiang, J.-M. A background study of the wetland ecosystem research station in the Ruoergai Plateau. *J. Sichuan For. Sci. Technol.* **2006**, *27*, 21–29.
11. Wang, R.; He, M.; Niu, Z.-G. Responses of Alpine Wetlands to Climate Changes on the Qinghai-Tibetan Plateau Based on Remote Sensing. *Chin. Geogr. Sci.* **2020**, *30*, 189–201. [CrossRef]
12. Kang, X.-M.; Hao, Y.-B.; Cui, X.-Y.; Chen, H.; Huang, S.-X.; Du, Y.-G.; Li, W.; Kardol, P.; Xiao, X.-M.; Cui, L.-J. Variability and Changes in Climate, Phenology, and Gross Primary Production of an Alpine Wetland Ecosystem. *Remote Sens.* **2016**, *8*, 391. [CrossRef]
13. Jin, X.-Y.; Qiang, H.-F.; Zhao, L.; Jiang, S.-Z.; Cui, N.-B.; Cao, Y.; Feng, Y. SPEI-based analysis of spatio-temporal variation characteristics for annual and seasonal drought in the Zoige Wetland, Southwest China from 1961 to 2016. *Theor. Appl. Climatol.* **2020**, *139*, 711–725. [CrossRef]
14. Xia, H.-M.; Zhao, W.; Li, A.-N.; Bian, J.H.; Zhang, Z.-J. Subpixel Inundation Mapping Using Landsat-8 OLI and UAV Data for a Wetland Region on the Zoige Plateau, China. *Remote Sens.* **2017**, *9*, 31. [CrossRef]
15. Shen, G.-Y.; Yang, X.-C.; Jin, Y.-X.; Xu, B.-Z.; Zhou, Q.-B. Remote sensing and evaluation of the wetland ecological degradation process of the Zoige Plateau Wetland in China. *Ecol. Indic.* **2019**, *104*, 48–58. [CrossRef]
16. Kuang, Q.; Yuan, Q.-Z.; Han, J.-C.; Leng, R.; Wang, Y.-S.; Zhu, K.-H.; Lin, S.; Ren, P. A remote sensing monitoring method for alpine grasslands desertification in the eastern Qinghai–Tibetan Plateau. *J. Mt. Sci.* **2020**, *17*, 1423–1437. [CrossRef]
17. Fei, Y.; Wang, J.-Y.; Wang, Z.-G. Dynamic changes of land desertification in Zoige Plateau. *J. Arid Land Resour. Environ.* **2019**, *33*, 146–152.
18. Ma, K.; Zhang, Y.; Tang, S.-X.; Liu, J.-G. Spatial distribution of soil organic carbon in the Zoige Alpine Wetland, Northeastern Qinghai-Tibet Plateau. *Catena* **2016**, *144*, 102–108. [CrossRef]
19. Ma, K.; Liu, J.-G.; Balkovič, J.; Skalský, R.; Azevedo, L.B.; Kraxner, F. Changes in soil organic carbon stocks of wetlands on China's Zoige plateau from 1980 to 2010. *Ecol. Model.* **2016**, *327*, 18–28. [CrossRef]
20. Li, J.-C.; Wang, W.-L.; Hu, G.-Y.; Wei, Z.-H. Changes in ecosystem service values in Zoige Plateau, China. *Agr. Ecosyst. Environ.* **2010**, *139*, 766–770. [CrossRef]

21. Wu, C.-Y.; Chen, W.; Cao, C.-X.; Tian, R.; Liu, D.; Bao, D.-M. Diagnosis of Wetland Ecosystem Health in the Zoige Wetland, Sichuan of China. *Wetlands* **2018**, *38*, 469–484. [CrossRef]
22. Lu, F.; Hu, H.-F.; Sun, W.-J.; Zhu, J.-J.; Liu, G.-B.; Zhou, W.-M.; Zhang, Q.-F.; Shi, P.-L.; Liu, X.-O.; Wu, X.; et al. Effects of national ecological restoration projects on carbon sequestration in China from 2001 to 2010. *Proc. Natl. Acad. Sci. USA* **2018**, *115*, 4039–4044. [CrossRef] [PubMed]
23. Zhang, X.; Wang, J.; Gao, Y.; Wang, L. Variations and controlling factors of vegetation dynamics on the Qingzang Plateau of China over the recent 20 years. *Geogr. Sustain.* **2021**, *2*, 74–85. [CrossRef]
24. Gao, J.-X.; Xu, M.-J.; Zou, C.-X. Development Achievement of Natural Conservation in 70 Years of New China. *Chin. J. Environ. Manag.* **2019**, *11*, 25–29.
25. Kuang, Q. Study on the Coupling Relationship between Development of Alpine Animal Husbandry and Restoration of Sandy Grassland in Northwest Sichuan: A Case Study of Zoige. Master's Thesis, Sichuan Normal University, Chengdu, China, 2020.
26. Dong, S.-L.; Zhang, T.-S. Ecological status quo and benefit analysis of protection measures in Zoige County. *Prataculture Anim. Husb.* **2012**, *8*, 30–32 + 53.
27. Hou, P.; Zhai, J.; Cao, W.; Yang, M.; Cai, M.-Y.; Li, J. Evaluation on ecosystem changes and protection of the national key ecological function zones in mountainous areas of central Hainan Island. *Acta Geogr. Sin.* **2018**, *73*, 429–441.
28. Gao, Y.-P.; Zhang, X.-X.; Fang, Y.; Zhang, X.; Guo, J.-B. Landscape fragmentation and wetland restoration priority in Zoige County of Sichuan Province of China. *J. China Agric. Univ.* **2019**, *24*, 145–158.
29. Cui, Y.; Zhang, X.-X.; Zhang, X.; Fang, Y.; Guo, C.-Q. Hydrological and geomorphologic ecological threshold of steady-state transformation of wetland in Zoige County. *Acta. Ecol. Sin.* **2020**, *40*, 8794–8804. [CrossRef]
30. Jiapaer, G.-L.; Chen, X.; Bao, A. A comparison of methods for estimating fractional vegetation cover in arid regions. *Agr. For. Meteorol.* **2011**, *151*, 1698–1710. [CrossRef]
31. Gutman, G.; Ignatov, A. The derivation of the green vegetation fraction from NOAA/AVHRR data for use in numerical weather prediction models. *Int. J. Remote Sens.* **1998**, *19*, 1533–1543. [CrossRef]
32. Ivits, E.; Cherlet, M.; Sommer, S.; Mehl, W. Addressing the complexity in non-linear evolution of vegetation phenological change with time-series of remote sensing images. *Ecol. Indic.* **2013**, *26*, 49–60. [CrossRef]
33. Song, W.-J.; Mu, X.-H.; Ruan, G.-Y.; Gao, Z. Estimating fractional vegetation cover and the vegetation index of bare soil and highly dense vegetation with a physically based method. *Int. J. Appl. Earth Obs.* **2017**, *58*, 168–176. [CrossRef]
34. Gao, J.-X.; Wang, Y.-C.; Hou, P.; Wan, H.-W.; Zhang, W.-G. Temporal and spatial variation characteristics of land surface water area in the Yellow River basin in recent 20 years. *J. Hydraul. Eng.* **2020**, *51*, 1157–1164. [CrossRef]
35. Pekel, J.F.; Cottam, A.; Gorelick, N.; Belward, A.S. High-resolution mapping of global surface water and its long-term changes. *Nature* **2016**, *540*, 418–422. [CrossRef] [PubMed]
36. Gong, S.; Yang, X.; Zheng, H.; Xiao, Y.; Ouyang, Z. Spatial patterns of ecosystem water conservation in China and its impact factors analysis. *Acta Ecol. Sin.* **2017**, *37*, 2455–2462. [CrossRef]
37. Wu, B.-F.; Yuan, Q.-Z.; Yan, C.-Z.; Wang, Z.-M.; Yu, X.-F.; Li, A.-N.; Ma, R.-H.; Huang, J.-L.; Chen, J.-S.; Chang, C.; et al. Land cover changes of China from 2000 to 2010. *Quat. Sci.* **2014**, *34*, 723–731.
38. Wu, D.-H.; Wu, H.; Zhao, X.; Zhou, T.; Tang, B.-J.; Zhao, W.-Q.; Jia, K. Evaluation of spatiotemporal variations of global fractional vegetation cover based on GIMMS NDVI data from 1982 to 2011. *Remote Sens.* **2014**, *6*, 4217–4239. [CrossRef]
39. Mann, H.-B. Nonparametric tests against trend. *Econometrica* **1945**, *13*, 245–259. [CrossRef]
40. Kendall, M.-G. *Rank Correlation Methods*; Griffin: London, UK, 1948.
41. Woolway, R.I.; Kraemer, B.M.; Lenters, J.D.; Merchant, C.J.; O'Reilly, C.M.; Sharma, S. Global lake responses to climate change. *Nat. Rev. Earth Environ.* **2020**, *1*, 388–403. [CrossRef]

diversity

Article

Detection of Large Herbivores in UAV Images: A New Method for Small Target Recognition in Large-Scale Images

Jiarong Ma [1], Zhuowei Hu [1,*], Quanqin Shao [2,3], Yongcai Wang [1], Yanqiong Zhou [1], Jiayan Liu [4] and Shuchao Liu [2,3]

1 College of Resource Environment and Tourism, Capital Normal University, Beijing 100048, China
2 Key Laboratory of Terrestrial Surface Pattern and Simulation, Institute of Geographic Sciences and Natural Resources Research, Chinese Academy of Sciences, Beijing 100101, China
3 University of Chinese Academy of Sciences, Beijing 100049, China
4 School of Environment, Beijing Jiaotong University, Beijing 100044, China
* Correspondence: huzhuowei@cnu.edu.cn

Abstract: Algorithm design and implementation for the detection of large herbivores from low-altitude (200 m–350 m) UAV remote sensing images faces two key problems: (1) the size of a single image from the UAV is too large, and the mainstream algorithm cannot adapt to it, and (2) the number of animals in the image is very small and densely distributed, which makes the model prone to missed detection. This paper proposes the following solutions: For the problem of animal size, we optimized the Faster-RCNN algorithm in terms of three aspects: selecting a HRNet feature extraction network that is more suitable for small target detection, using K-means clustering to obtain the anchor frame size that matches the experimental object, and using NMS to eliminate detection frames that have sizes inconsistent with the size range of the detection target after the algorithm generates the target detection frames. For image size, bisection segmentation was used when training the model, and when using the model to detect the whole image, we propose the use of a new overlapping segmentation detection method. The experimental results obtained for detecting yaks, Tibetan sheep (Tibetana folia), and the Tibetan wild ass in remote sensing images of low-altitude UAV from Maduo County, the source region of the Yellow River, show that the mean average precision (mAP) and average recall (AR) of the optimized Faster-RCNN algorithm are 97.2% and 98.2%, respectively, which are 9.5% and 12.1% higher than the values obtained by the original Faster-RCNN. In addition, the results obtained from applying the new overlap segmentation method to the whole UAV image detection process also show that the new overlap segmentation method can effectively solve the problems of the detection frames not fitting the target, missing detection, and creating false alarms due to bisection segmentation.

Keywords: HRNet; MMdetection; large herbivores; UAV remote sensing image; overlapping segmentation

Citation: Ma, J.; Hu, Z.; Shao, Q.; Wang, Y.; Zhou, Y.; Liu, J.; Liu, S. Detection of Large Herbivores in UAV Images: A New Method for Small Target Recognition in Large-Scale Images. *Diversity* **2022**, *14*, 624. https://doi.org/10.3390/d14080624

Academic Editors: Luc Legal and Christoph Meyer

Received: 31 May 2022
Accepted: 4 August 2022
Published: 5 August 2022

1. Introduction

The term 'herbivores' refers to animals that live on the roots, stems, leaves, and seeds of grass [1]. Large herbivores have a large body size and consume huge amounts of grass, and their interaction with grassland plays a crucial role in maintaining the grassland ecosystem [2]. With the increase in protection efforts, the population of wild herbivores in the source region of the Yellow River has gradually increased, and the phenomenon of competition between wild herbivores and farm livestock for pasture has become increasingly prominent, even having some impact on the local grassland ecosystem and animal husbandry production [3]. Traditional grassland animal husbandry is facing development bottlenecks such as overgrazing, grassland degradation, and seasonal imbalance, which seriously threaten the service function of grassland ecosystems [4]. The key to solving these problems is to: find out the quantity and distribution of various large herbivores over time so as to provide a scientific basis for maintaining the balance between grassland

and livestock, including wild animals and farm livestock; formulate scientific and effective grassland resource utilization plans; and maintain the cycle of the grassland ecosystem [5]. Animal survey is of great significance. In this paper, the large herbivores, including the large wild herbivores and the large domestic herbivores, were detected.

Animal survey methods include ground survey methods, satellite remote sensing survey methods, and aerial remote sensing survey methods (including manned aerial vehicle survey methods and unmanned aerial survey vehicle methods) [1]. The ground survey method takes a long time, and some animals are difficult for humans to approach. The satellite remote sensing survey method has the problems of the flight height being impossible to set according to real-time needs, the data resolution being low, and it being unable to identify a single animal. The manned aerial vehicle survey method has the problems of a high cost and high noise. Compared with these, the unmanned aerial vehicle survey method (UAV survey method) has the advantages of fast image acquisition, flexible and convenient operation, and low cost and risk, providing an effective and reliable tool for regional wildlife surveys [6]. Therefore, we detected large herbivores from the images obtained by UAV.

In recent years, machine learning, especially deep learning, has made many breakthroughs in target recognition, and the related detection algorithms have been gradually applied to the target detection task of images [7]. At present, target detection algorithms with good detection effects include Faster-RCNN, SSD, YOLO, and R-FCN [8]. Faster-RCNN [9] and R-FCN [10] generally accept 1000 × 600 pixel VOC data set images or 1388 × 800 pixel COCO data set images, SSD [11] usually uses 300 × 300 or 512 × 512 pixel input images, and YOLO [12] uses 416 × 416 or 544 × 544 pixel input images. The standard size of the UAV remote sensing images obtained in this paper was 6000 × 4000 pixel, meaning that the UAV images to be detected did not match the image size required by the model. Compared with the YOLO algorithm and the SSD algorithm, Faster-RCNN has a higher accuracy in small-target detection and the stages of feature extraction, detection, and classification are separated, meaning that they can be improved and optimized separately [13]. Therefore, we improved the Faster-RCNN algorithm in this paper. Aiming at the problem of the size of a single image being too large, a new solution is proposed: equal segmentation in training, overlapping segmentation in model detection, discarding the detection frame close to the edge, and then non-maximum suppression (NMS).

In addition, the proportion of large herbivores in UAV images is often very small, and the detection accuracy of small targets is often low due to the small proportion of images, the insignificant number of texture features, the insufficient semantic information of shallow features, the lack of high-level feature information, and other reasons [1]. We replace the feature extraction network in the Faster-RCNN algorithm with a feature extraction network that is more suitable for small-target detection. The feature extraction network used in the target detection algorithm is used to reduce the data dimension and extract effective features for subsequent use. Classical feature extraction networks include the following: the LeNet network [14], which applied the convolutional neural network to practical tasks for the first time; AlexNet [15], which used dropout to prevent over fitting and proposed a ReLU activation function; the VGG network [16], which used the idea of modularization to build the network model and applied a convolution layer using multiple smaller convolution filters to replace a convolution layer with a larger convolution filter; the ResNet network [17], which used a residual network to avoid the problems of gradient explosion and gradient disappearance caused by model deepening, meaning that the number of network layers can become very deep; GoogLeNet [18], which proposed the Inception module and used multiple branches and convolution kernels; the ResNext network [19], which combined Inception and ResNet; and SENet [20], which proposed a channel domain attention mechanism. HRNet [21] maintains the high-resolution representation by connecting high-resolution to low-resolution convolution in parallel and enhances high-resolution representation by repeating multi-scale fusion across parallel convolution, making it more suitable for the detection of small targets. Therefore, we selected the HRNet network as the feature extraction network in this paper. In addition, we used K-means clustering to obtain an anchor frame size that matches the experimental object and filtered out detection frames with sizes that do not fall within the range after the algorithm generated the detection frames so as to further improve the detection accuracy of large herbivores in UAV images.

2. Data and Relevant Technical Principles

2.1. Data Sources

Maduo County, the study area, belongs to the Golog Tibetan Autonomous Prefecture and is located in the south of the Qinghai Province at the north foot of Bayankala mountain. The land cover type of Maduo County is mainly grassland, and there are small areas of swamps, lakes, bare rock gravel, sandy land, rivers, etc. [22]. Maduo County is located in the Qinghai Tibet Plateau and belongs to the Qinghai Tibet Plateau climate system. It is a typical continental climate type on the plateau. The large wild herbivores in the study area include the Tibetan wild donkey, Tibetan gazelle, and rock sheep, whereas the large domestic herbivores in the area include the domestic yak, domestic Tibetan sheep, and horse [23]. In this paper, the Tibetan wild donkey was selected as a representative species of large wild herbivores, whereas the yak and Tibetan sheep were selected as representative species of large domestic herbivores to detect large herbivores in UAV images.

The UAV remote sensing image data used in this paper were obtained and provided by Shao Quanqin research group of the Institute of Key Laboratory of Terrestrial Surface Pattern and Simulation, Institute of Geographic Sciences and Natural Resources Research, Chinese Academy of Sciences. The remote sensing image acquisition platform of the UAV was an electric fixed-wing UAV (Figure 1). The fixed-wing UAV has the advantages of strong wind resistance, fast flight speed, high operation efficiency, and long endurance time, and it is suitable for information collection in large areas. The camera carried is a Sony ILCE-5100. A dual-camera system was used for aerial photography. The images were taken in Maduo County from 9–18 April 2017. The aerial shooting time was 7:00–11:00 every day. Those days were sunny or cloudy. There were 16 aerial sorties and 14 effective sorties. The shooting height was 200–350 m and the resolution was 4–7 cm. In total, 23,784 images were obtained, with an effective shooting area of 356 km^2 (high resolution, large heading, and lateral overlap), and the effective utilization area was 326.6 km^2 (excluding corners). In this paper, images taken on 9 April and 15 April were used to detect large herbivores including yaks, Tibetan sheep, and Tibetan wild asses.

Figure 1. Figure of an electric fixed-wing UAV.

2.2. Data Preprocessing

In this paper, the images obtained by the UAV were used for detection. With the help of python, each UAV remote sensing image was divided into 100 small images with heights of 600 pixels and widths of 400 pixels. Bisection segmentation may cause a target to be divided into two halves; in order to ensure the quality of model training, when selecting the segmented images for training samples, images containing half an animal were not selected. In order to enhance the robustness, 50 small images of Tibetan sheep, 50 small images of Tibetan wild donkeys, and 50 small images of yaks without half an animal were randomly selected to create the training samples. These images included 308 Tibetan sheep, 210 Tibetan wild asses, and 130 yaks. The production and selection methods applied to the test samples used for evaluation were the same as those applied to the training samples. In order to ensure the robustness of the evaluation, all test images were randomly selected and none of the test images or training images were the same image.

After randomly selecting all images that did not contain half an animal from the cropped images, they needed to be labeled to generate the training data set. The software tool LabelImg was used to label Tibetan sheep, yaks, and Tibetan wild donkeys. The labeling process is shown in Figure 2. In this process, a corresponding XML format file was established for each image, in which the category of each target animal in the image and its corresponding location information were recorded.

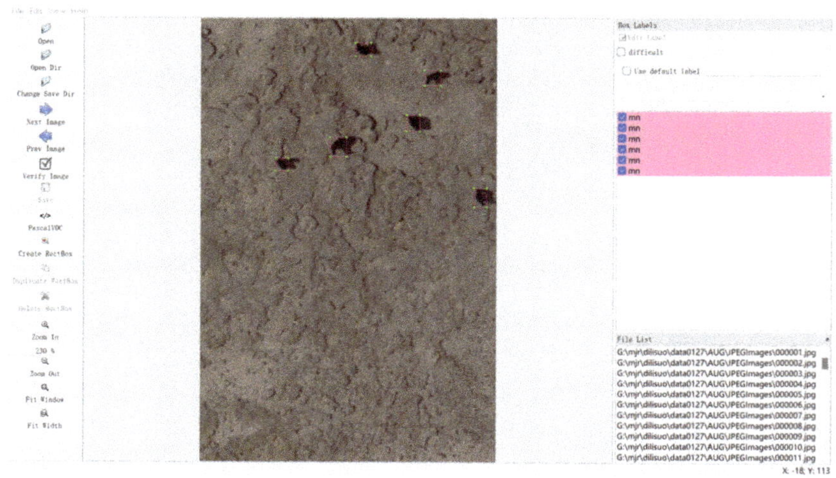

Figure 2. Diagram of the training sample creation process (the boxes composed of green points and red lines are the ground truth frames of large herbivores we annotated).

In the training process of a deep learning model, the more model parameters and the less data there are, the more likely the model is to experience the problem of over fitting [24]. Increasing the amount of data is the most effective way to solve the problem of over fitting. We observed targets on the ground from the angle of the UAV, where the observation angle was almost vertical. Therefore, we could use data enhancement to simulate the posture of various animals from an overhead view to expand the data. By observing the posture and behavior of large herbivores, we: expanded the training samples to 9 times the original through 90-, 180-, and 270-degree rotation; horizontally, vertically, and diagonally mirrored the image; and changed light and shade so as to improve the generalization ability of the model. In order to reduce the workload of annotation, after the randomly selected training samples were annotated, we performed data enhancement operations on the image and its corresponding XML annotation file at the same time, and then converted the data into the required COCO format and input them into the network for training.

2.3. *Faster-RCNN Algorithm*

Faster-RCNN is a two-stage target detection algorithm where VGG16 is used as the feature extraction network. VGG16 includes 13 convolution layers and 3 full connection layers, and the convolution kernel size is 3 × 3 [25]. The VGG network conducts several down sampling processes in the process of extracting feature information, resulting in the small targets in the feature map having basically only single digit pixel sizes, resulting in a poor detection effect for small targets. The reasoning speed reaches 5 fps on GPU (including all steps) [10]. The algorithm process can be divided into four steps. (1) Feature extraction network. We input the image into the feature extraction network to extract feature maps. (2) Region Proposal Network (RPN). The RPN generated 9 anchors with 3 box areas of 128^2, 256^2, and 512^2 pixels and 3 aspect ratios of 1:1, 1:2, and 2:1 at each position of the feature map [26] and sampled 256 samples in all anchor boxes to train the RPN network, in which the ratio of positive to negative samples was 1:1. On the one hand, it judged whether the category of anchor boxes belonged to the foreground or background through softmax; on the other hand, it calculated the regression offset of the bounding box to obtain an accurate candidate box and eliminate candidate boxes that were too small or beyond the boundary. After screening, the RPN finally retained 2000 candidate frames and projected them onto the feature map to obtain the corresponding feature matrices. (3) Region of interest pooling (ROI pooling). The feature matrices obtained from the RPN layer were mapped into uniform size feature vectors by ROI pooling. (4) Classification and regression. The feature vectors obtained by ROI pooling were transmitted to the classifier through the full connection layer to judge the category. Additionally, the more accurate positions of the detection frames were regressed by the regressor. The flow chart is shown in Figure 3.

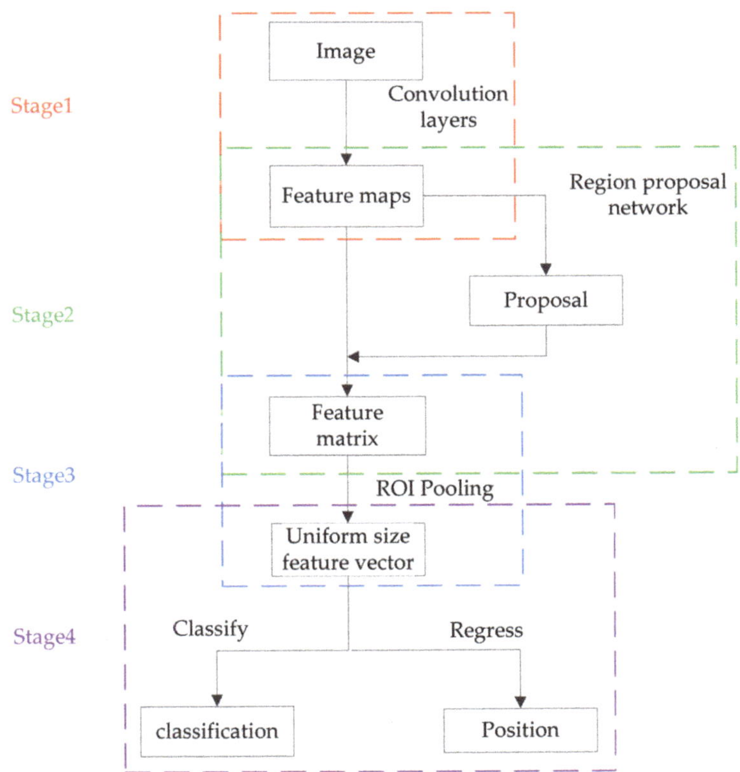

Figure 3. Flow chart of the Faster-RCNN algorithm.

2.4. *MMdetection*

We used the MMdetection toolbox to optimize the Faster-RCNN to complete the detection of large herbivores in UAV images. MMdetection [27] is an open-source deep learning target detection toolbox based on pytorch from Shangtang technology. It adopts modular design to connect different components, and can easily build a customized target detection framework. It supports many classical target detection algorithm models, such as RPN, Fast-RCNN, Faster-RCNN, SSD, RetinaNet, etc., as well as toolkits of various modules. All basic box and mask operations now run on GPU, which has the characteristics of a slightly higher performance, faster training speed, and less video memory, which is conducive to experiments. Therefore, this paper improved the Faster-RCNN algorithm with the help of the MMdetection toolbox.

3. Improvement of Faster-RCNN Algorithm

In view of the problems mentioned above (in that the original Faster-RCNN model has low small target detection accuracy, the UAV single image is too large, and its size does not match the size required by the Faster-RCNN), this paper mainly improved the Faster-RCNN target detection algorithm in terms of the following aspects: (1) Replacing the feature extraction network: We selected HRNet, which is more suitable for small target detection, as the feature extraction network. The design details are shown in Section 3.1. (2) Optimizing the Region Proposal Network: We used the K-means algorithm to cluster the target size and set the anchor frame size and proportions suitable for large herbivores to improve the Region Proposal Network. The design details are shown in Section 3.2. (3) Post-processing optimization: After the detection frames were generated by the Faster-RCNN algorithm for classification and regression, post-processing was carried out, and NMS was used to eliminate the detection frames with scores lower than 0.5. By observing the results, we found that there were some cases where the background was incorrectly detected as a target, and the size of these backgrounds was usually different from the target. In this paper, we optimized the post-processing process by eliminating the detection frames that were inconsistent with the size range of the detection target. The details are shown in Section 3.3. (4) New detection method of overlap segmentation: When using the trained model to detect a whole UAV image, if a whole remote sensing image is evenly divided into small images directly, it may cause the problem of false alarm or missed detection. In this paper, we proposed a method of overlapping segmentation, removing the detection frame close to the edge, and then NMS. The details are shown in Section 3.4.

3.1. Replacing the Feature Extraction Network

The convolutional neural network near the input can extract the edges and corners, then the later convolutional neural network can extract the local details of the object, and the convolutional neural network near the output can extract the abstract object structure. The convolution layer usually contains five pooling layers to alleviate the problem of over fitting and reduce the amount of computation. Five down sampling will reduce the resolution of the feature map by 32 times. However, the length and width of the large herbivores to be detected only account for 5–50 pixels. After multiple instances of down sampling, the animals will disappear in the feature map or only have a very simple feature representation. Therefore, the conventional feature extraction network is not suitable for too-small targets.

In this paper, HRNet (High-Resolution Net) was used as the feature extraction network. HRNet was originally proposed by Sun K and used in human posture detection tasks [21]. It tends to generate high-resolution feature maps. Its advantage is that it not only maintains high-resolution feature maps, but also maintains the information interaction between different resolution feature maps in a unique parallel way. In this way, a high-quality feature map is generated, which not only covers the semantic information of feature maps at different levels, but also maintains the original resolution. The main structure of the network is shown in Figure 4. It can be seen from the figure that the network is divided into four stages, of which the third stage has three sizes of feature maps and the fourth stage has four sizes of feature maps. Each stage has a feature map of one size greater than the previous stage, and the feature maps of different levels repeatedly interact with each other in different stages. After reaching the last stage, there will be convolution streams with four resolutions. HRNetV1 only outputs the highest resolution feature map, which is often used for human posture estimation. Its output representation is shown in Figure 5a. HRNetV2 connects all low-resolution feature maps with the highest resolution feature map after up sampling, which is commonly used for semantic segmentation and face key point detection. Its output representation is shown in Figure 5b. HRNetV2p [28] is a new feature pyramid formed by down sampling the output representation of HRNetV2. It is predicted separately on each scale and is commonly used for target detection. Its output expression is shown in Figure 5c. In this paper, HRNetV2p was used as the feature extraction network, and the HRNet mentioned in this paper used as the feature extraction network refers to HRNetV2p. The down sampling shown in Figures 4 and 5 was realized by many 3 × 3 convolution filters with stride 2, and the up sampling was realized by the nearest neighbor algorithm and 1 × 1 convolution, as shown in Figure 6.

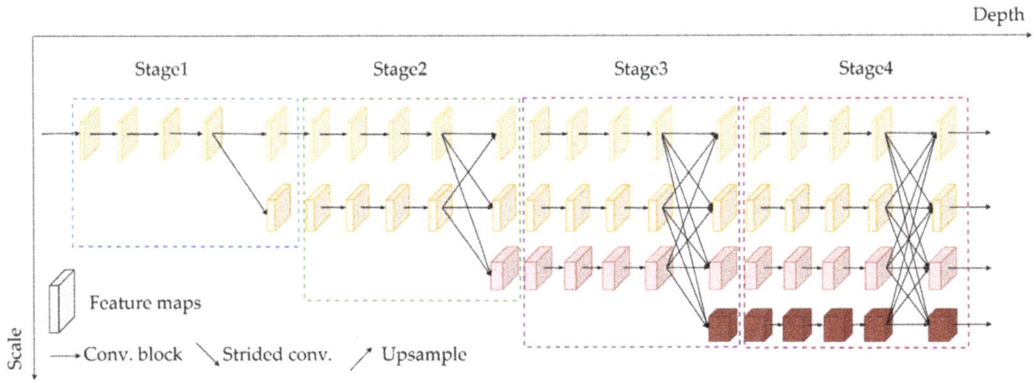

Figure 4. Schematic diagram of the HRNet network.

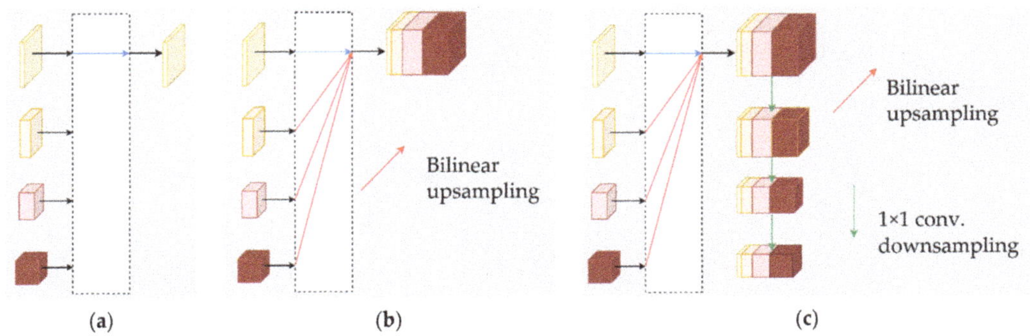

Figure 5. Output representation of different versions of HRNet: (**a**) output representation of HRNetV1; (**b**) output representation of HRNetV2; (**c**) output representation of HRNetV2p.

As can be seen from Figures 4–6, HRNet improves the resolution of feature maps and enriches the semantic information contained in feature maps by means of feature map fusion at different levels, which is similar to the idea of Feature Pyramid Networks (FPNs). FPNs [29] improve the detection effect of multi-scale targets by building feature pyramids and integrating feature maps of different levels, but HRNet was realized through a unique parallel mode. The high-resolution subnet of HRNet started as the first stage, gradually adding high-resolution to low-resolution subnets to form more stages and connecting the multi-resolution subnets in parallel. Through multiple multi-scale fusion, each high-resolution to low-resolution feature map repeatedly received information from other parallel representations so as to form rich high-resolution feature maps. Therefore, HRNet is more conducive for the detection of smaller targets. In this paper, HRNetV2p was used as the feature extraction network.

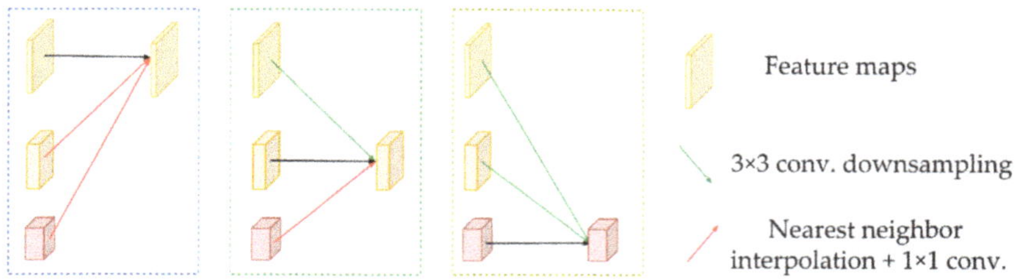

Figure 6. Up sampling and down sampling methods.

3.2. Optimizing the Region Proposal Network

The Region Proposal Network module of the original Faster-RCNN generates 9 anchor frames with 3 box areas of 128^2, 256^2, and 512^2 pixels and 3 aspect ratios of 1:1, 1:2, and 2:1 at each position of the feature map [27]. We selected the correct detection boxes from these anchor frames and adjusted their position and size to finally complete the detection of the target. The size of large herbivores in the UAV images ranged from 5 pixels to 50 pixels. The anchor frame size generated by the original algorithm was not suitable for large herbivores. In this paper, the size of each training sample was calculated by reading the coordinates (xmin) and (ymin) in the lower left corner and (xmax) and (ymax) in the upper right corner of the target in the annotation file. Then, the K-means algorithm was used to cluster the sizes of all training samples to obtain the sizes of six clustering centers. The clustering results are shown in Figure 7, in which the abscissa is the height of the annotation box and the ordinate is the width of the annotation box.

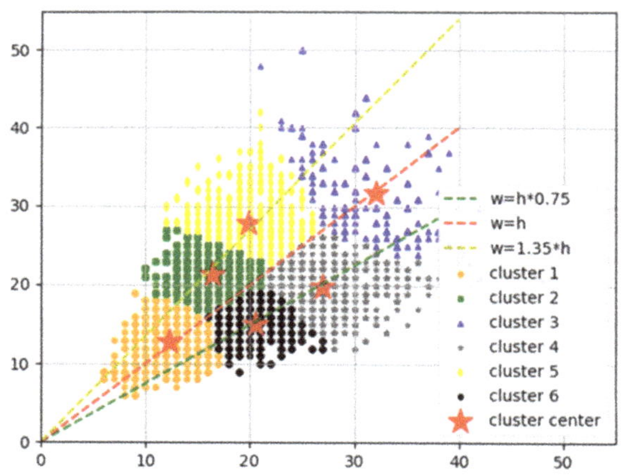

Figure 7. Results of K-means clustering on the size of large herbivores.

The red pentagram is the final six cluster centers, with coordinates of [21.25, 16.27], [15.03, 20.54], [12.95, 12.38], [19.83, 26.99], [32.08, 32.08], and [26.72, 19.88]. The aspect ratios were 1.31, 0.73, 1.05, 0.96, 0.73, 1, and 1.39, respectively. It can be seen from the results of K-means clustering on the size of large herbivores that the aspect ratio was roughly clustered around 0.73, 1, and 1.35. We set the scale of detection frame to 0.75, 1.0, and 1.35 and set the basic size of anchor frame to 4 and the step sizes to 4, 8, 16, 32, and 64, so that the algorithm generated 15 anchor frames at each position on the feature map generated by HRNet.

3.3. *Post-Processing Optimization*

In the classification and regression phase of the original Faster-RCNN algorithm, after the detection frames were generated, the detection frames needed to be filtered again to eliminate some wrong detection frames. By observing the detection results, it was found that no matter whether the feature extraction network was replaced or the anchor frame size was modified, there were false detections in the detection results where snow was recognized as Tibetan sheep, rock, or shadow or where other backgrounds were recognized as Tibetan wild donkeys. Additionally, the size of these false detected backgrounds was usually inconsistent with the target. As shown in Figure 8, too-small and too-large detection frames were usually formed by the false detection of some confusing backgrounds. After screening, the detection accuracy could be improved. Therefore, we optimized the post-processing stage. With the help of python, we calculated that the size range of the target was between 5 pixels and 50 pixels. So, we eliminated the detection frames with a score of less than 0.5 and the detection frames with lengths or widths less than 3 pixels and lengths or widths greater than 55 pixels at the same time, so as to reduce the number of false alarms during detection. After screening, the detection accuracy could be improved.

Figure 8. An example of where a detection frame obtained by mistakenly detecting snow as Tibetan sheep needs to be removed (the red boxes in the figure are the detection result boxes of the original Faster-RCNN model).

3.4. *New Detection Method of Overlap Segmentation*

As the whole UAV image was too large, its size did not match the size required by the model. During training, we divided the image equally for training. When using the model to detect the whole UAV image, it was also necessary to divide the image into small blocks. After the detection frames of each block were obtained, these detection frames were mapped to the original image. The method of bisection segmentation was usually used for detection, but if a whole remote sensing image was evenly divided into small images for detection, this may cause the problem of false alarms or missed detection. As shown in Figure 9, if a yak is divided into two halves in the segmentation, both parts of the body may be detected or not detected, resulting in false alarms or missed detection.

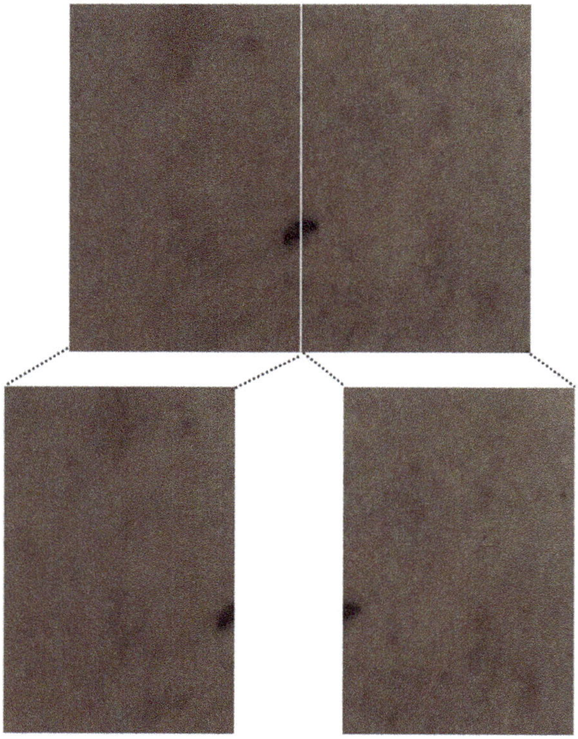

Figure 9. Example diagram of bisection segmentation that may cause false detection or false alarms.

Aiming at the problem of the detection method of bisection segmentation causing false alarms or missed detection, Adam van Etten proposed overlapping segmentation when improving the YOLO algorithm for satellite image detection [30]. Large-scale remote sensing images were overlapped and segmented, and then the NMS algorithm was used to eliminate the detection frame of repeated detection of a target. However, in some cases, this method would still detect a target repeatedly. As shown in Figure 10, a yak cut into a small half was detected in the upper right corner of the left figure, and the whole yak was detected in the right figure. It can be seen from the figure that the intersection of the two frames was relatively small. The NMS algorithm cannot suppress the redundant detection frames in the left figure, resulting in the problem of the multiple detection of a single target.

In order to further reduce the incidence of false alarms and missed detection, we proposed a detection method of overlapping segmentation and then discarding the detection frame close to the edge. The slice size was 600 pixels high and 400 pixels wide. For two adjacent slices, we made them overlap 100 pixels. The naming method of the slice was: Upper left row coordinates_Upper left column coordinates_Slice height_Slice width.jpg. For example, 1000_1800_600_400.jpg means that the slice is located in the third row and seventh column of the original image, and the slice size was 600 × 400 pixels. After each slice was detected, the detection frame within 50 pixels of the edge of each slice was discarded first, meaning that there will be no false alarm caused by a target being cut in half that cannot be filtered out by NMS. As shown in Figure 11, the detection frame with the center point in the orange area was directly eliminated. However, if the detection frame within 50 pixels of the edge of all slices was directly screened out, the detection frame of the edge of the original UAV image would also be removed. In order to avoid this situation, the edge detection frame was retained according to the slice name. For example, if the row coordinate of the upper left corner was 0 and the column coordinate of the upper left corner was 0, all detection frames with a center row coordinate between 0 and 350 pixels and a column coordinate between 0 and 550 pixels would be retained. Then, the position of the detection frame in the original image was obtained according to the slice name, and then the target detection frame of the original UAV image was obtained and mapped to the original image after NMS for all the detection frames.

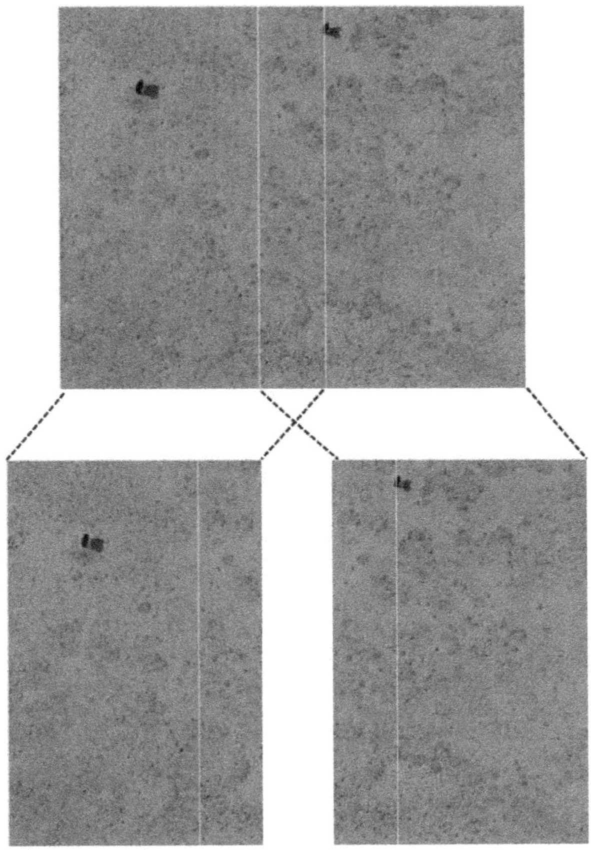

Figure 10. Conventional overlap segmentation method (the green boxes in the figure are the detection result boxes of the original Faster-RCNN model using the original overlapping segmentation method).

Figure 11. Overlapping segmentation method proposed in this paper (the green boxes in the figure are the detection result boxes of the original Faster-RCNN model using the original overlapping segmentation method).

4. Experimental Verification and Analysis

4.1. Experimental Configuration

All experiments described in this paper were carried out on the Linux platform, using the Ubuntu 20.04 system. The computer was configured as RTX3080, the V2 version of the MMdetection toolbox was used, and the programming language used was python. Data clipping, data enhancement, K-means clustering, size statistics of large herbivores, the elimination of detection frames that do not meet the range, the elimination of detection frames with the center within 50 pixels of the edge after overlapping segmentation, and NMS were all realized using python. The optimizer selected was SGD (Stochastic Gradient Descent). By default, the Faster-RCNN algorithm in the MMdetection toolbox uses eight GPUs, where each GPU processes two pictures, and the learning rate is 0.02. In this paper, two GPUs were used, each GPU processed two pictures, and the corresponding setting of the learning rate was 0.005. In total, 24 epochs were trained. The learning rate increased linearly in the first 500 iterations, the learning rate reduced at 18 to 22 epochs, and the momentum was set to 0.9. The weight attenuation was generally two to three orders of magnitude different from the learning rate. In this paper, the weight attenuation was set to 0.0001 [31]. After the image input, the size of the resize was set to 600 × 400—that is, the size of the input image was not changed.

4.2. Experimental Design and Evaluation Index

In order to verify the effectiveness of several improved methods used in this paper for the detection of large herbivores in UAV images, we designed four groups of experiments. In experiment 1, VGG16, ResNet50, ResNet101, and HRNet were, respectively, used as feature extraction networks for model training in the framework of Faster-RCNN so as to verify the impact of different feature extraction networks on the model training accuracy and the superiority of HRNet as the feature extraction network in small-target detection. In experiment 2, the Faster-RCNN algorithm was used as the framework, and HRNet was used as the feature extraction network. Different anchor frame sizes and proportions were used to train the model and the effect of modifying the anchor frame size according to the K-means clustering results was compared and verified. In experiment 3, Faster-RCNN was used as the framework and HRNet was used as the feature extraction network. NMS was used to eliminate detection frames that did not meet the scope under the conditions of modifying the anchor frame size and not modifying the anchor frame size in order to verify the impact of post-processing optimization on the experimental effect. In experiment 4, the whole UAV image was detected using bisection segmentation, original overlap segmentation, and the new overlap segmentation method proposed in this paper, and the experimental results were compared and analyzed.

The commonly used evaluation indicators for target detection are precision and recall. Precision refers to the proportion of the number of correct targets in the detection results, as shown in Formula (1). The recall rate refers to the proportion of the number of correct test results in the total number of targets to be tested, as shown in Formula (2) [32].

$$\text{Precision} \qquad P = \frac{TP}{TP + FP} \tag{1}$$

$$\text{Recall} \qquad R = \frac{TP}{TP + FN} \tag{2}$$

In this paper, TP represents the number of large herbivores that were correctly detected, FP represents the number of large herbivores that were incorrectly detected, and FN represents the number of large herbivores that were not detected. However, the precision and recall are in conflict under certain circumstances. This paper used each category of AP (average precision), the whole category of AR (average recall), and mAP (mean average precision) as evaluation indicators [33]. AP is the area under the curve of the precision–recall curve, AR is the maximum recall of a given number of detection results per image, and mAP is the average of all categories of AP.

When evaluating the accuracy of a target detection algorithm, different IoU values are usually selected to judge the correctness of the detection frame predicted by the algorithm. IoU represents the intersection and union ratio between the detection frame predicted by the algorithm and the ground truth, as shown in Formula (3).

$$\text{IoU} = \frac{area\ (B_{op} \cap B_{gt})}{area\ (B_{op} \cup B_{gt})} \tag{3}$$

where B_{op} represents the prediction box, B_{gt} represents the ground truth, $area\ (B_{op} \cap B_{gt})$ represents the intersection of the prediction box and the ground truth, and $area\ (B_{op} \cup B_{gt})$ represents the union of the prediction box and the ground truth. The higher the IoU value is, the more accurate the detected box will be and the smaller the difference between the box and the ground truth will be. In this paper, PR curves constructed under different IoU values were compared. Finally, a threshold with an IoU of 0.5 was selected. If the IoU between the detection frame and the ground truth was greater than 0.5, the detection frame was considered to be the correct prediction frame.

4.3. *Comparative Experimental Results and Analysis of Improved Faster-RCNN Algorithm*

4.3.1. Experimental Results and Analysis of the Original Faster-RCNN Algorithm Using Different Feature Extraction Networks

A feature extraction network is used to extract or sort out effective features for subsequent use and plays an important role in the accuracy of a model. In order to verify the impact of different feature extraction networks on the accuracy of the model: we used the

MMdetection toolbox; took Faster-RCNN as the framework; and used VGG16, ResNet50, ResNet101, and HRNet as the feature extraction networks, respectively, to train the model. The precision–recall curves (PR curve) of the four feature extraction network experiments are shown in Figure 12.

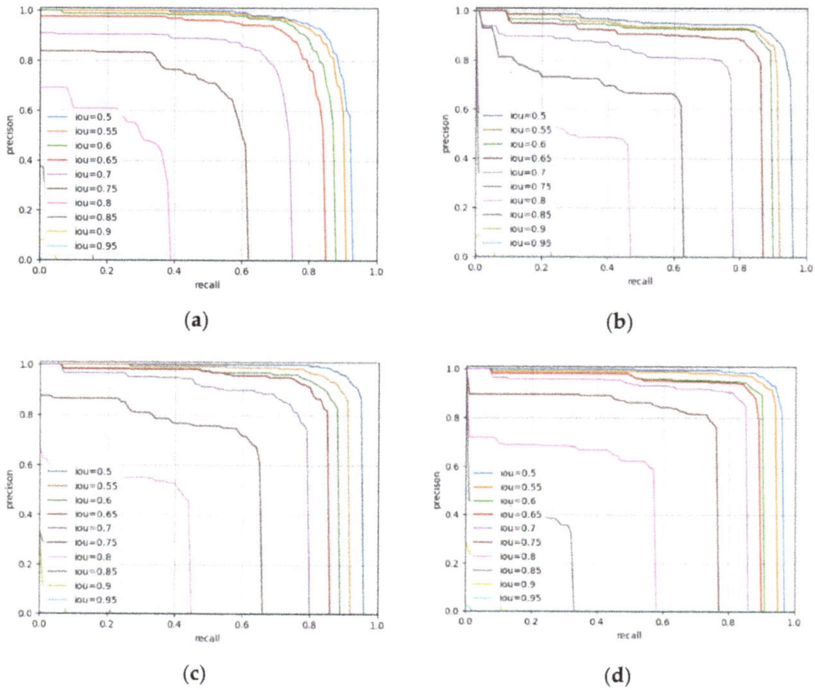

Figure 12. The PR curves of the original Faster-RCNN algorithm constructed using different feature extraction networks: (**a**) the PR curve of the VGG16 feature extraction network experiment; (**b**) the PR curve of the ResNet50 feature extraction network experiment; (**c**) the PR curve of the ResNet101 feature extraction network experiment; (**d**) the PR curve of the HRNet feature extraction network experiment.

From the PR curves of the four feature extraction network experiments, it can be seen that, when taking the same IoU, the accuracy and recall rate of the HRNet feature extraction network were generally higher than those of the VGG16 feature extraction network, the ResNet50 feature extraction network, and the ResNet101 feature extraction network. The precision and recall of the four feature extraction networks were the highest when IOU was 0.5. The category AP, full category AR, and mAP of the four feature extraction networks when IoU was 0.5 are shown in Table 1.

Table 1. Experimental results of the original Faster-RCNN algorithm using different feature extraction networks when IoU was 0.5.

Feature Extraction Network	AP			AR	mAP
	Yak	Tibetan Wild Ass	Tibetan Sheep		
VGG16	0.893	0.903	0.756	0.887	0.851
ResNet50	0.908	0.951	0.832	0.921	0.897
ResNet101	0.944	0.96	0.811	0.925	0.905
HRNet	0.952	0.973	0.868	0.942	0.931

By comparing the results of the four feature extraction networks, it was found that: the VGG16 network used by the original Faster-RCNN had a low accuracy; ResNet50, ResNet101, and HRNet had higher accuracies than VGG16; and the AP and AR of ResNet101 were slightly higher than those of ResNet50. It was found that the overall accuracy of HRNet was higher than that of VGG16, ResNet50, and ResNet101. The VGG network and ResNet performed five down sampling operations on the feature map when extracting features, which seriously affected the detection accuracy of the algorithm for small targets and overlapping targets [34], whereas HRNet realized the efficient detection of small targets and overlapping targets by generating a high-resolution feature map with a large amount of semantic information so as to achieve a high accuracy.

4.3.2. Experimental Results and Analysis of Optimizing Region Proposal Network

In the RPN module of the original Faster-RCNN algorithm, the anchor frame size was preset. In this paper, the objects needing to be detected were all small targets, so the original size and proportions were not necessarily suitable for the data set used in this paper. Therefore, in this paper we used K-means clustering to obtain the anchor frame size matching the detection target. To verify whether modifying the anchor frame size according to the results of K-means clustering can improve the experimental accuracy, we selected HRNet as the feature extraction network and set different anchor frame sizes according to the results obtained in Section 3.3. We set the anchor frame sizes to 8 and 4 and the anchor frame ratios to 0.5, 1.0, 2.0, and 0.75, 1.0, 1.35, respectively. The strides used in the four groups were 4, 8, 16, 32, and 64, and the PR curves of the four groups are shown in Figure 13.

It can be seen from the PR curves of the four groups of experiments that the precision and recall of the model were improved no matter whether the size or proportions of the anchor frame were modified. When the size and proportions of the anchor frame were modified at the same time, the precision and recall of the model were the highest, whereas the precision and recall of the four groups of experiments were the highest when the IoU was 0.5. The AP of each category and the AR and mAP of all categories in the four groups of experiments when IoU was 0.5 are shown in Table 2.

It can be seen from the experimental results that modifying the size and proportions of the anchor frame can improve the experimental accuracy. By modifying the size of the anchor frame, mAP was increased by 0.012; by modifying the proportions of the anchor frame, mAP was increased by 0.035; and by modifying the size and proportions of the anchor frame at the same time, mAP was increased by 0.037.

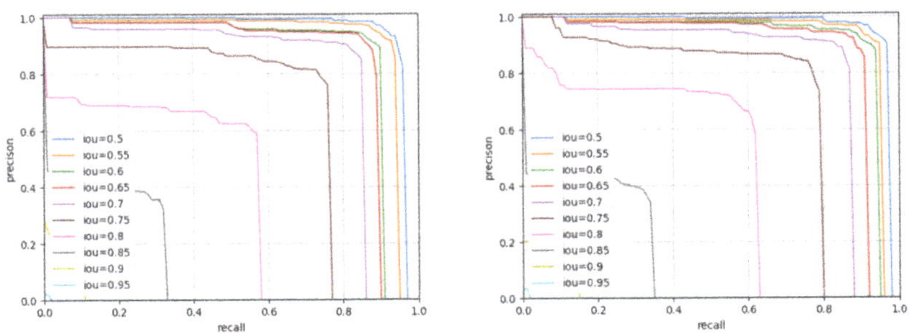

(a) (b)

Figure 13. *Cont.*

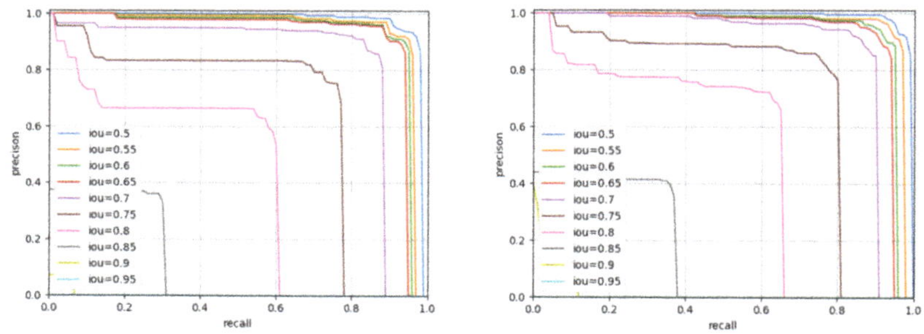

(c) (d)

Figure 13. Modified anchor frame size experimental PR curves: (**a**) the PR curve of the experiment with anchor frame foundation size of 8, and proportions of 0.5, 1, and 2; (**b**) the PR curve of the experiment with anchor frame foundation size of 8, and proportions of 0.75, 1, and 1.35; (**c**) the PR curve of the experiment with anchor frame foundation size of 4, and proportions of 0.5, 1, and 2; (**d**) the PR curve of the experiment with anchor frame foundation size of 4, and proportions of 0.75, 1, and 1.35.

Table 2. Test results before and after modifying the size and proportions of the anchor frame when IoU was 0.5.

Anchor Frame Size	Anchor Frame Scale	AP			AR	mAP
		Yak	Tibetan Wild Ass	Tibetan Sheep		
8	0.5, 1.0, 2.0	0.952	0.973	0.868	0.942	0.931
8	0.75, 1.0, 1.35	0.962	0.966	0.902	0.947	0.943
4	0.5, 1.0, 2.0	0.972	0.977	0.948	0.974	0.966
4	0.75, 1.0, 1.35	0.985	0.963	0.955	0.980	0.968

4.3.3. Test Results and Analysis of Post-Processing Optimization

After the algorithm generates the predicted detection frames through classification and regression, it is necessary to filter the detection frames and eliminate some wrong detection frames. The original Faster-RCNN algorithm will eliminate the detection frames with scores of less than 0.5. In the UAV images used in this paper, the size of large herbivores was between 5 pixels and 50 pixels, and the detection frames with too small or large lengths or widths generally presented a confusing background. Therefore, on the basis of eliminating the detection frames with scores of less than 0.5, we also eliminated detection frames with lengths or widths less than 3 pixels or with lengths or widths greater than 55 pixels. As for whether using NMS to eliminate anchor frames that do not fall within the range can improve the experimental accuracy, we used the HRNet feature extraction network to perform four groups of experiments under the conditions of modifying the anchor frame size and not modifying the anchor frame size. The PR curves of the four groups of experiments are shown in Figure 14.

From the PR curves of the four groups of experiments, it can be seen that the experimental accuracy can be improved by using NMS to eliminate detection frames that do not fall within the range, whether or not the anchor frame size was modified. The experimental accuracy of the four groups of experiments was the highest when IoU was 0.5. The AP of each category and the AR and mAP of all categories in the four groups of experiments when IoU was 0.5 are shown in Table 3.

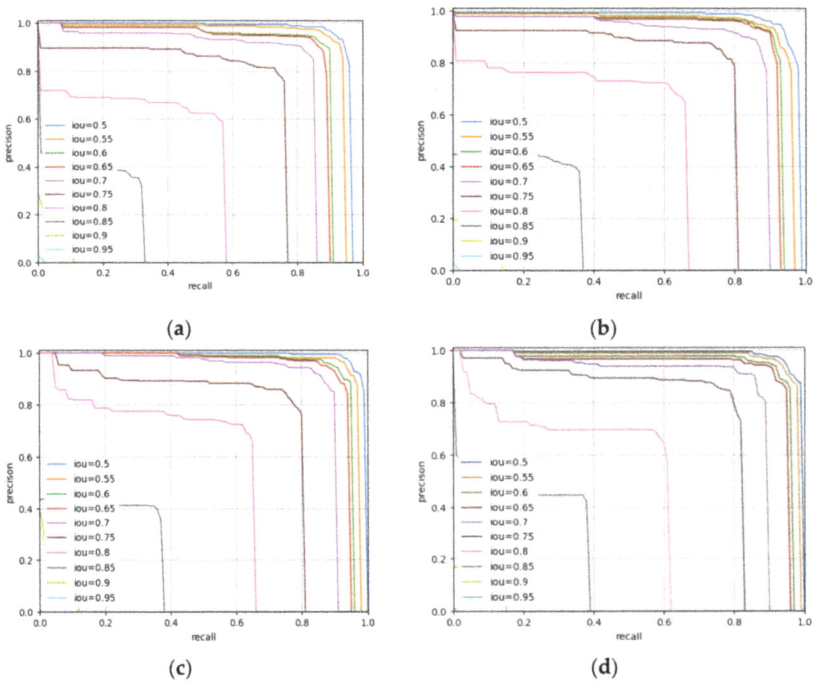

Figure 14. The PR curves of the experiment of using NMS to eliminate the detection frames with sizes that did not meet the range: (**a**) the PR curve of the experiment performed without eliminating the non-conforming detection frame by NMS, with the anchor frame foundation size of 8 and proportions of 0.5, 1, 2; (**b**) the PR curve of the experiment performed eliminating the non-conforming detection frames by NMS, with the anchor frame foundation size of 8 and proportions of 0.5, 1, 2; (**c**) the PR curve of the experiment performed without eliminating the non-conforming detection frame by NMS, with the anchor frame foundation size of 4 and proportions of 0.75, 1, 1.35; (**d**) the PR curve of the experiment performed eliminating the non-conforming detection frames by NMS, with the anchor frame foundation size of 4 and proportions of 0.75, 1, 1.35.

Table 3. Test results before and after using NMS to eliminate the detection frames that did not fall within the scope when IoU was 0.5.

NMS	Anchor Frame Size	Anchor Frame Scale	AP			AR	mAP
			Yak	Tibetan Wild Ass	Tibetan Sheep		
No	8	0.5, 1.0, 2.0	0.952	0.973	0.868	0.942	0.931
Yes	8	0.5, 1.0, 2.0	0.964	0.962	0.902	0.958	0.942
No	4	0.7, 1.0, 1.35	0.985	0.963	0.955	0.980	0.968
Yes	4	0.7, 1.0, 1.35	0.971	0.978	0.967	0.982	0.972

It can be seen from the experimental results that whether the anchor frame size was the original size or the modified size, using NMS to eliminate the detection frames that did not fall within the range can improve the experimental accuracy. Among these, when the size and proportions of the anchor frames were modified at the same time and NMS was used to eliminate the detection frames that did not fall within the scope, the detection effect was the best. The mAP was 0.041 higher than that achieved before the improvement, and the APs of yaks, Tibetan wild donkeys, and Tibetan sheep were also improved.

4.3.4. Experimental Results and Analysis of Different Segmentation Methods

When training the model, we divided the image equally. When using the model to detect the whole UAV image, we also needed to divide the image into small blocks, and map back to the original image after obtaining the detection frames of each block. In order to verify the effectiveness of the proposed new overlapping segmentation method, we used the best model trained by modifying the size and proportions of the anchor frame and used NMS to eliminate the detection frames that did not fall within the range to detect the whole UAV image. The bisection segmentation, the original overlapping segmentation, and the overlapping segmentation method proposed in this paper were used to test the image. Figure 15 shows the result of detection conducted using bisection segmentation, Figure 16 shows the result of detection conducted using the original overlapping segmentation method, and Figure 17 shows the result of detection conducted using the new overlapping segmentation method.

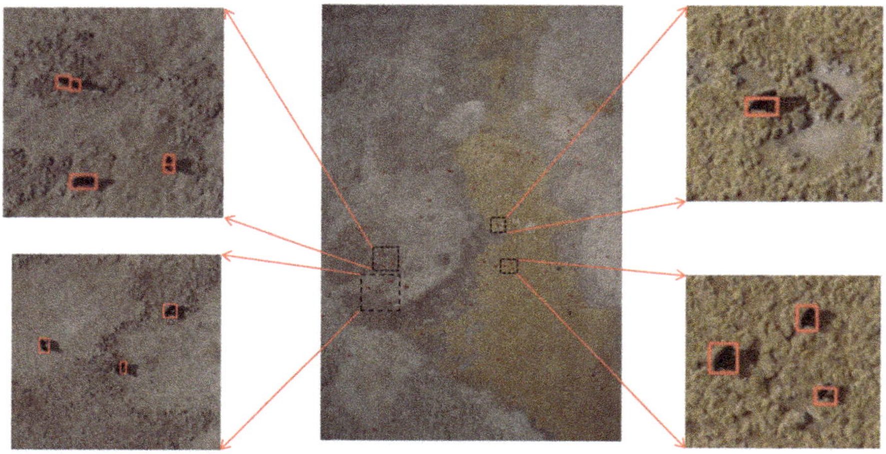

Figure 15. Result diagram of detection performed using bisection segmentation (the red boxes in the figure are the detection result boxes detected by the improved Faster-RCNN model using the bisection segmentation method).

It can be seen that after the image was detected by bisection segmentation, there were a large number of cases where the detection frame did not fit the target because the animal was divided into two halves, and there were also cases where the same target was detected twice. Using the original overlap segmentation method for detection reduced the incidence of situations where the detection frame did not fit the target and false alarm, but there were still cases where the target was detected twice because the coincidence degree was not high and it could not be suppressed by NMS, as shown in Figure 16. The overlapping segmentation detection method proposed in this paper can completely avoid the false alarms and missed detection events caused by the segmentation of animals, and it can prevent cases where the detection frame does not fit the target. In the detection result diagram obtained using the new overlap detection method, except that two backgrounds were mistakenly detected as yaks and one background was mistakenly detected as Tibetan sheep, the remaining large herbivores were correctly detected.

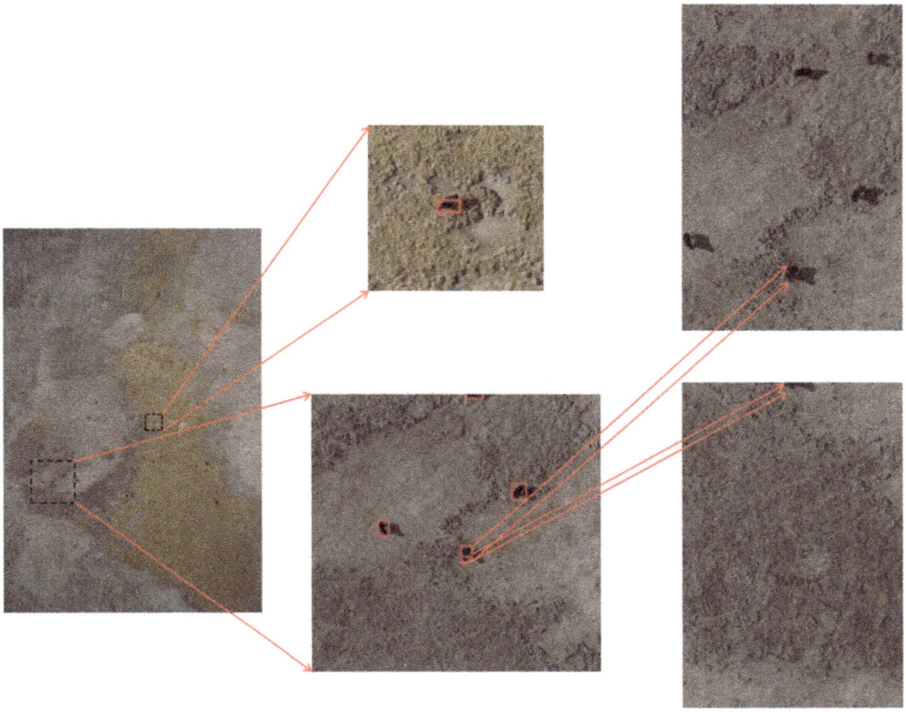

Figure 16. Result diagram of detection performed using original overlapping segmentation (the red boxes in the figure are the detection result boxes detected by the improved Faster-RCNN model using the original overlapping segmentation method).

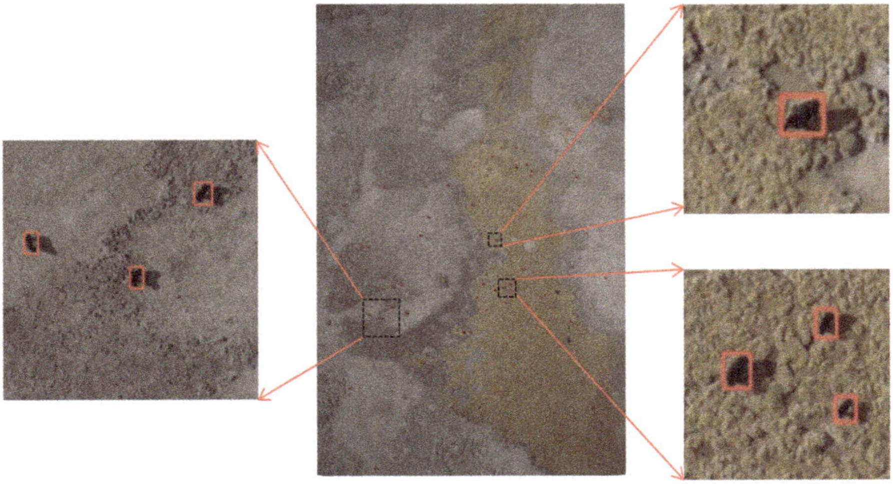

Figure 17. Result diagram of detection performed using new overlapping segmentation (the red boxes in the figure are the detection result boxes detected by the improved Faster-RCNN model using the new overlapping segmentation method proposed in this paper).

5. Conclusions

The size of UAV images is large and the proportion of large herbivores is very small. The current mainstream target detection algorithms cannot be directly used to identify large herbivores in UAV images. With the help of the MMdetection toolbox, after comparing different feature extraction networks, we selected HRNet, which has excellent performance in small- and medium-sized target detection, as the feature extraction network. Then we set a suitable anchor frame size and proportions for large herbivores according to the results of K-means clustering, and used NMS to eliminate the detection frames that did not fall within the range to improve the Faster-RCNN algorithm. The following conclusions were drawn:

(1) Compared with the VGG16 network, ResNet50 network, and ResNet101 network, it was proven that the HRNet feature extraction network is more suitable for the detection of large herbivores in UAV images.
(2) According to the results of K-means clustering, the size and proportions of the anchor frame were adjusted. AR was increased by 0.038 and mAP was increased by 0.037. This shows that setting sizes and proportions of anchor frames that are suitable for the target according to the results of K-means clustering can improve the accuracy.
(3) Using the results of K-means clustering to adjust the size and proportions of anchor frames and using NMS to eliminate the detection frames that did not fall within the range at the same time, the AP of yaks, Tibetan wild donkeys, and Tibetan sheep reached 0.971, 0.978, and 0.967, respectively, values which were 0.019, 0.005, and 0.099 higher than those obtained before the two improvements, whereas the mAP reached a value that was 0.972, and 0.041 higher than that obtained before the two improvements.
(4) We used the detection method of overlapping segmentation first, removing the detection frame within 50 pixels of the edge, and then NMS could realize the high-precision detection of the whole UAV image, and there were no cases where the detection frame did not fit the target or where false alarms or missed detection were caused by the animals being divided into two halves.

In this paper, after using the improved Faster-RCNN algorithm, the mean average precision (mAP) and average recall (AR) were improved to varying degrees. Several optimized methods used in this paper can be used to detect small targets in other large-scale images. The optimized model reached a high level of precision and could be used to: more accurately count the number of large herbivores; estimate the proportions of wild herbivores and farm livestock; and further provide a scientific basis for maintaining the balance between grass and livestock, including wild animals and farm livestock. However, the anchor frame size designed according to this data set may not be suitable for use in other data sets. Then, we can further optimize the algorithm by adaptively calculating the K-means clustering results according to the data set to modify the anchor box size, and adaptively calculate the size range of the ground truth frames, and eliminate too-large and too-small detection frames. In addition, this method can be applied to more complex scenes, such as UAV images in different phases, more target types, and more easily confused distractors, in order to check the generalization ability of the scheme.

Author Contributions: Conceptualization, Z.H.; methodology, Z.H. and J.M.; validation, J.M., Y.Z. and Y.W.; formal analysis, Z.H.; resources, Q.S. and S.L.; data curation, J.M. and S.L.; writing—original draft preparation, J.M.; writing—review and editing, Z.H., Y.Z. and Y.W.; visualization, J.M.; supervision, J.L.; project administration, Z.H.; funding acquisition, Z.H. and Q.S. All authors have read and agreed to the published version of the manuscript.

Funding: This research was financially supported by the National Natural Science Foundation of China (No. 42071289) and the National Key Research and Development Program of China (grant no. 2018YFC1508902).

Institutional Review Board Statement: Ethical review and approval were waived for this study due to the UAV aerial photography taking place at an altitude of 200 to 350 m, meaning that it was far from the ground, with low noise, causing little or no interference to wildlife, and with none of the personnel involved in the aerial photography having direct contact with animals. We guarantee that the aerial photography was conducted under the guidance of relevant regulations, such as the "technical regulations for the second national terrestrial wildlife resources investigation" and "technical specifications for national terrestrial wildlife resources investigation and monitoring" of the State Forestry Administration.

Informed Consent Statement: Not applicable.

Data Availability Statement: Data are available from the corresponding authors upon request, because the data is provided by the Shao Quanqin research group of the Institute of Key Laboratory of Terrestrial Surface Pattern and Simulation, Institute of Geographic Sciences and Natural Resources Research, Chinese Academy of Sciences, and the data provider does not want the data to be made public.

Acknowledgments: We would like to thank the Shao Quanqin research group of the Institute of Key Laboratory of Terrestrial Surface Pattern and Simulation, Institute of Geographic Sciences and Natural Resources Research, Chinese Academy of Sciences, for obtaining and providing the UAV remote sensing image data.

Conflicts of Interest: The authors declare no conflict of interest. The funders had no role in the design of the study; in the collection, analysis, or interpretation of data; in the writing of the manuscript; or in the decision to publish the results.

References

1. Song, Q.J. Unmanned Aircraft System for Surveying Large Herbivores—A Case Study in Longbao Wetland National Nature Reserve, Qinghai Province, China. Master's Thesis, Lanzhou University, Lanzhou, China, 26 May 2018.
2. Lu, W.J. Mechanisms of Carbon Transformation in Grassland Ecosystems Affected by Grazing. Master's Thesis, China Agricultural University, Beijing, China, 4 December 2017.
3. Zhang, S.; Shao, Q.Q.; Liu, J.Y.; Xu, X.L. Grassland Cover Change near the Source of the Yellow River: Case Study of Madoi County, Qinghai Province. *Resour. Sci.* **2008**, *30*, 1547–1554.
4. Liu, J.Y.; Xu, X.L.; Shao, Q.Q. Grassland degradation in the "Three-River Headwaters" region, Qinghai Province. *J. Geogr. Sci.* **2008**, *18*, 259–273. [CrossRef]
5. Yang, F.; Shao, Q.Q.; Guo, X.J.; Li, Y.Z.; Wang, D.L.; Zhang, Y.X.; Wang, Y.C.; Liu, J.Y.; Fan, J.W. Effects of wild large herbivore populations on the grassland-livestock balance in Maduo County. *Acta Prataculturae Sin.* **2018**, *27*, 1–13.
6. Luo, W.; Shao, Q.Q.; Wang, D.L.; Wang, Y.C. An Object-Oriented Classification Method for Detection of Large Wild Herbivores: A Case Study in the Source Region of Three Rivers in Qinghai. *Chin. J. Wildl.* **2017**, *38*, 561–564.
7. Wang, N.; Li, Z.; Liang, X.L.; Wu, X.N.; Lü, Z.H. Research on Application of UAV Autonomous Target Recognition and Positioning. *Comput. Meas. Control* **2021**, *29*, 215–219.
8. Shi, L.; Jing, M.E.; Fan, Y.B.; Zeng, X.Y. Segmentation Detection Algorithm Based on R-CNN Algorithm. *J. Fudan Univ. Nat. Sci.* **2020**, *59*, 412–418.
9. Ren, S.Q.; He, K.M.; Girshick, R.B.; Sun, J. Faster R-CNN: Towards Real-Time Object Detection with Region Proposal Networks. *IEEE Trans. Pattern Anal. Mach. Intell.* **2017**, *39*, 1137–1149. [CrossRef] [PubMed]
10. Dai, J.F.; Yi, L.; He, K.M.; Sun, J. R-FCN: Object Detection via Region-based Fully Convolutional Networks. In Proceedings of the Thirtieth Conference on Neural Information Processing Systems, Centre Convencions Internacional Barcelona, Barcelona, Spain, 5 December 2016.
11. Liu, W.; Anguelov, D.; Erhan, D.; Szegedy, C.; Reed, S.E.; Fu, C.Y.; Berg, A.C. SSD: Single Shot MultiBox Detector. *IEICE Trans. Fundam. Electron. Commun. Comput. Sci.* **2015**, *9905*, 21–37.
12. Redmon, J.; Divvala, S.K.; Girshick, R.; Farhadi, A. You Only Look Once: Unified, Real-Time Object Detection. In Proceedings of the IEEE Conference on Computer Vision and Pattern Recognition, Las Vegas, NV, USA, 12 December 2016.
13. Peng, H.; Li, X.M. Small target detection model based on improved Faster R-CNN. *Electron. Meas. Technol.* **2021**, *44*, 122–127.
14. Lecun, Y.; Bottou, L. Gradient-based learning applied to document recognition. *Proc. IEEE* **1998**, *86*, 2278–2324. [CrossRef]
15. Krizhevsky, A.; Sutskever, I.; Hinton, G. ImageNet Classification with Deep Convolutional Neural Networks. In Proceedings of the 26th Annual Conference on Neural Information Processing Systems 2012, Lake Tahoe, NV, USA, 3–6 December 2012.
16. Simonyan, K.; Zisserman, A. Very deep convolutional networks for large-scale image recognition. *Comput. Sci. Name* **2014**, arXiv:1409.1556.
17. He, K.M.; Zhang, X.Y.; Ren, S.Q.; Sun, J. Deep Residual Learning for Image Recognition. In Proceedings of the IEEE Conference on Computer Vision and Pattern Recognition (CVPR), Las Vegas, NV, USA, 10 April 2016.
18. Szegedy, C.; Liu, W.; Jia, Y.Q.; Reed, S.; Anguelov, D.; Erhan, D.; Vanhoucke, V.; Sermanet, P.; Rabinovich, A. Going Deeper with Convolutions. In Proceedings of the IEEE Conference on Computer Vision and Pattern Recognition (CVPR), Boston, MA, USA, 7–12 June 2015.

19. Xie, S.N.; Girshick, R.; Dollar, P.; Tu, Z.W.; He, K.M. Aggregated Residual Transformations for Deep Neural Networks. In Proceedings of the IEEE Conference on Computer Vision and Pattern Recognition (CVPR), Honolulu, HI, USA, 25 July 2017.
20. Hu, J.; Shen, L.; Sun, G. Squeeze-and-Excitation Networks. In Proceedings of the IEEE Conference on Computer Vision and Pattern Recognition (CVPR), Salt Lake City, UT, USA, 21 June 2018.
21. Sun, K.; Xiao, B.; Liu, D.; Wang, J.D. Deep High-Resolution Representation Learning for Human Pose Estimation. In Proceedings of the IEEE/CVF Conference on Computer Vision and Pattern Recognition, Long Beach, CA, USA, 19 June 2019.
22. Zhang, S. Analysis of Land Use/Cover Change and Driving Forces in Maduo County, the Source Region of the Yellow River. Master's Thesis, Institute of Geographic Sciences and Natural Resources Research, CAS, Beijing, China, 6 July 2007.
23. Shao, Q.Q.; Guo, X.J.; Li, Y.Z.; Wang, Y.C.; Wang, D.L.; Liu, J.Y.; Fan, J.W.; Yang, J.W. Using UAV remote sensing to analyze the population and distribution of large wild herbivores. *J. Remote Sens.* **2018**, *22*, 497–507.
24. Xie, X.H.; Lu, J.B.; Li, W.T.; Liu, C.X.; Huang, H.M. Classification Model of Clothing Image Based on Migration Learning. *Comput. Appl. Softw.* **2020**, *37*, 88–93.
25. Huang, Y.X.; Fang, L.M.; Huang, S.Q.; Gao, H.L.; Yang, L.B.; Lou, S.X. Research on Crown Extraction Based on lmproved Faster R-CNN Model. *For. Resour. Manag.* **2021**, *1*, 173–179.
26. Cheng, Z.A. Automatic Recognition of Terrestrial Wildlife in Inner Mongolia based on Deep Convolution Neural Network. Master's Thesis, Beijing Forestry University, Beijing, China, 12 June 2019.
27. Chen, X.E.; Li, H.M.; Chen, D.T. Research and Implementation of Dish Recognition Based on Convolutional Neural Network. *Mod. Comput.* **2021**, *27*, 95–98.
28. Sun, K.; Zhao, Y.; Jiang, B.; Cheng, T.H.; Xiao, B.; Liu, D.; Mu, Y.D.; Wang, X.G.; Liu, W.Y.; Wang, J.D. High-Resolution Representations for Labeling Pixels and Regions. *arXiv* **2019**, arXiv:1904.04514.
29. Lin, T.Y.; Dollar, P.; Girshick, R.; He, K.M.; Hariharan, B.; Belongie, S. Feature pyramid networks for object detection. In Proceedings of the IEEE Conference on Computer Vision and Pattern Recognition (CVPR), Honolulu, HI, USA, 9 November 2017.
30. Van Etten, A. You Only Look Twice: Rapid Multi-Scale Object Detection In Satellite Imagery. *arXiv* **2018**, arXiv:1805.09512.
31. Tan, T. The Optimization Algorithm Research of Stochastic Gradient Descent Based on Convolutional Neural Network. Master's Thesis, Southwest University, Chongqing, China, 18 April 2020.
32. Wang, J.; Zhang, D.Y.; Kang, X.Y. Improved Detection Method of Low Altitude Small UAV by Faster-RCNN. *J. Shenyang Ligong Univ.* **2021**, *40*, 23–28.
33. Long, J.H.; Guo, W.Z.; Lin, S.; Wen, C.W.; Zhang, Y.; Zhao, C.J. Strawberry Growth Period Recognition Method Under Greenhouse Environment Based on Improved YOLOv4. *Smart Agric.* **2021**, *3*, 99–110.
34. Chen, Y.Z. Research on Object Detection algorithm Based on Feature fusion and Adaptive Anchor Framework. Master's Thesis, Henan University, Kaifeng, China, 24 June 2020.

diversity

MDPI

Article

Hyperspectral Inversion of Soil Carbon and Nutrient Contents in the Yellow River Delta Wetland

Leichao Nie [1,2,3], Zhiguo Dou [1,2,3], Lijuan Cui [1,2], Xiying Tang [1,2,3], Xiajie Zhai [1,2,3], Xinsheng Zhao [1,2,3], Yinru Lei [1,2,3], Jing Li [1,2,3], Jinzhi Wang [1,2,3] and Wei Li [1,2,3,*]

1 Institute of Wetland Research, Chinese Academy of Forestry, Beijing 100091, China
2 Beijing Key Laboratory of Wetland Services and Restoration, Beijing 100091, China
3 Institute of Ecological Conservation and Restoration, Chinese Academy of Forestry, Beijing 100091, China
* Correspondence: wetlands207@caf.ac.cn

Abstract: Hyperspectral inversion techniques can facilitate soil quality monitoring and evaluation. In this study, the Yellow River Delta Wetland Nature Reserve was used as the study area. By measuring and analyzing soil samples under different vegetation types and collecting soil reflectance spectra, the relationships between vegetation types, soil depth, and the changes in soil total carbon (TC), total nitrogen (TN), and total phosphorus (TP) contents were assessed. The spectral data set was changed by spectral first derivative processing and division of the sample set according to vegetation type. The correlation between soil carbon, nitrogen, and phosphorus contents, and soil spectra was also analyzed, sensitive bands were selected, and the partial least-squares (PLS) method, support vector machine (SVM) method, and random forest (RF) model were used to establish the inversion model based on the characteristic bands. The optimal combination of spectral transformation, sample set partitioning, and inversion model was explored. The results showed significant differences ($p < 0.05$) in soil TC, TN, and TP contents under reed and saline alkali poncho vegetation, but not between soil element contents under different stratifications of the same plant species. The first derivative reflectance had higher correlation coefficients with soil TC, TN, and TP contents compared with the original reflectance, while the sensitive bands and quantities of the three elements differed. The division of the sample sets according to vegetation type and the first derivative treatment can improve the prediction accuracy of the model. The best combination of sample set plus FD plus RF for TC, TN, and TP in reed soil and sample set plus FD plus SVM for TC, TN, and TP in saline alkali pine soil provides technical support to further improve the prediction accuracy of TC, TN, and TP in wetland soil.

Keywords: soil nutrient; hyperspectral; inversion model

Citation: Nie, L.; Dou, Z.; Cui, L.; Tang, X.; Zhai, X.; Zhao, X.; Lei, Y.; Li, J.; Wang, J.; Li, W. Hyperspectral Inversion of Soil Carbon and Nutrient Contents in the Yellow River Delta Wetland. *Diversity* **2022**, *14*, 862. https://doi.org/10.3390/d14100862

Academic Editor: Giovanni Bacaro

Received: 17 July 2022
Accepted: 6 October 2022
Published: 11 October 2022

1. Introduction

Soils are unparalleled in terms of their complexity and dynamics, and they contain minerals, organic matter, innumerable microorganisms, and varying amounts of air, water, and essential nutrients [1–3]. Soil carbon, nitrogen, and phosphorus are important elements required for plant physiological processes in terrestrial ecosystems, and they have a great impact on the structure and function of ecosystems, along with being important indicators of soil nutrient levels [4–7]. The soil whole carbon content is an important indicator of the soil carbon pool, while soil nitrogen and phosphorus are indicators of soil nutrient elements [8–10].

To understand the key role played by soils in global material cycling, a quantitative assessment of the soil carbon, nitrogen, and phosphorus contents and their management is needed [11]. However, the traditional methods of soil element quantification are laborious and expensive, and a large number of samples is required to maintain the statistical robustness of the analysis [12,13]. Therefore, traditional estimation methods pose critical analytical and environmental challenges. Reflectance spectroscopy techniques serve as alternatives to laboratory practices that require more analysis time and use large amounts of hazardous reagents [14]. The principles of reflectance spectroscopy in soil science are related to the variability of material surfaces and their optically active components [15]. For example, soil elements, such as carbon, nitrogen, and phosphorus, have a significant impact on the form and nature of soil reflectance spectra and can be estimated quickly [16–19]. Recently, the use of hyperspectral techniques to obtain information on soil elemental content has gained popularity and become a reliable method for exploring soil-related issues [20–22]. In the process of modelling inversion, scholars have found that different models perform differently because of the differences in computational principles; hence, it is necessary to construct different models to compare the inversion effects to determine the best inversion model [23–25]. Owing to the redundancy of hyperspectral data, a mathematical transformation of spectral data or of the extraction of sensitive bands via principal component analysis and the correlation coefficient method can improve the modelling accuracy [26–28]. Naveen et al. collected soil hyperspectral data from mangrove and salt marsh wetlands and established a partial least-squares regression model between the spectral information and soil carbon and nitrogen variables in an attempt to determine the best band for soil variable inversion [29]. Meanwhile, Zhang et al. used several mathematical transformation methods to screen out the sensitive bands of soil carbon and nitrogen. Based on the sensitive bands, the authors established a hyperspectral inversion model of the coastal wetland surface soil carbon and nitrogen contents and achieved a better prediction accuracy [30]. In addition, the partitioning of the hyperspectral sample also has an important impact on the predictive power of the model [31,32]. Taking coastal wetland soil as the study object, Wei et al. established partial least-squares regression and support vector machine (SVM) prediction models based on three different sample set division methods and found that different sample set partitioning methods also impact modelling accuracy [33].

Previous studies have focused on different models and different data processing methods to carry out research, and although some studies have used the division of sample sets, there remains a lack of research that incorporates the influence of surface vegetation into the division factors of sample sets. Surface vegetation is the primary factor influencing soil elemental carbon, nitrogen, and phosphorus contents, while apoplastic material and root secretion during vegetation growth cause differences in soil physicochemical properties under different vegetation types, and the level of soil carbon, nitrogen, and phosphorus contents causes differences in soil spectral properties. Therefore, when targeting samples of soil carbon, nitrogen, and phosphorus contents of different vegetation types, this study, through different data processing methods and models, proposes a strategy to divide the sample set according to the surface vegetation, and it determines whether dividing the sample pool according to different types of vegetation types could improve the model's prediction accuracy. The objectives of this study are to: (1) investigate and analyze the differences in soil total carbon (TC), total nitrogen (TN), and total phosphorus (TP) contents under reed and saline alkali pong communities in coastal wetlands of the Yellow River Delta; (2) compare and analyze the effects of different data processing methods, different sample set division methods, and different models on soil carbon, nitrogen, and content prediction; and (3) evaluate the reliability of soil carbon, nitrogen, and phosphorus content prediction using hyperspectral techniques.

2. Materials and Methods

2.1. Study Area

The study area is located in the Yellow River Delta Wetland Nature Reserve in Shandong Province (37°35′ N–38°12′ N, 118°33′ E–119°20′ E) (Figure 1). The Yellow River Delta is the largest estuarine delta nature reserve in China. It is a representative example of estuarine wetland ecosystems worldwide and has been included in the list of internationally important wetlands by the Ramsar Convention [34]. The terrain of this area is flat, with an altitude of 2.0–15.0 m. The total research area is approximately 2902 km^2, and the land use types mainly include cultivated land, wetlands, and saline-alkali land [35]. The Yellow

River Delta belongs to a warm, temperate, semi-humid, continental monsoon climate zone, with significant temperature differences between the four seasons. The annual average temperature is 11.7–12.6 °C, the extreme maximum temperature is 41.9 °C, and the extreme minimum temperature is −23.3 °C; the frost-free period is 211 d, and the average annual rainfall is 530–630 mm [36].

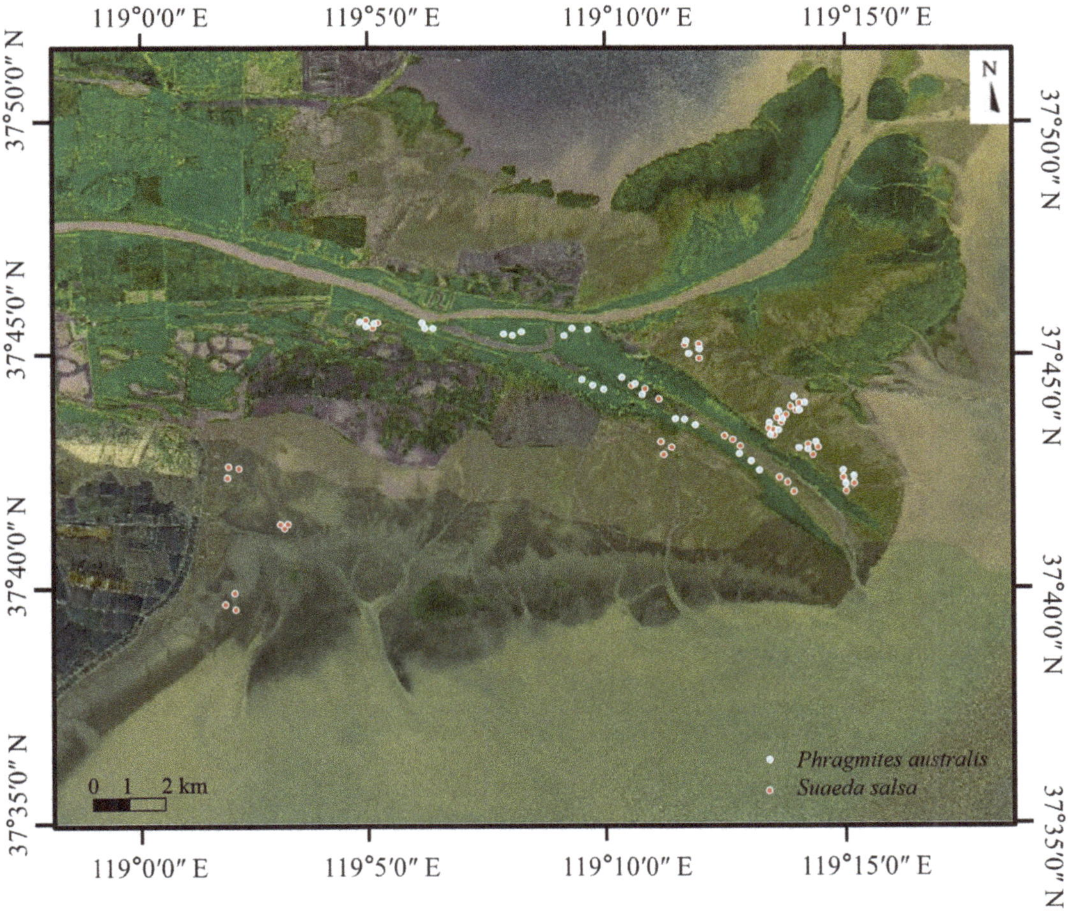

Figure 1. Location of the study area and sampling sites.

2.2. Data Collection

Sampling was conducted in October 2021, and the sampling sites were randomly distributed within the study area. A total of 80 sampling sites were selected, of which 42 were *Suaeda salsa* ponies and 38 were *Phragmites australis*. Three layers of soil were collected from each sampling site, and the sampling depth was 0–60 cm, divided into three layers (0–20 cm, 20–40 cm, and 40–60 cm). A total of 240 soil samples were collected, of which 126 were collected from the *P. australis* ponies and 114 were collected from the *S. salsa*. The minimum quantity of each soil sample was 500 g. The soil samples were dried naturally in a cool, dry, and ventilated area and then ground and sieved through a 100-mesh sieve after removing impurities (e.g., plant roots and stones). The screened soil was divided into two parts: one part was used for the determination of soil carbon, nitrogen, and phosphorus using traditional chemical methods. The soil organic carbon (SOC) content was determined via potassium dichromate-ferric sulfate titration, soil TN content was determined using the semi-micro Kjeldahl method, and soil TP content was measured using the sulfuric acid-perchloric acid digestion-molybdenum antimony colorimetric method. Soil spectral reflectance data were collected using an ASD FS4 portable geospectrometer (Analytical Spectral Devices, Inc., Boulder, CO, USA) equipped with a soil spectral probe in the wavelength range of 350–2500 nm with a sampling interval of 1 nm. The soil samples were sieved and placed in a 1.0 cm deep glass Petri dish such that the soil surface was flat and the probe was kept perpendicular to the soil surface during measurements. Reference white plate calibration was performed before each spectral test.

2.3. Data Set Division

Firstly, the significant difference analysis of soil carbon, nitrogen, and phosphorus contents between different vegetation types and soil layers was conducted to determine whether the classification of the sample set was reasonable based on the results. Different classification criteria of the sample data sets will cause differences in modelling effects [37]. Therefore, in this study, all soil spectra obtained were classified into two categories: the soil spectra of P. australis and the soil spectra of S. salsa, based on the different surface vegetation. From this, three sample libraries could be established: a sample library of P. australis soil spectra containing 114 spectral data, a sample library of S. salsa soil spectra containing 126 spectral data, and a total sample library of 240 soil spectra not classified according to surface vegetation. Based on the three sample libraries, subsequent pre-processing and modelling validation were conducted to investigate the effect of dividing the soil sample set according to the surface samples on the modelling effect.

2.4. Pre-Processing Methods

Viewspec Pro software was used to extract the spectral data. First, the spectral curve was modified through the parabola correction function to avoid the jumping of connection points in spectral acquisition, and then the Savitzky–Golay smoothing filter with 10 points was used to smooth the spectral reflectance curve in order to eliminate the reflectance error caused by background noise during spectral data acquisition.

To highlight the correlation between the spectral reflectance and soil elements, two spectral mathematical transformations, original spectral reflectance (OR) and first derivative reflectance (FD), were used. The first derivative processing of the spectrum can decompose the overlapping mixed spectrum, expand the spectral characteristic difference between samples, and facilitate the determination of the spectral sensitive band (SB) [38]. The first derivative (FD) conversion formula is:

$$\mathrm{FDR}(\lambda_i) = \frac{R(\lambda_{i+1}) - R(\lambda_{i-1})}{\Delta\lambda} \tag{1}$$

where λ_i is the wavelength of each band, $\mathrm{FDR}(\lambda_i)$ is the first derivative spectral value at wavelength λ_i, and $\Delta\lambda$ is the wavelength value from the band I to band i plus 1.

In addition, owing to data redundancy in many hyperspectral data bands, to improve the model's accuracy, the original spectral reflectance (OR) and first derivative reflectance (FD) were used as independent variables, while Pearson correlation analysis was performed with soil carbon, nitrogen, and phosphorus contents, separately; this process was implemented based on R software. Since the correlation between the raw spectral reflectance and soil carbon, nitrogen, and phosphorus contents was poorly calculated, the band with the first derivative spectral reflectance correlation coefficient of > 0.3 was selected as the sensitive band.

2.5. Model Establishment and Verification

The three soil spectral sample libraries were divided into two groups: one group was the sample set for the model building construction, while the other was the validation set for verifying the accuracy of the model built. The ratio of the number of samples in the modelling set to the number of samples in the validation set for each sample library was 2:1 by soil. The hybrid sample library was modelled and validated based on the original spectral reflectance and first derivative spectral reflectance, respectively; the *P. australis* and *S. salsa* soil sample libraries were modelled and validated based on the original spectral reflectance, first derivative spectral reflectance, and sensitive band spectral reflectance. In this study, three models—random forest, support vector machine, and partial least-squares regression—were selected for hyperspectral inversion of soil carbon, nitrogen, and phosphorus contents. SVM is a popular machine learning technique with relevant learning algorithms for the analysis, classification, and regression analysis of the data provided. PLSR is an operational method based on principal component analysis that facilitates data dimensionality reduction. RF is an integrated learning algorithm for classification and regression and is constructed by combining the results of various decision trees and bagging the original dataset to select samples. The model construction in this study was implemented using Wake 3.8 software, where the model was first trained by a modelling set and then tested for accuracy using a validation set. The accuracy and stability of the models were assessed using the coefficient of determination (R^2), root-mean-square error (RMSE), and residual prediction deviation (RPD). The larger the R^2, the smaller the RMSE, indicating a higher model estimation accuracy; otherwise, the accuracy of model estimation was poor [39]. The RPD values reflect the calibration model's ability to predict the chemical data. Regarding the RPD statistic, an RPD of <1.4 indicates that it is insufficient for applications, values ranging from 1.4–2.0 indicate approximately quantitative predictions, and a value of >2.0 indicates excellent prediction [40].

3. Results

3.1. Characterization of Carbon, Nitrogen, and Phosphorus Contents of Soils in the Yellow River Delta

One-way ANOVA implemented in SPSS was used to test the significance of the carbon, nitrogen, and phosphorus contents in the soil under the two coastal wetland plants and under different soil layers, respectively. The results in Figure 2 show a significant difference ($p < 0.05$) between the soil TC and TP contents under *P. australis* and *S. salsa*, while no significant difference was observed in soil TN. Uniformly, there were no significant differences in the soil carbon, nitrogen, and phosphorus contents between the soil layers for either *P. australis* or *S. salsa*. The trend of the soil carbon, nitrogen, and phosphorus contents of *S. salsa* basically followed the pattern of decreasing with the deepening of the soil layer, while the second soil layer (20–40 cm) of *P. australis* had the lowest TC and TN, and TP increased with the deepening of the soil layer.

Based on the results of the significance tests of the soil TC, TN, and TP under different vegetation types, *P. australis* and *S. salsa* can be divided into separate layers for their respective modelling predictions, while there is no significant difference between the carbon, nitrogen, and phosphorus contents of the different soil layers, and hence, the three layers of soil spectral data can be mixed for analysis and processing.

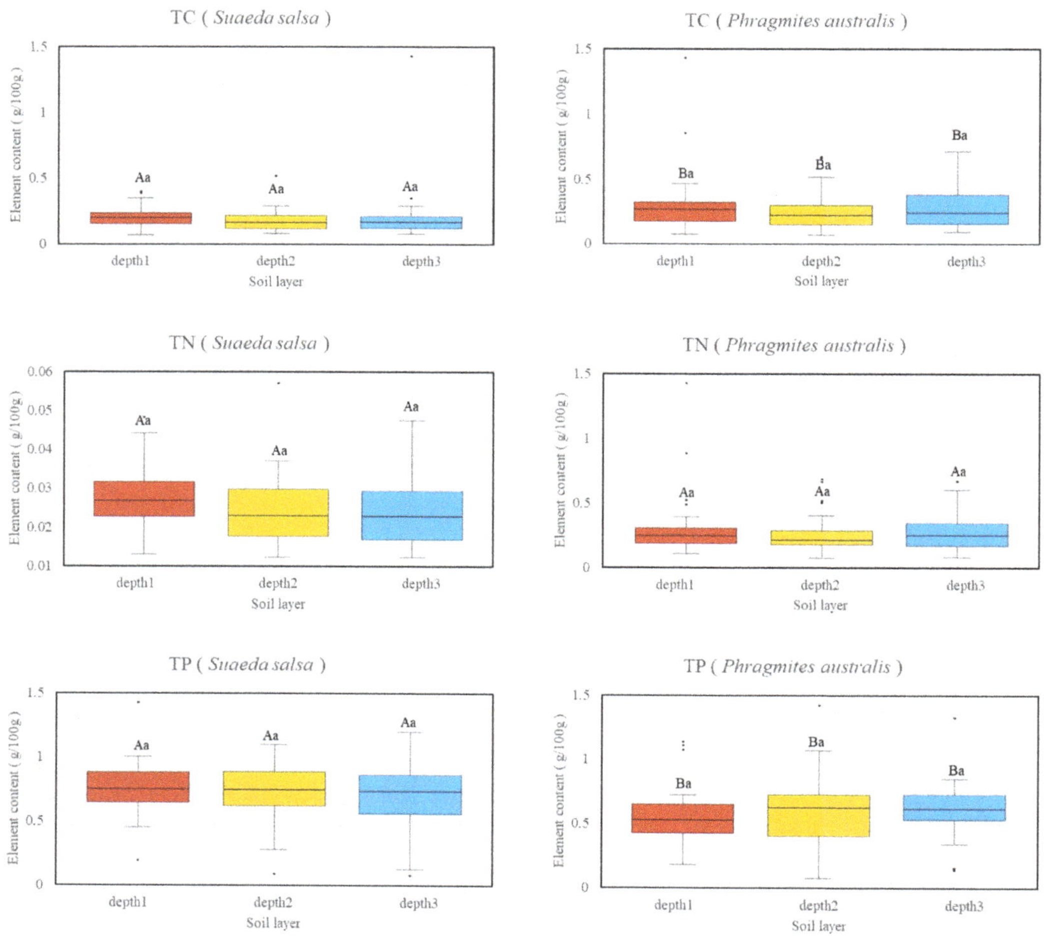

Figure 2. Total carbon (TC), total nitrogen (TN), and total phosphorus (TP) contents of the inter-root soils of two plants, *S. salsa* and *P. australis* (depth 1 = 0–20 cm, depth 2 = 20–40 cm, and depth 3 = 40–60 cm). Notes: different capital letters (A, B) represent significant differences in the soil elements under the different vegetation types, and lowercase letters (a) represent nosignificant differences in the soil elements in the different soil layers of the same vegetation type.

3.2. Spectral Data of the Two Wetland Plant Soils

The spectral reflectance curves of all soil samples under the two vegetation types in the wavelength range of 350–2500 nm are shown in Figure 3, where the general trends of the measured spectral reflectance curves of the soil samples under the different vegetation types are the same. The soil spectra under both vegetation types showed distinct soil spectral absorption peaks near 1400, 1750, and 2300 nm, but the depths and areas of the absorption peaks were different. Comparing the spectral reflectance curves of *S. salsa* and *P. australis*, the reflectance curves of *P. australis* soils were more concentrated, indicating that the structural components of the *P. australis* root soils were relatively stable.

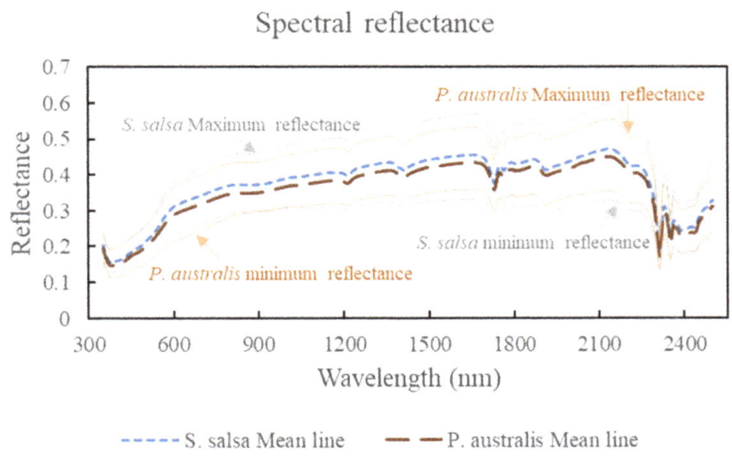

Figure 3. Spectral reflectance curves of *P. australis* and *S. salsa*.

3.3. Correlation Analysis

The correlation coefficients were calculated between the soil TC, TN, and TP contents and the soil original spectral reflectance and first derivative reflectance, respectively (Figure 4). The soil correlation coefficient curves of the two vegetation types showed similar trends, but the maximum correlation coefficients differed.

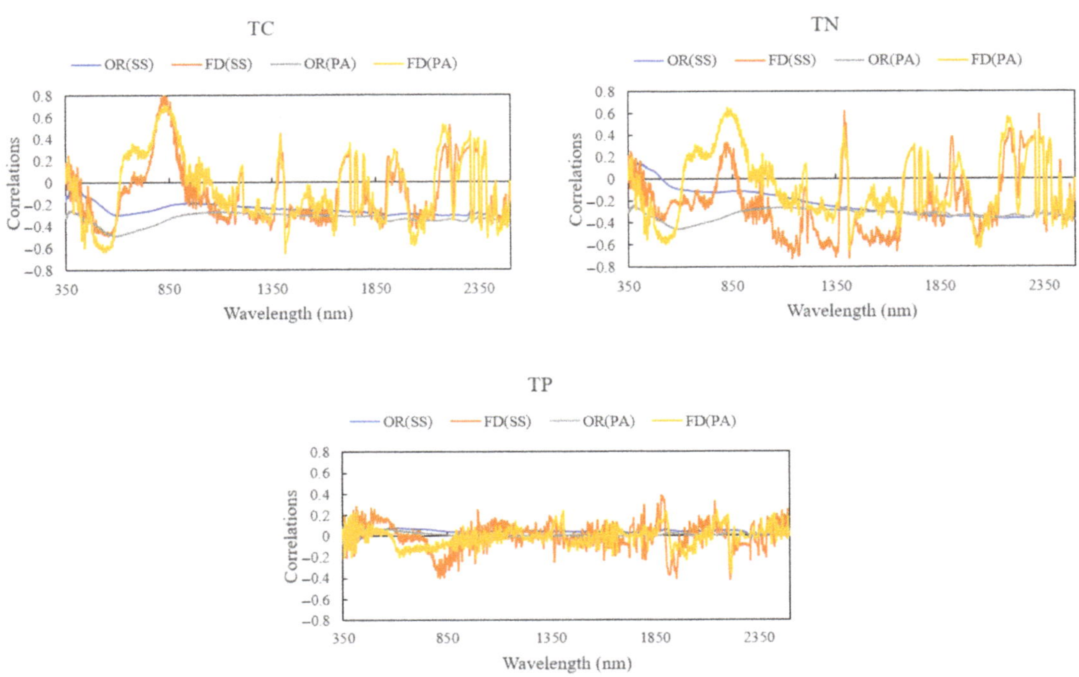

Figure 4. Correlation between the original reflectance (OR) and first derivative reflectance (FD) of *P. australis* (PA) and *S. salsa* (SS) and the soil total carbon (TC), total nitrogen (TN), and total phosphorus (TP) contents.

Among them, the soil TC and TN are negatively correlated with spectral reflectance, while the soil TP is positively correlated with spectral reflectance. Compared with the original spectral reflectance, the correlation coefficients of the first derivative reflectance were higher, while those between the soil TC, TN, and TP contents and the first derivative reflectance showed a positive and negative crossover, with more peaks and valleys, and the maximum correlation coefficients were greatly improved compared with the original spectra. As the correlation coefficients between the first derivative reflectance and the soil carbon, nitrogen, and phosphorus were much higher than the original reflectance, the first derivative reflectance of each element was used to establish a new database of sensitive bands for each element.

3.4. Mixed Species Modelling Effects

The *P. australis* and *S. salsa* sample data were mixed and then divided into sample and validation sets. Prediction models based on the original spectral reflectance and first derivative spectral reflectance data were established using PLSR, RF, and SVM models. Comparing the models obtained using these three methods (Figure 5) revealed that the modelling accuracies of the PLSR and RF were relatively higher than that of the SVM. Additionally, the R^2 values of the PLSR models were >0.82, with most values being >0.90. However, the modelling accuracy of SVM was relatively low. Except for the R^2 of the TN prediction model based on first derivative processing, which reached 0.99, the other R^2 values were <0.80. The prediction model based on the first derivative processing of the spectral reflectance was significantly more accurate than the prediction model based on the original spectral reflectance data, except for the SVM models for TC and TN.

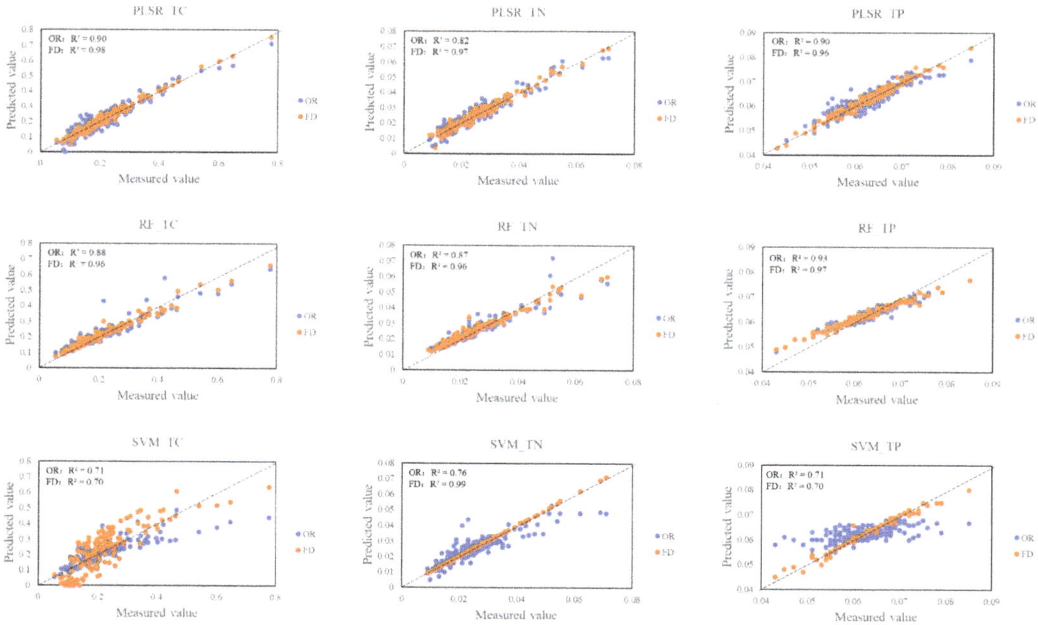

Figure 5. Modelling results of the carbon, nitrogen, and phosphorus levels in inverted wetland soils using three models: partial least-squares regression (PLSR), random forest (RF), and support vector machine (SVM). OR: original reflectance; FD: first derivative reflectance.

Among the three modelling methods, the PLSR and RF models had the best modelling and prediction accuracies, whereas the SVM model had a relatively poor prediction accuracy.

3.5. Effect of Sample Set Division and Sensitive Band on Modelling

3.5.1. Random Forest Regression Models

The RF model had a high prediction accuracy for the soil TC, TN, and TP (Figure 6). The modelling and verification of these three models were great, with high stability and accuracy. The prediction accuracy of the soil TC, TN, and TP in the models was high. The R^2 of the TC model was at least 0.84, and the RMSE was 0.053 g/kg. The accuracy of soil TN prediction in the model was slightly better than that of TC, with a lowest R^2 of 0.90 and an RMSE of 0.006. When modelling the soil TP, the model accuracy was the highest, with the highest R^2 (reaching 0.92), while the RMSE was 0.002. Considering model accuracy, the RF model was reliable and excellent in predicting the results of the soil TC, TN, and TP.

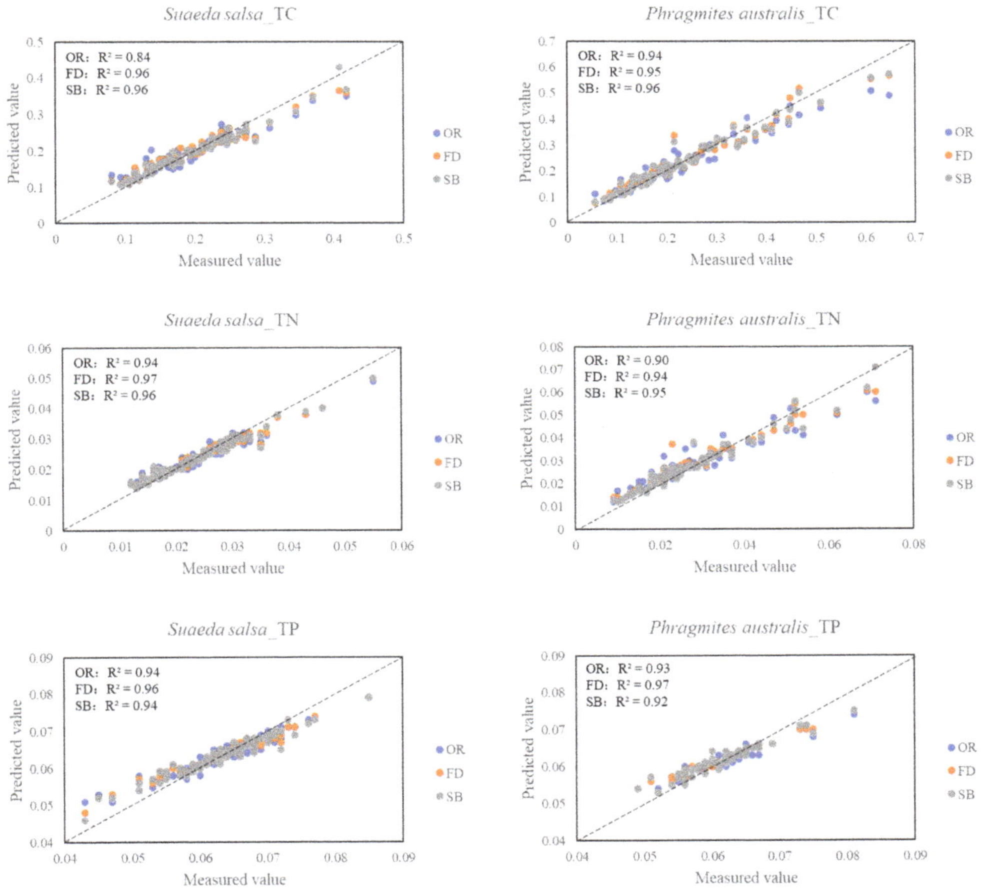

Figure 6. RF modelling results for *P. australis* and *S. salsa*. OR: original reflectance; FD: first derivative reflectance; SB: sensitive band.

3.5.2. Partial Least-Squares Regression Models

The PLSR model could accurately model and predict the soil carbon, nitrogen, and phosphorus contents (Figure 7). The modelling accuracy of the PLSR model based on the original spectral reflectance and first derivative spectral reflectance was very high, with an R^2 of >0.95. The PLSR model based on the sensitive bands performed equally well in most cases, but the R^2 value of the PLSR model for the *P. australis* rhizosphere soil TP was only 0.37; thus, the model was not reliable. Comparing the models constructed using the three spectral data types, the model accuracy of the PLSR model based on the first derivative for both *P. australis* and *S. salsa* was very high, with an R^2 of 0.99. The PLSR model based on the sensitive bands was less stable than the PLSR model based on the other two spectral data types.

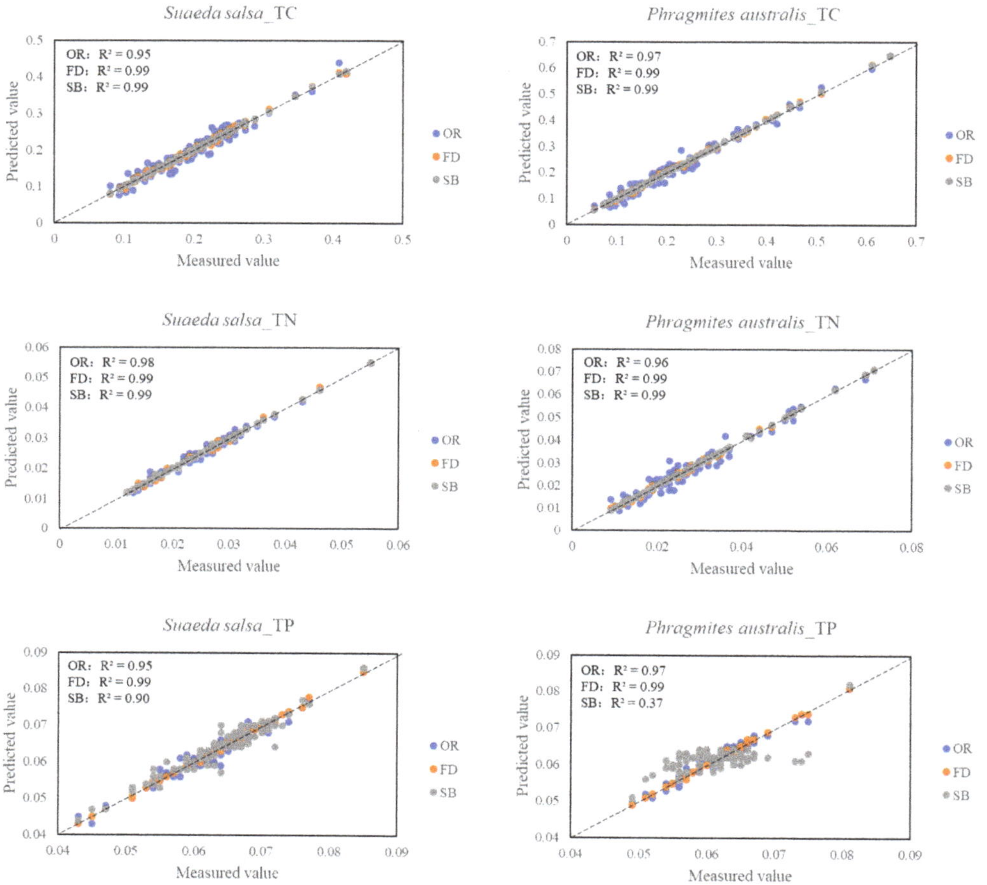

Figure 7. PLSR modelling results for *P. australis* and *S. salsa*. OR: original reflectance; FD: first derivative reflectance; SB: sensitive band.

3.5.3. Support Vector Machine Regression Models

The SVM model had good prediction effects for the soil TC, TN, and TP. The prediction accuracy among the different elements decreased in the following order: TC, TN, and TP (Figure 8). The prediction effect for the soil TC was the best, where the highest R^2 was 0.90 and the lowest was 0.82. The prediction of the soil TN was slightly less accurate, where the highest R^2 was 0.82 and the lowest was 0.71. The prediction of the soil TP was the worst, with the highest R^2 being 0.78 and the lowest being 0.33. Consequently, the reliability of the model predictions was low. Comparing the models of the three spectral data, we found that the R^2 of the SVM model based on the sensitive bands was improved compared with that of the SVM model based on the original spectral reflectance, and it was significantly improved compared with that of the SVM model based on the first derivative spectral reflectance. However, the usage of the sensitive bands reduced the R^2 of the model when modelling the contents of TC and TN in the *P. australis* root soil.

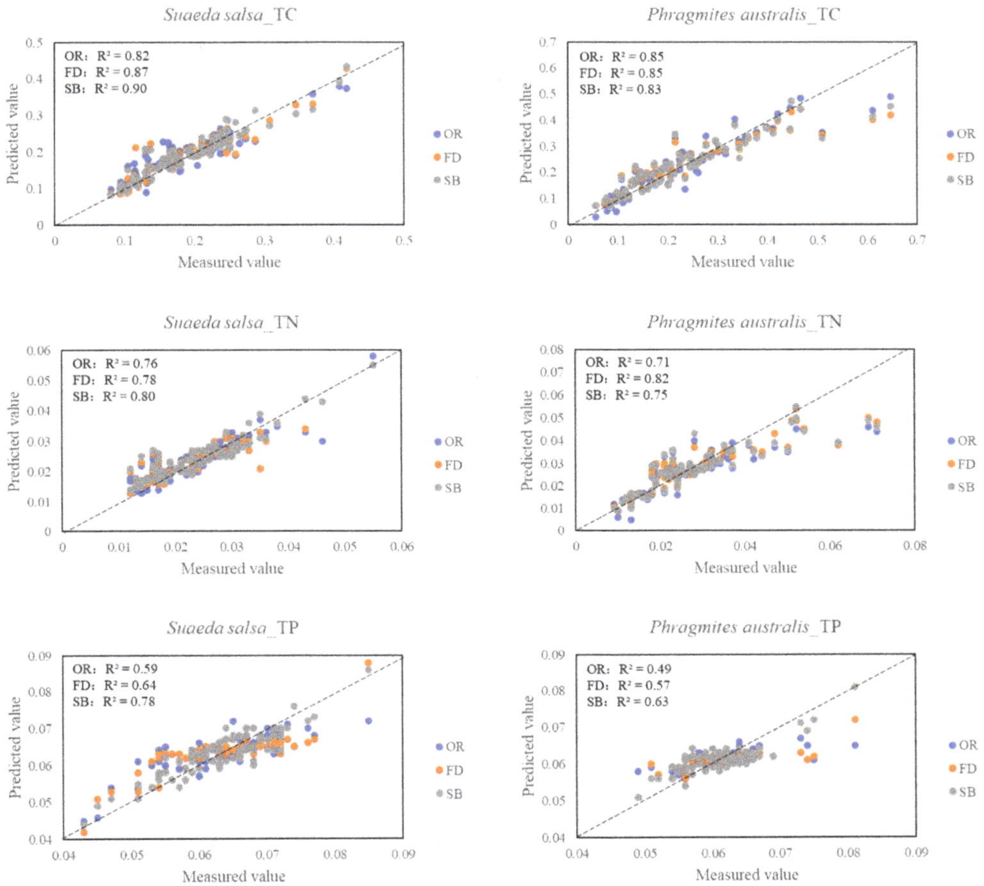

Figure 8. SVM modelling results for *P. australis* and *S. salsa*. OR: original reflectance; FD: first derivative reflectance; SB: sensitive band.

3.6. Accuracy of the Prediction Models

In this study, the R^2, RMSE, and RPD were used to evaluate the stability and accuracy of retrieving the soil carbon, nitrogen, and phosphorus contents of two dominant plant species using different spectral processing methods and different models. As shown in Table 1, among the three models, the RF model had the best inversion results for the soil carbon, nitrogen, and phosphorus contents. Both the modelling and verification effects were better than those of the other two models, while the PLSR model was slightly better than the SVM model in the other two models. Among the inversion models for soil TC, TN, and TP, the model for TC had the highest prediction accuracy, where its R^2 was the highest (0.57) and its RPD was 1.46, indicating that the prediction results of the model were reliable. The prediction effect of TN was slightly worse than that of TC, where the lowest R^2 value was 0.47. The inversion effect of each model was largely reliable. The inversion effect of TP was the worst: its lowest R^2 was 0.29 and its RPD was less than 1.4. However, some of these prediction models are unreliable. By comparing the prediction effects on different plant root systems, it was found that the prediction effect for the *P. australis* root soil nutrient elements was slightly better than that for the *S. salsa* soil.

Table 1. Accuracy of the prediction models for *P. australis* (PA) and *S. salsa* (SS) total carbon (TC), total nitrogen (TN), and total phosphorus (TP).

Index			PLSR					RF					SVM				
			Model		Test		RPD	Model		Test		RPD	Model		Test		RPD
			R^2	RMSE	R^2	RMSE		R^2	RMSE	R^2	RMSE		R^2	RMSE	R^2	RMSE	
TC	PA	OR	0.97	0.020	0.79	0.078	2.11	0.94	0.051	0.87	0.054	2.69	0.85	0.094	0.77	0.074	1.37
		FD	0.99	0.004	0.86	0.064	2.31	0.95	0.045	0.91	0.046	2.90	0.85	0.102	0.80	0.084	0.98
		SB	0.99	0.001	0.89	0.052	2.95	0.96	0.042	0.92	0.043	3.19	0.83	0.098	0.79	0.083	1.09
	SS	OR	0.95	0.016	0.57	0.060	1.46	0.84	0.053	0.67	0.047	1.56	0.82	0.099	0.79	0.039	1.78
		FD	0.99	0.015	0.62	0.056	1.58	0.96	0.023	0.81	0.037	1.93	0.87	0.024	0.85	0.033	2.37
		SB	0.99	0.005	0.63	0.057	1.55	0.96	0.010	0.87	0.031	2.27	0.90	0.023	0.88	0.028	2.76
	MIX	OR	0.90	0.035	0.67	0.119	1.12	0.88	0.050	0.73	0.109	0.91	0.71	0.089	0.64	0.104	0.76
		FD	0.98	0.002	0.65	0.122	1.04	0.96	0.027	0.86	0.092	0.97	0.70	0.096	0.61	0.113	1.37
TN	PA	OR	0.96	0.003	0.58	0.008	1.99	0.90	0.006	0.85	0.006	2.26	0.71	0.012	0.61	0.011	0.82
		FD	0.99	0.001	0.76	0.008	1.88	0.94	0.005	0.92	0.005	2.92	0.82	0.011	0.68	0.010	1.04
		SB	0.99	0.001	0.87	0.006	2.74	0.95	0.004	0.94	0.004	3.43	0.75	0.011	0.64	0.010	0.91
	SS	OR	0.98	0.001	0.47	0.006	1.40	0.94	0.003	0.64	0.005	1.23	0.76	0.004	0.71	0.004	1.22
		FD	0.99	0.001	0.62	0.005	1.34	0.97	0.001	0.76	0.004	1.80	0.78	0.002	0.73	0.004	1.58
		SB	0.99	0.001	0.66	0.005	1.46	0.96	0.001	0.78	0.004	1.85	0.80	0.002	0.76	0.004	1.68
	MIX	OR	0.82	0.003	0.71	0.010	1.36	0.87	0.005	0.48	0.012	0.74	0.76	0.008	0.72	0.010	1.29
		FD	0.97	0.001	0.64	0.011	3.33	0.96	0.002	0.70	0.009	1.03	0.99	0.004	0.78	0.008	1.36
TP	PA	OR	0.97	0.001	0.52	0.004	1.41	0.93	0.002	0.62	0.003	1.33	0.49	0.004	0.44	0.004	1.10
		FD	0.99	0.001	0.60	0.004	1.49	0.97	0.002	0.70	0.003	1.18	0.57	0.004	0.50	0.004	1.17
		SB	0.37	0.004	0.29	0.005	0.75	0.92	0.002	0.76	0.003	1.73	0.63	0.003	0.53	0.004	1.20
	SS	OR	0.95	0.002	0.45	0.006	1.18	0.94	0.003	0.52	0.005	1.00	0.59	0.005	0.52	0.005	1.19
		FD	0.99	0.001	0.53	0.005	1.21	0.96	0.001	0.64	0.004	1.34	0.64	0.003	0.61	0.004	1.11
		SB	0.90	0.002	0.69	0.004	1.74	0.94	0.001	0.73	0.004	1.81	0.78	0.003	0.74	0.004	1.95
	MIX	OR	0.90	0.002	0.47	0.007	1.00	0.93	0.002	0.40	0.006	0.47	0.71	0.002	0.39	0.005	0.48
		FD	0.96	0.001	0.49	0.006	0.93	0.97	0.001	0.31	0.006	0.30	0.70	0.002	0.33	0.006	0.72

4. Discussion

4.1. Soil Nutrient Differences

Changes in the soil nutrient contents are the result of a combination of environmental and biological factors, with the former dominated by the altitude gradient, temperature, and soil texture and the latter dominated by vegetation type and soil animal activity. Soil nutrient elements vary in their concentrations and forms of existence, among which the soil organic carbon content is primarily influenced by the decomposition and transformation rate of apoplastic matter, and it is also highly affected by surface vegetation growth [41]. The soil TN content is also closely related to the soil organic carbon content, whereas soil nitrogen is mainly fixed by rhizobia. The soil TP content varied greatly among the sample types. A likely cause of such variation is that soil phosphorus content is affected by multiple soil ecochemical processes, such as weathering and the leaching enrichment of phosphorus-bearing ores [42].

As shown in Figure 2, the distribution trends of the root soil nutrients displayed differences between the *P. australis* and *S. salsa* samples. The C, N, and P contents in the root soil of *S. salsa* in saline land decreased with increasing soil depth; however, the root soil of *P. australis* showed the opposite trend. Noticeably, the C, N, and P contents in the deepest soil (20–40 cm) were the highest. This phenomenon may have been caused by the combined effects of the degree of soil flooding and the morphological differences between *P. australis* and *S. salsa* [43]. The C content of *P. australis* root soil was higher, which was likely due to the fact that *P. australis* has much more biomass than *S. salsa* and that the residual organic matter in the soil is higher. The N and P contents in the root soil system of *S. salsa* were higher than those in *P. australis*, which may be caused by the long-term flooding of the soil near the sea and the high contents of N and P in the soil pore water.

4.2. Differences in the Element Contents Retrieved from the Hyperspectral Data

Hyperspectral technology is used to identify substances and determine their chemical compositions and relative contents according to their spectral reflectance. Earlier studies have reported that the visible spectrum is created by the outer electronic transition, whereas the near-infrared spectrum is mainly affected by molecular vibrations, which potentially reflect the compositions and structures of molecules [44]. This basic principle is applied during the quantitative analysis of target substances using hyperspectral technology. While no two substances have the same spectral characteristics, this approach ensures that similar substances will have similar spectral characteristics. Noticeably, during the collection of the soil spectral information, the presence of soil moisture and clay minerals was the main cause for the occurrence of spectral absorption peaks [45,46]. The complex structure and composition of soil often leads to a large number of interference factors in the spectral information [47]. To avoid these unwanted spectral peaks, unified drying and grinding steps are used, which reduce the impacts of soil moisture and soil structure on the spectral reflectance, thus improving the signal-to-noise ratio and contributing to modelling and inversion in the next step.

The contents of soil C and N are considered soil properties that have a direct impact on reflectance [48]. The multiplicity and ensemble frequencies of molecular vibrations are the main sources of differences in spectral reflectance; therefore, spectral analysis is commonly used for the analysis of organic matter containing C-H, N-H, O-H, and other groups. The vast majority of soil N is in the organic-bound state and displays a strong correlation with the C content. Therefore, hyperspectral technology can be used to establish a model of the C and N soil contents for rapid estimation, as well as to achieve a high prediction accuracy. In our study, the prediction accuracy of the TP content prediction models for all types of samples and treated soils was lower than that for soil TC and soil TN. This situation was consistent with the results of the correlation analysis. The poor accuracy of the prediction models and correlation analysis was likely caused by the low P content in the soil, which is known to increase the difficulty of prediction [49].

4.3. Preprocessing Transformations

Hyperspectral data have very high spectral resolutions; thus, while they provide information, most are redundant. Therefore, it is necessary to remove this redundant information to reduce its impact on the establishment of soil TC, TN, and TP content prediction models. In our study, three types of soil spectral reflectance data, namely original spectral reflectance, first derivative spectral reflectance, and first derivative sensitive band spectral reflectance, were selected. The soil nutrient contents of TC, TN, and TP were estimated using three modelling methods—PLSR, RF, and SVM, respectively—and the model prediction accuracy was also improved or largely similar in this study compared with the results of the previous study (Table 2). The correlation analysis indicated that the first derivative processing of spectral reflectance significantly improved the correlation between the spectral reflectance and soil nutrient content compared with the original spectral reflectance, and this result is consistent with the conclusions of other recent studies [50]. Table 1 and Figures 4–6 show the prediction results of the prediction models based on the three spectral reflectance data. In general, the accuracy of the prediction model based on the sensitive band was higher than that of those based on the original spectral reflectance and first derivative spectral reflectance. However, there were exceptions, such as the prediction model of soil TP content, which displayed a decline in accuracy. This was likely caused by the spectral reduction in the signal-to-noise ratio [51]. The process of extracting sensitive bands removed some information related to soil phosphorus. In addition, the prediction model based on the first derivative spectral reflectance had better accuracy than the prediction model based on the original spectral reflectance, which suggests that the conversion of spectral variables could effectively eliminate the impacts of environment, soil, and other factors on the spectral information.

Table 2. Comparison of soil carbon, nitrogen, and phosphorus between other research results and this paper.

Element	Accuracy	Spectral Data	Model	Author	Accuracy in the Present Study
C	R^2 = 0.95	First derivative	RF	Wang S. et al., 2022 [52]	R^2 = 0.91
C	R^2 = 0.44	Untransformed	PLSR	Mondal B. P. et al., 2019 [53]	R^2 = 0.67
C	R^2 = 0.81	Smoothed reflectance	PLSR	Ribeiro S. G. et al., 2021 [49]	R^2 = 0.89
C	RPD = 2.52	Untransformed	PLSR	Anna P. et al., 2020 [54]	RPD = 2.11
N	R^2 = 0.76	Untransformed	RF	Lin X. L. et al., 2022 [55]	R^2 = 0.85
N	R^2 = 0.35	Untransformed	PLSR	Pechanec V. et al., 2021 [56]	R^2 = 0.71
N	R^2 = 0.94	Untransformed	SVM	Xu S. X. et al., 2021 [23]	R^2 = 0.72
N	R^2 = 0.81	Smoothed reflectance	PLSR	Li H. Y. et al., 2019 [57]	R^2 = 0.87
P	R^2 = 0.34	Untransformed	PLSR	Malmir M. et al., 2019 [58]	R^2 = 0.47
P	R^2 = 0.54	Untransformed	PLSR	Lu P. et al., 2013 [59]	R^2 = 0.47

4.4. Differences in the Prediction Accuracy of the Different Models

RF is an integral machine learning algorithm that is used for classification and regression. It was constructed by combining the results of various decision trees and bagging the original dataset to select samples. SVM is a popular machine learning technology. This supervised learning model contains related learning algorithms that are used to analyze, classify, and conduct regression analyses on the supplied data. PLSR integrates various analyses, such as correlation, principal component analysis, and multiple linear regression, to identify the primary control factors affecting the dependent variable (soil C, N, and P contents) from the high-dimensional data while reducing the dimensionality of spectral analysis, which makes the constructed model more robust [60].

By comparing the results in Table 1 and Figures 5–8, we found that the best inversion model for the soil nutrient content was the RF model. Compared with the other two models, the modelling R^2 was higher, and the RMSE was lower. The R^2 of the PLSR model was much higher than that of the SVM model during modelling; however, there was little difference between the PLSR and SVM models during validation. This could have been caused by overfitting during modelling by the PLSR, resulting in high modelling but low validation accuracy.

5. Conclusions

To explore the spatial distribution characteristics of soil carbon, nitrogen, and phosphorus contents of different plant roots in the Yellow River Delta and to predict soil carbon, nitrogen, and phosphorus potential using hyperspectral techniques, stratified soil sampling was conducted for two vegetation types. In addition, soil samples were subjected to traditional chemical measurements and spectral data collection. While studying the distribution characteristics of TC, TN, and TP soil nutrients based on the measured values, a rapid inversion of soil total carbon, nitrogen, and phosphorus contents and real-time monitoring of the soil quality were established based on the original spectral reflectance and first spectral reflectance of the soil. Our findings are as follows:

(1) There was a significant difference in soil total carbon and phosphorus contents between the two vegetation types, and no significant difference in soil carbon, nitrogen, and phosphorus contents between the different strata under the same vegetation type was observed. Therefore, the influence of vegetation type should be considered prior to making modelling predictions.

(2) The trends of soil correlation coefficient curves were similar for different vegetation types, while the maximum correlation coefficients differed. The first derivative reflectance had a large increase in correlation coefficients compared with the correlation coefficients between the original spectral reflectance and soil carbon, nitrogen, and phosphorus contents.

(3) The first derivative treatment and division of sample sets according to vegetation types improved the modelling accuracy. The best prediction method for the TC, TN, and TP contents of the *P. australis* soils was to divide the sample set plus FD plus RF, while that of the *S. salsa* soils was to divide the sample set plus FD plus SVM.

Author Contributions: Conceptualization, L.N. and W.L.; data curation, L.N. and Z.D.; funding acquisition, W.L.; methodology, X.T.; project administration, Y.L.; resources, L.C.; software, X.Z. (Xiajie Zhai) and X.Z. (Xinsheng Zhao); supervision, J.W.; visualization, J.L. and W.L.; writing—original draft, L.N.; writing—review and editing, W.L. All authors have read and agreed to the published version of the manuscript.

Funding: This work was supported by China's Special Fund for Basic Scientific Research Business of Central Public Research Institutes (CAFYBB2021ZB003) and the National Key R&D Program of China (2017YFC0506200).

Institutional Review Board Statement: Not applicable.

Data Availability Statement: Data is available from the corresponding author upon request.

Conflicts of Interest: The authors declare no conflict of interest.

References

1. Bartholomeus, H.; Kooistra, L.; Stevens, A.; van Leeuwen, M.; van Wesemael, B.; Ben-Dor, E.; Tychon, B. Soil Organic Carbon mapping of partially vegetated agricultural fields with imaging spectroscopy. *Int. J. Appl. Earth Obs. Geoinf.* **2011**, *13*, 81–88. [CrossRef]
2. Rahman, M.M.; Kamal, M.Z.U.; Ranamukhaarachchi, S.; Alam, M.S.; Alam, M.K.; Khan, M.A.R.; Islam, M.M.; Alam, M.A.; Jiban, S.I.; Al Mamun, M.A.; et al. Effects of Organic Amendments on Soil Aggregate Stability, Carbon Sequestration, and Energy Use Efficiency in Wetland Paddy Cultivation. *Sustainability* **2022**, *14*, 4475. [CrossRef]
3. Ledford, K.; Schmidt, S.A.; Ahn, C. Assessing carbon storage potential of forested wetland soils in two physiographic provinces of northern virginia, USA. *Sustainability* **2022**, *14*, 2048. [CrossRef]
4. Reich, P.B.; Tjoelker, M.G.; Machado, J.; Oleksyn, J. Universal scaling of respiratory metabolism, size and nitrogen in plants. *Nature* **2006**, *439*, 7075. [CrossRef] [PubMed]
5. Lu, M.; Liu, K.; Zhang, L.; Zeng, F.; Song, T.; Peng, W.; Du, H. Stoichiometric Variation in Soil Carbon, Nitrogen, and Phosphorus Following Cropland Conversion to Forest in Southwest China. *Forests* **2022**, *13*, 1155. [CrossRef]
6. Spohn, M.; Stendahl, J. Carbon, nitrogen, and phosphorus stoichiometry of organic matter in Swedish forest soils and its relationship with climate, tree species, and soil texture. *Biogeosciences* **2022**, *19*, 2171–2186. [CrossRef]
7. Wang, X.W.; Sun, X.X.; Sun, L.; Chen, N.; Du, Y. Small-Scale Variability of Soil Quality in Permafrost Peatland of the Great Hing'an Mountains, Northeast China. *Water* **2022**, *14*, 2597. [CrossRef]
8. Serrani, D.; Cocco, S.; Cardelli, V.; D'Ottavio, P.; Borguete, A.R.R.; Feniasse, D.; Vilanculos, A.; Fernández-Marcos, M.L.; Giosué, C.; Tittarelli, F.; et al. Soil fertility in slash and burn agricultural systems in central Mozambique. *J. Environ. Manag.* **2022**, *322*, 116031. [CrossRef]
9. Tedontsah, V.P.L.; Mbog, M.B.; Ngon, G.F.N.; Edzoa, R.C.; Tassongwa, B.; Bitom, D.; Etame, J. Physicochemical Properties and Fertility Assessment of Soils in Foumban (West Cameroon). *Appl. Environ. Soil Sci.* **2022**, *2022*, 7889430. [CrossRef]
10. Chen, M.Y.; Yang, X.; Shao, M.A.; Wei, X.R.; Li, T.C. Changes in soil C-N-P stoichiometry after 20 years of typical artificial vegetation restoration in semiarid continental climate zones. *Sci. Total Environ.* **2022**, *852*, 158380. [CrossRef]
11. Sweetman, A.J.; Valle, M.D.; Prevedouros, K.; Jones, K.C. The role of soil organic carbon in the global cycling of persistent organic pollutants (POPs): Interpreting and modelling field data. *Chemosphere* **2005**, *60*, 959–972. [CrossRef] [PubMed]
12. Amol, D.V.; Karbhari, V.K.; Sandeep, V.G. Estimation of soil nitrogen in agricultural regions by VNIR reflectance spectroscopy. *SN Appl. Sci.* **2020**, *2*, 21–56.
13. Zheng, W.; Lu, X.; Li, Y. Hyperspectral identification of chlorophyll fluorescence parameters of *Suaeda salsa* in coastal wetlands. *Remote Sens.* **2021**, *13*, 2066. [CrossRef]
14. Lin, C.; Hu, Y.; Liu, Z.; Peng, Y.; Wang, L.; Peng, D. Estimation of cultivated land quality based on soil hyperspectral data. *Agriculture* **2022**, *12*, 93–98. [CrossRef]
15. Dhawale, N.M.; Adamchuk, V.I.; Prasher, S.O.; Rossel, R.A.V.; Ismail, A.A. Evaluation of two portable hyperspectral-sensor-based instruments to predict key soil properties in canadian soils. *Sensors* **2022**, *22*, 2556. [CrossRef] [PubMed]
16. Yu, S.y.; Bu, H.R.; Dong, W.C.; Jiang, Z.; Zhang, L.X.; Xia, Y.Q. Construction and Evaluation of Prediction Model of Main Soil Nutrients Based on Spectral Information. *Appl. Sci.* **2022**, *12*, 6298. [CrossRef]
17. Odebiri, O.; Mutanga, O.; Odindi, J.; Naicker, R.; Masemola, C.; Sibanda, M. Deep learning approaches in remote sensing of soil organic carbon: A review of utility, challenges, and prospects. *Environ. Monit. Assess.* **2021**, *193*, 802. [CrossRef]

18. Misbah, K.; Laamrani, A.; Khechba, K.; Dhiba, D.; Chehbouni, A. Multi-Sensors remote sensing applications for assessing, monitoring, and mapping npk content in soil and crops in african agricultural land. *Remote Sens.* **2021**, *14*, 81. [CrossRef]
19. Chandan, G.; Naorem, J.S.; Bijoy, K.H. Hyperspectral spectroscopic study of soil properties a review. *Int. J. Plant Soil Sci.* **2020**, *23*, 14–25.
20. Ewing, J.; Oommen, T.; Jayakumar, P.; Alger, R. Utilizing hyperspectral remote sensing for soil gradation. *Remote Sens.* **2020**, *12*, 3312. [CrossRef]
21. Mônica, S.C.; Marcos, R.N.; Marcelo, L.C.; Renato, H.F.; Everson, C.; Roney, B.D.O. Hyperspectral remote detection as an alternative to correlate data of soil constituents. *Remote Sens. Appl. Soc. Environ.* **2019**, *16*, 100270.
22. Michael, V.; Marie, L.; Sören, T.; Bernard, L. Quantification of soil properties with hyperspectral data: Selecting spectral variables with different methods to improve accuracies and analyze prediction mechanisms. *Remote Sens.* **2017**, *9*, 1103.
23. Xu, S.X.; Wang, M.Y.; Shi, X.Z.; Yu, Q.B.; Zhang, Z.Q. Integrating hyperspectral imaging with machine learning techniques for the high-resolution mapping of soil nitrogen fractions in soil profiles. *Sci. Total Environ.* **2021**, *754*, 142135. [CrossRef] [PubMed]
24. Dhawale, N.M.; Adamchuk, V.I.; Prasher, S.O.; Rossel, R.A.V. Evaluating the Precision and Accuracy of Proximal Soil vis–NIR Sensors for Estimating Soil Organic Matter and Texture. *Soil Syst.* **2021**, *5*, 48. [CrossRef]
25. Guo, H.L.; Zhang, R.R.; Dai, W.H.; Zhou, X.W.; Zhang, D.J.; Yang, Y.H.; Cui, J. Mapping Soil Organic Matter Content Based on Feature Band Selection with ZY1-02D Hyperspectral Satellite Data in the Agricultural Region. *Agronomy* **2022**, *12*, 2111. [CrossRef]
26. Zhu, J.J.; Jin, X.; Li, S.W.; Han, Y.L.; Zheng, W.R. Prediction of Soil Available Boron Content in Visible-Near-Infrared Hyperspectral Based on Different Preprocessing Transformations and Characteristic Wavelengths Modeling. *Comput. Intell. Neurosci.* **2022**, *2022*, 9748257. [CrossRef]
27. Tang, R.N.; Li, X.W.; Li, C.; Jiang, K.X.; Hu, W.F.; Wu, J.J. Estimation of Total Nitrogen Content in Rubber Plantation Soil Based on Hyperspectral and Fractional Order Derivative. *Electronics* **2022**, *11*, 1956. [CrossRef]
28. Zhang, Z.H.; Guo, F.; Xu, Z.; Yang, X.Y.; Wu, K.Z. On retrieving the chromium and zinc concentrations in the arable soil by the hyperspectral reflectance based on the deep forest. *Ecol. Indic.* **2022**, *144*, 109440. [CrossRef]
29. Naveen, J.P.A.; Amr, H.A.; David, B.L.; Nicole, A.H. Modeling soil parameters using hyperspectral image reflectance in subtropical coastal wetlands. *Int. J. Appl. Earth Obs. Geoinf.* **2014**, *33*, 47–56.
30. Zhang, S.; Lu, X.; Zhang, Y.; Nie, G.; Li, Y. Estimation of soil organic matter, total nitrogen and total carbon in sustainable coastal wetlands. *Sustainability* **2019**, *11*, 667. [CrossRef]
31. Zhao, D.M.; Wang, J.Z.; Miao, J.; Zhen, J.N.; Wang, J.J.; Gao, C.J.; Jiang, J.C.; Wu, G.F. Spectral features of Fe and organic carbon in estimating low and moderate concentration of heavy metals in mangrove sediments across different regions and habitat types. *Geoderma* **2022**, *426*, 116093. [CrossRef]
32. Chen, Z.Y.; Ren, S.J.; Qin, R.M.; Nie, P.C. Rapid Detection of Different Types of Soil Nitrogen Using Near-Infrared Hyperspectral Imaging. *Molecules* **2022**, *27*, 2017. [CrossRef] [PubMed]
33. Wei, D.; Zheng, G. Estimation of soil total phosphorus content in coastal areas based on hyperspectral reflectance. *Spectrosc. Spectr. Anal.* **2022**, *42*, 158–163.
34. Gao, Y.; Liu, L.; Zhu, P.; Yang, S.; Guo, W.; Yu, X. Patterns and dynamics of the soil microbial community with gradual vegetation succession in The Yellow River Delta, China. *Wetlands* **2021**, *41*, 9. [CrossRef]
35. Li, F.; Xie, Y.; Chen, X.; Hou, Z.; Li, X.; Deng, Z.; Liu, Y.; Hu, J.; Liu, N. Succession of aquatic macrophytes in the Modern Yellow River Delta after 150 years of alluviation. *Wetl. Ecol. Manag.* **2013**, *21*, 219–228. [CrossRef]
36. Liu, Q.; Liu, G.; Huang, C.; Li, H. Soil physicochemical properties associated with quasi-circular vegetation patches in the Yellow River Delta, China. *Geoderma* **2019**, *337*, 202–214. [CrossRef]
37. Li, G.; Ma, S.; Li, K.; Zhou, M.; Lin, L. Heterogeneity classification based on hyperspectral transmission imaging and multivariate data analysis. *Infrared Phys. Technol.* **2022**, *123*, 104180. [CrossRef]
38. Chen, Y.; Wang, J.; Liu, G. Hyperspectral estimation model of forest soil organic matter in northwest Yunnan Province, China. *Forests* **2019**, *10*, 217–233. [CrossRef]
39. Nikolaos, T.; Nikolaos, T.; Yaron, O.; Eleni, K.; Eyal, B.; John, T.; George, Z. An integrated methodology using open soil spectral libraries and Earth Observation data for soil organic carbon estimations in support of soil-related SDGs. *Remote Sens. Environ.* **2020**, *244*, 111793.
40. Wang, J.J.; Xu, Y.; Wu, G.F. The integration of species information and soil properties for hyperspectral estimation of leaf biochemical parameters in mangrove forest. *Ecol. Indic.* **2020**, *115*, 106467. [CrossRef]
41. Cui, L.; Dou, Z.; Liu, Z.; Zuo, X.; Lei, Y.; Li, J.; Zhao, X.; Zhai, X.; Pan, X.; Li, W. Hyperspectral inversion of *phragmites australis* carbon, nitrogen, and phosphorus stoichiometry using three models. *Remote Sens.* **2020**, *12*, 1998. [CrossRef]
42. Lin, Y.; Li, Z.; Lv, S.; Huang, H.; Hu, J. Detection of soil total nitrogen, phosphorus and potassium content based on the spectral information of citrus canopy. *Am. J. Biochem. Biotechnol.* **2020**, *16*, 177–183.
43. Sun, W.; Shuo, L.; Xia, Z.; Yao, L. Estimation of soil organic matter content using selected spectral subset of hyperspectral data. *Geoderma* **2022**, *409*, 115653. [CrossRef]
44. Danesh, M.; Bahrami, H.A. Modeling of Soil Sand Particles Using Spectroscopy Technology. *Commun. Soil Sci. Plant Anal.* **2022**, *53*, 2216–2228. [CrossRef]
45. Divyesh, M.V.; Ajay, K.M.; Onkar, D. Development of Spectral Indexes in Hyperspectral Imagery for Land Cover Assessment. *IETE Tech. Rev.* **2019**, *36*, 216–230.

46. Andrea, C.; Frédéric, A.; Francesca, G.; Gilles, G.; Sébastien, L.; Leandro, C.; Sandro, M. Improved estimation of soil clay content by the fusion of remote hyperspectral and proximal geophysical sensing. *J. Appl. Geophys.* **2015**, *116*, 135–145.
47. Nikolaos, T.; Nikolaos, T.; Eyal, B.; John, T.; George, Z. Employing a Multi-Input Deep Convolutional Neural Network to Derive Soil Clay Content from a Synergy of Multi-Temporal Optical and Radar Imagery Data. *Remote Sens.* **2020**, *12*, 1389.
48. Michael, S.K.; David, S.; Kevin, T.; Court, S. The Effects of tidal inundation on the reflectance characteristics of coastal marsh vegetation. *J. Coast. Res.* **2009**, *25*, 1177–1186.
49. Ribeiro, S.G.; Teixeira, A.D.S.; de Oliveira, M.R.R.; Costa, M.C.G.; Araújo, I.C.D.S.; Moreira, L.C.J.; Lopes, F.B. Soil Organic Carbon Content Prediction Using Soil-Reflected Spectra: A Comparison of Two Regression Methods. *Remote Sens.* **2021**, *13*, 4752. [CrossRef]
50. Gao, L.L.; Zhu, X.C.; Han, Z.Y.; Wang, L.; Zhao, G.X.; Jiang, Y.M. Spectroscopy-Based Soil Organic Matter Estimation in Brown Forest Soil Areas of the Shandong Peninsula, China. *Pedosphere* **2019**, *29*, 810–818. [CrossRef]
51. Xu, X.; Chen, S.; Xu, Z.; Yu, Y.; Zhang, S.; Dai, R. Exploring appropriate preprocessing techniques for hyperspectral soil organic matter content estimation in black soil area. *Remote Sens.* **2020**, *12*, 3765. [CrossRef]
52. Wang, S.; Guan, K.Y.; Zhang, C.H.; Lee, D.; Margenot, A.J.; Ge, Y.F.; Peng, J.; Zhou, W.; Zhou, Q.; Huang, Y.Z. Using soil library hyperspectral reflectance and machine learning to predict soil organic carbon: Assessing potential of airborne and spaceborne optical soil sensing. *Remote Sens. Environ.* **2022**, *271*, 112914. [CrossRef]
53. Mondal, B.P.; Sekhon, B.S.; Sahoo, R.N.; Paul, P. Vis-nir reflectance spectroscopy for assessment of soil organic carbon in a rice-wheat field of ludhiana district of punjab. *ISPRS-Int. Arch. Photogramm. Remote Sens. Spat. Inf. Sci.* **2019**, *43*, 417–422. [CrossRef]
54. Anna, P.; Marcin, C.; Jakub, R.; Tadeusz, U. Application of FT-NIR spectroscopy and NIR hyperspectral imaging to predict nitrogen and organic carbon contents in mine soils. *Measurement* **2020**, *164*, 108117.
55. Lin, L.X.; Liu, X.X. Mixture-based weight learning improves the random forest method for hyperspectral estimation of soil total nitrogen. *Comput. Electron. Agric.* **2022**, *192*, 106634. [CrossRef]
56. Pechanec, V.; Mráz, A.; Rozkošný, L.; Vyvlečka, P. Usage of airborne hyperspectral imaging data for identifying spatial variability of soil nitrogen content. *ISPRS Int. J. Geo-Inf.* **2021**, *10*, 355. [CrossRef]
57. Li, H.Y.; Jia, S.Y.; Le, Z.C. Quantitative analysis of soil total nitrogen using hyperspectral imaging technology with extreme learning machine. *Sensors* **2019**, *19*, 4355. [CrossRef]
58. Malmir, M.; Tahmasbian, I.; Xu, Z.H.; Farrar, M.B.; Bai, S.H. Prediction of soil macro- and micro-elements in sieved and ground air-dried soils using laboratory-based hyperspectral imaging technique. *Geoderma* **2019**, *340*, 70–80. [CrossRef]
59. Lu, P.; Wang, L.; Niu, Z.; Li, L.H.; Zhang, W.H. Prediction of soil properties using laboratory VIS–NIR spectroscopy and Hyperion imagery. *J. Geochem. Explor.* **2013**, *132*, 26–33. [CrossRef]
60. Li, W.; Dou, Z.; Cui, L.; Wang, R.; Zhao, Z.; Cui, S.; Zhai, X. Suitability of hyperspectral data for monitoring nitrogen and phophorus content in constructed wetlands. *Remote Sens. Lett.* **2020**, *11*, 495–504. [CrossRef]

diversity

MDPI

Article

Evaluation of Production–Living–Ecological Functions in Support of SDG Target 11.a: Case Study of the Guangxi Beibu Gulf Urban Agglomeration, China

Ziyan Ling [1,2], Weiguo Jiang [2,*], Chaoming Liao [3], Yanshun Li [4], Yurong Ling [1], Kaifeng Peng [2] and Yawen Deng [2]

1 College of Geography and Planning, Nanning Normal University, Nanning 530001, China; lingziyan@nnnu.edu.cn (Z.L.); lingyurong@nnnu.edu.cn (Y.L.)
2 Beijing Key Laboratory for Remote Sensing of Environment and Digital Cities, Faculty of Geographical Science, Beijing Normal University, Beijing 100875, China; pengkaifeng@mail.bnu.edu.cn (K.P.); dengyawen@mail.bnu.edu.cn (Y.D.)
3 College of Natural Resources and Surveying, Nanning Normal University, Nanning 530001, China; liaochaoming@nnnu.edu.cn
4 Manpower Logistics Academy, Guangxi Vocational and Technical College, Nanning 530226, China; smlydh@gmail.com
* Correspondence: jiangweiguo@bnu.edu.cn

Abstract: Sustainable Development Goals (SDGs) target 11.a is a good vision for the coordinated development of the economy, society and environment in urban agglomerations. However, there was an extreme lack of indicators, data or case studies for SDG target 11.a, since it is a vague "process target", which is not conducive to the implementation of SDG target 11.a. It is important to propose a quantitative, convenient, and local policies relevant method to promote the realization or to test the implementation effects of SDG target 11.a. Combined with socio-economic data and land use data, this study uses the methods of comprehensive evaluation model, coupling and coordination degree, and comparative advantage degree methods to study the pattern evolution, coordination characteristics and advantageous areas of production–living–ecological (PLE) functions in the Guangxi Beibu Gulf Urban Agglomeration (GBG_UA) from 1995 to 2019. The results showed that, (1) considering the spatiotemporal distribution of PLE functions, the study area has a relatively stable ecological function as well as fluctuating production and living functions. Considering the coordination characteristics of PLE functions, high–high and low–low clustering effects were observed, and primary coordination maintained the highest proportion, accounting from 55.26% in 1995 to 71.05% in 2019, indicating the SDG target 11.a level in the GBG_UA was poor. Considering the advantageous areas for PLE functions, the region mostly comprises single-function advantageous areas and a few multifunction advantageous areas, including 20 single-function advantage counties (accounting for 52%), 15 dual-function advantage counties (accounting for 39%), and three multi-function advantage counties (accounting for 7.8%), which indicates the lack of diversified land use structures in this region. (2) Optimization suggestions for the coordinated development and realization of SDG target 11.a for the GBG_UA were provided. Suggestions were made based on the radiation and driving role of Nanning city to guide the coordinated development of surrounding counties (districts). Suggestions were also made to improve the design of the integrated transportation network as well as to optimize allocation according to the resource endowment of land and to realize an upgraded ecology as well as agricultural products and services. (3) The evaluation of PLE functions is a quantitative and convenient method that can optimize national and regional development planning and test the implementation effects of SDG target 11.a. This study offers foundational knowledge for the realization of SDG target 11.a in the GBG_UA and provides a reference for the research and implementation of SDG target 11.a in other regions around the world.

Keywords: production–living–ecological (PLE) functions; coordination characteristics; SDG target 11.a; Guangxi Beibu Gulf urban agglomeration (GBG_UA)

Citation: Ling, Z.; Jiang, W.; Liao, C.; Li, Y.; Ling, Y.; Peng, K.; Deng, Y. Evaluation of Production–Living–Ecological Functions in Support of SDG Target 11.a: Case Study of the Guangxi Beibu Gulf Urban Agglomeration, China. *Diversity* **2022**, *14*, 469. https://doi.org/10.3390/d14060469

Academic Editor: Raúl Romero-Calcerrada

Received: 10 April 2022
Accepted: 9 June 2022
Published: 11 June 2022

Publisher's Note: MDPI stays neutral with regard to jurisdictional claims in published maps and institutional affiliations.

1. Introduction

The Sustainable Development Goals (SDGs) were first proposed at the 2012 Rio Earth Summit [1]. In total, 193 countries around the world jointly signed *"Changing our future: 2030 Agenda for Sustainable Development"* in September 2015. The SDGs framework could be regarded as the blueprint that promotes sustainable development of Member States, which commit to the harmony of ecological environment and social economy. They address the global challenges we face, including poverty, inequality, climate change, environmental degradation, peace and justice, and the SDGs have become a global research hotspot in recent years. In July 2017, the United Nations General Assembly adopted the global indicator framework, which included 17 goals, 169 targets and 232 indicators [2], in which Goal 11 proposed to *"Make cities inclusive, safe, resilient and sustainable"*. Urban agglomerations include urban, peri urban, and rural areas that, are gathering areas for human activities. For example, comprising 29.12% of the national land area, Chinese urban agglomerations concentrate 75.19% of the total population, 80.05% of the GDP, 82.37% of the total social fixed asset investment and 91.19% of the national fiscal revenue [3]. Nevertheless, there are still many problems regarding the development of urban agglomerations, such as rapid economic development, the rapid expansion of urban construction, the crowding out of ecological space, prominent resource and environmental problems, and the deterioration of the human living environment [4–6], as such, SDG target 11.a proposed to *"Support positive economic, social and environmental links between urban, peri urban and rural areas by strengthening national and regional development planning"*, and SDG target 11.a is a good vision for the coordinated development of economy, society, and environment in urban agglomerations [7].

There have been many studies on the goal, targets and indicators of SDG11. For example, in terms of theory, Caprotti et al. discussed important policy and practical opportunities as well as challenges of the new urban agenda [8]; Mccarton et al. explored the key components needed to achieve safe, resilient and sustainable cities and communities in the EU [9]; and Lawanson et al. revealed that paucity of data, weak institutional capacity as well as poor governance strategies are major impediments to mainstreaming SDG11 in Lagos, Nigeria [10]. In terms of indicators and data, Cochran et al. took EnviroAtlas as an example to show and analyze how earth observation indicators can help fill the gaps in SDG monitoring data [11]. Ni et al. constructed an indicator system for the SDG 11.1–11.7 targets in the urban dimensions in China; however, SDG target 11.a−11.c targets were not included [12]. In terms of case studies, Patel took Cape Town as an example to explore the role of urban experimentation in helping cities cope with the data and governance challenges faced in the implementation of SDG 11 [13]. Abubakar et al. assessed the implementation of SDG11 in Nigeria at the national level [14]. However, there is an extreme lack of indicators, data or case studies for SDG target 11.a, which is not conducive to the implementation of SDG target 11.a.

There are two kinds of targets that are included in SDG 11: one kind are the so-called "outcome targets", which are marked by numbers, e.g., 11.1, 11.2, 11.3, etc., the others are so-called "process targets", which are marked with letters, e.g., 11.a, 11.b, etc. [15]. The indicators of "outcome targets" are clear and quantifiable, but those of "process targets" are not, they are vague. Klopp and Petretta investigated the relationship between indicators, complexity, and the politics of measuring cities, emphasizing the need to reduce the vagueness of indicators to avoid fuzziness in local implementation [16]. Thus far, only Erblin et al. have developed a set of SDG target 11.a indicators to assess the quality of spatial governance and planning in Europe [15]. However, this method requires many indicators that are difficult to obtain, and this method is based on the European context. Hansson et al. suggested that domestic actors should be allowed to select indicators "that fulfil the criteria of easy measurement or collection, appropriateness, convenience and relevance to current conditions and national and local development policies, priorities and programmes" [17]. As a result, it is urgent to find quantitative, convenient, and local policies relevant evaluation method and case study for SDG target 11.a.

Land is the carrier of all human activities, and land use is multifunctional [18]. Optimizing management options from the perspective of multifunctional land use can promote sustainable land management [19]. In the European project "Sustainability Impact Assessment: Tools for Environmental Social and Effects of Multifunctional Land Use in Europe Regions (SENSOR)", land-use functions are classified into three main functions: economic, social and environmental functions [20]. In China, it has been proposed that all human land relations are embodied and included in the utilization of production function, living function, and ecological function [21], which are called production–living–ecological (PLE) functions. China wanted to *"promote intensive and efficient production space, appropriate living space and beautiful ecological space"* in 2012, and further emphasized that policies should *"firmly follow the civilized development path of production development, affluent living and good ecology"* in 2017. This represents the planning framework for the coordinated development of PLE functions, which means supporting a positive production function (economy), living function (social), and ecological function (environmental) links by land spatial planning at the national, provincial, prefectural, district, and county levels. Therefore, the coordination of PLE functions is consistent with the national and local development policies, and it can be quantitatively evaluated and has strong operability. Therefore, evaluating of PLE functions is of great significance when formulating reasonable local development planning locally, constructing positive economic, social and environmental links and serving SDG target 11.a.

In recent years, great progress has been made in the evaluation of PLE functions. Since pattern evolution can grasp the spatial distribution patterns and development trends of PLE functions, the coordination characteristics reveal the degree of interaction and game process of PLE functions, and advantageous areas show the natural resources and economic social development of each space unit, PLE functions can be comprehensively evaluated by these indicators. The comprehensive evaluation model [18,22,23], coupling and coordination degree [24–26] and comparative advantage degree [27] were used to study the function value, pattern evolution, coordination characteristics and advantageous areas of PLE functions, and the results showed that these methods were very effective, which provided a good technical basis for our study. Although PLE functions play an important role in the SDGs, the existing studies on PLE functions are lack of connection with the SDGs, which is not conducive to providing decision-making services for sustainable development.

Overall, the objectives of this research were as follows:

(1) Propose a quantitative, convenient, and local policies relevant evaluation method for SDG target 11.a based on evaluating of PLE functions.
(2) Take the Guangxi Beibu Gulf Urban Agglomeration (GBG_UA), which is one of the new urban agglomerations constructed in China as an example, and analyze the pattern evolution, coordination characteristics, and advantageous areas of PLE functions from 1995 to 2019 and offer foundational knowledge for the development planning and realization of SDG11.a.
(3) Put forward the optimization of the development planning of PLE functions in the GBG_UA and promote the realization of SDG target 11.a locally. At the same time, this study aims to provide a reference for the research and implementation of SDG target 11.a. in other regions around the world.

2. Materials and Methods

2.1. Study Area

The GBG_UA is located between 20°26′ N and 24°02′ N, and between 106°33′ E and 110°53′ E (Figure 1), and is one of the new urban agglomerations in the south of China [28], with 6 prefecture-level cities and 38 counties (districts) being included in the area. The terrain is high in the west, north, and east; inclines in the middle and south; and there are karst mountains in the west. Complex hills and small basins have formed in the middle of continuous mountains, and piedmont plains, river alluvial plains and deltas are formed in the south. The GBG_UA has a large amount of forest land and farmland, so it has a high ecological level and a good agricultural industrial foundation, and it has great potential to provide ecology, agricultural products and services [29]. The development intensity of the GBG_UA is low, and there is a relatively large stock of land resources that can be developed into construction land. Located in the tropical and subtropical zone, the study area is affected by high temperatures, abundant heat, and rich rainfall. The agglomeration is in the largest bay in southern China and ranks first in China for the quality of its ecological environment. It has plenty of ports, a long coastline, and oil and gas, agricultural, forestry, and tourism resources. The region has a flat terrain, a large environmental capacity, and a strong population and economy carrying capacity. The cities in the GBG_UA have profound historical and humanistic origins. In recent years, major planning, resource development and utilization, industrial layout, and public services have been integrated, and the integrated development of the GBG_UA has a solid foundation, strong momentum, and huge potential.

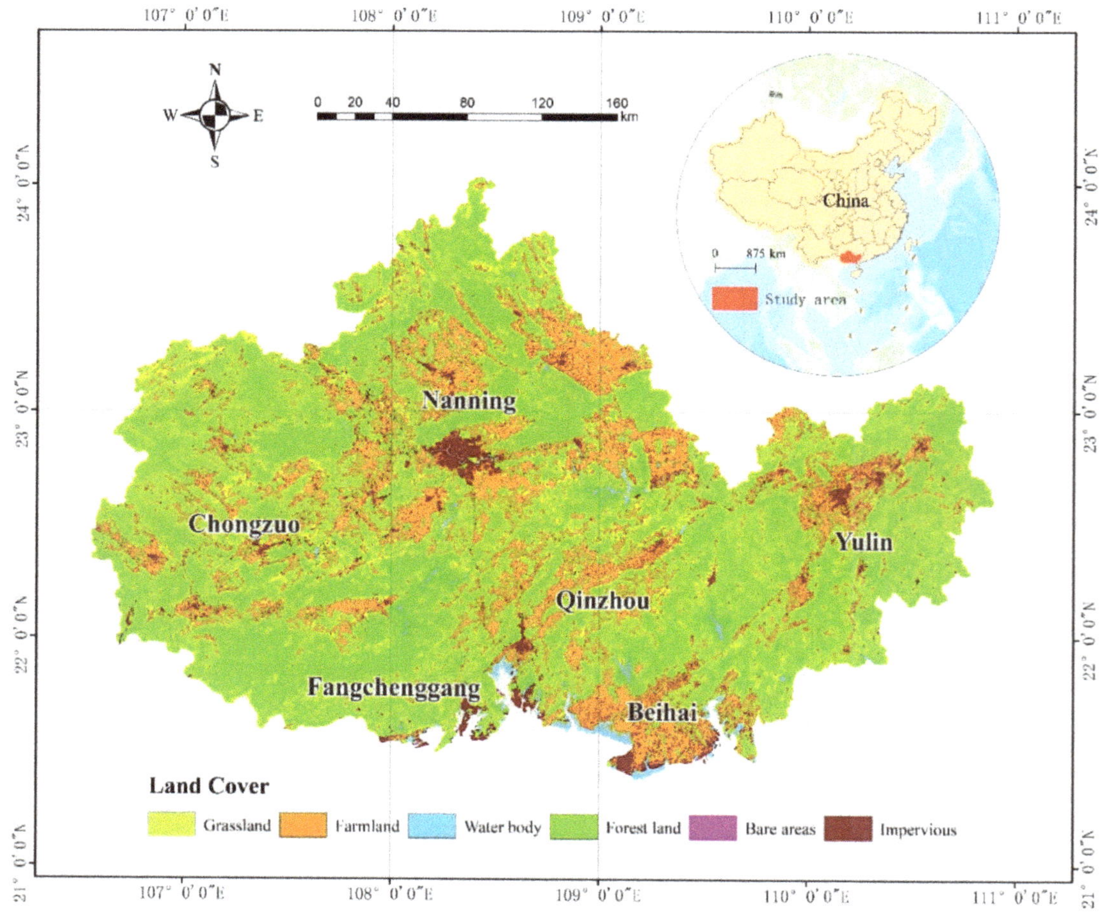

Figure 1. The land cover of the Guangxi Beibu Gulf urban agglomeration.

2.2. Data Sources

The data used in this paper mainly include land use maps and socioeconomic data, which are used to establish the evaluation system for the PLE functions in the GBG_UA. The details are as follows. (1) Land Use Data: The land use data for the study area from 1995 to 2019 were derived from remote sensing monitoring data such as Landsat

TM/ETM+ and HJ-1A/1B with a spatial resolution 30 m × 30 m. There are six first-class types including farmland, forest land, grassland, water body, impervious and bare areas, and 25 s-class land use types, including forest land, shrub forest, sparse forest land, other forest land and grassland with high, medium, and low coverage in the land use data, and its classification accuracy degree is greater than 90% [30]. The data are from the Resource and Environmental Science and Data Center of Chinese Academy of Sciences (https://www.resdc.cn, accessed on 1 June 2021), the Ministry of Ecology and Environment's Center for Satellite Application on Ecology and Environment, and the China National Environmental Monitoring Center. (2) Socioeconomic Data: The socioeconomic data for the study area from 1995 to 2019 are mainly derived from the Statistical Yearbook of Guangxi Zhuang Autonomous Region, the Statistical Yearbook of Nanning City, the Statistical Yearbook of Beihai City, the Statistical Yearbook of Qinzhou City, the Statistical Yearbook of Fangchenggang City, the Statistical Yearbook of Chongzuo City, the Statistical Yearbook of Yulin City, and from the socioeconomic statistical communiques of cities and counties. (3) Administrative Division Data: The administrative division data are derived from the National Geomatics Center of China (http://www.ngcc.cn/ngcc, accessed on 1 June 2021).

2.3. Methods

2.3.1. Theoretical Foundation Establishment

SDG target 11.a demands coordinated development of economy, society, and environment, which guarantees human land relations safe and sustainable. All human land relations are embodied and included in the utilization of PLE functions [21]. Production function supports the development of regional industries and provides industrial products, agricultural products, and service products [21,31]. Living function supports residence, consumption, leisure and entertainment [21,31]. Ecological function involves climate regulation, soil conservation, and guaranteeing regional ecological security [21,31]. The interrelations of PLE functions are tradeoffs and synergies [32]. The production function provides economic support for promoting quality of life and maintenance of ecology, the ecological function is the beautiful and healthy foundation for living and production, and the living function supports labor for production. However, the excessive development of living and production function will destroy the ecological environment, and then deteriorate the living function, leading to a vicious circle.

In summary, PLE functions should develop in a coordinated way, specifically, production is efficient and intensive, living is rich and comfortable, and ecology is beautiful and healthy, and achieve SDG target 11.a finally (Figure 2). This study establishes the relationship table of PLE functions coordination levels corresponding to the implementation stage of SDG target 11.a (Table 1).

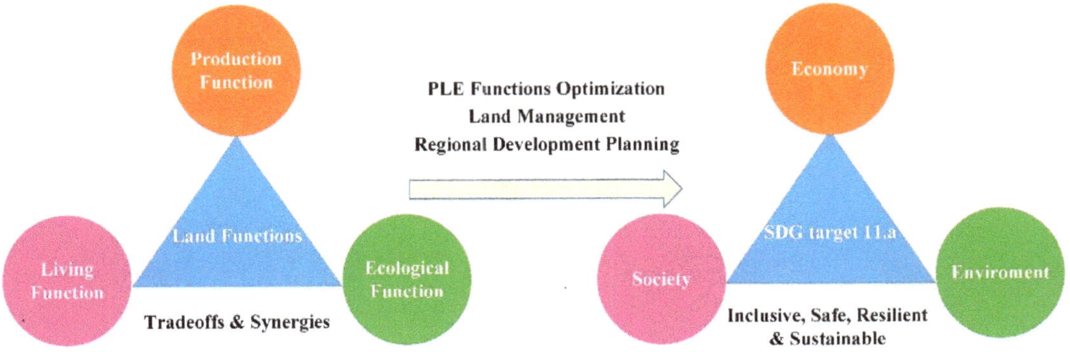

Figure 2. Schematic diagram of research ideas.

Table 1. Link between PLE functions coordination and SDG target 11.a.

Coordination Level of PLE Functions	SDG Target 11.a Characteristic
Imbalance	Extremely poor production, or the over development of production function has led to the serious extrusion of other functions, such as poor living conditions or serious ecological pollution.
Primary Coordination	Production has initially developed, life has gradually improved, or the ecology is fragile.
Intermediate Coordination	Gradually transformed into an intensive and efficient production mode, and began to pay attention to repair the ecological problems caused by production or living activities.
Good Coordination	Production developed and ecological restoration has achieved good results, and the overall living environment has been greatly improved.
High-quality Coordination	PLE functions promote each other, there is a positive link betweem PLE functions, and they realize the orderly sustainable development of urban agglomerations.

2.3.2. Building the Evaluation Index System of PLE Functions

The rationality for index system construction is the basis of function evaluation. The principles for index selection include: (a) Typicality and Comparability: there are great differences in the development level of each county in the GBG_UA. When selecting the indices, they should not only be typical, but also comparable. The indices should highlight the differences in the characteristics of PLE functions between each county, so that the evaluation results can be comparable and reflect the differences in the PLE functions of the counties. (b) Operability and Accessibility: As there are many indices that affect the PLE functions, in order to avoid a one-sided pursuit of diversity and comprehensiveness, operable, available and representative indices should be selected.

In this study, 20 indices were selected in terms of land use and socioeconomic types based on the actual use of PLE functions in the GBG_UA. Thus, a function evaluation index system was built for the PLE functions (Table 2).

Table 2. Evaluation index system for the production–ecological–living spaces functions in the Guangxi Beibu Gulf urban agglomeration.

Target Layer	Guideline Layer	Index Layer (Unit)	Weight
Production function	Economic development	Regional GDP (CNY ten thousand)	0.0932
		Fiscal revenue (CNY ten thousand)	0.0758
		Total fixed investment (CNY ten thousand)	0.0937
		Percentages of the output value of secondary and tertiary industries (%)	0.1034
	Agricultural production	Output values of agriculture, forestry, animal husbandry and fishery (CNY ten thousand)	0.1227
		Cultivated land area (km^2)	0.1236
		Grain output (tons)	0.0745
	Industrial production	Industrial and mining production space area (km^2)	0.1102
		Total output value of industries above designated scale (CNY ten thousand)	0.0997
		Number of designated-scale industrial and mining enterprises (EA)	0.1031
Living function	Living carrying	GDP per capita (CNY per person)	0.1598
		Total retail sales of consumer goods per capita (CNY per person)	0.1515
		Residents' saving balance per capita (CNY per person)	0.1490
	Living service	Living space area (km^2)	0.1571
		Urbanization rate (%)	0.1083
		Number of schools (EA)	0.1463
		Number of medical beds (EA)	0.1280
Ecological function	Ecological supply	Ecological space area (km^2)	0.1712
		Forest coverage rate (%)	0.1793
		Percentage of waters area (%)	0.1721
	Ecological maintenance	Proportion of days with excellent air quality (%)	0.1625
		Water quality compliance rate (%)	0.1691
		Harmless treatment rate of domestic waste (%)	0.1458

Note: The space area indices in relation to the PLE functions described in the table are derived from land use types (where the living space area includes urban living spaces and rural residential land area, whereas the ecological space area covers forest land, grassland, and water body). Other data are calculated based on the statistical yearbooks.

2.3.3. The Evaluation Model for PLE Functions

There are many methods that can be implemented to measure the weight of indices, such as the analytic hierarchy process, expert scoring method and other subjective weight determination methods [33,34]. Generally, the subjective weight determination method is greatly affected by the subjectivity of the evaluator or consultor and has limitations. The objective weighting law mainly analyzes the importance of the indices in the whole system through the correlation between indices, to determine the weight of the indices, such as through the entropy method and principal component analysis [35,36]. This paper uses the entropy method to measure the weight of the indices.

First, the indices are subject to dimensionless standardization. The entropy method was then used to measure the weight of each index for target layer [37,38]. Finally, a comprehensive evaluation model was used to evaluate the functions at the target layer. Assume that there are m counties under evaluation (including all counties at different times) and n evaluation indices. The matrix formed by the original data is $X = (X_{ij})_{m\times n}$, where x_{ij} is the initial value of the jth index of evaluation object i. The formulas are as follows:

$$P_{ij} = \{(x_{ij} - \min x_j)/(\max x_j - x_{ij})\} \tag{1}$$

$$f_{ij} = p_{ij}/\sum_{i=1}^{n} p_{ij} \tag{2}$$

$$H_j = -k\sum_{i=1}^{n}(f_{ij} \times \ln f_{ij}) \tag{3}$$

$$w_j = e_j/\sum_{j}^{m} e_j \tag{4}$$

$$V = \sum_{i=1}^{m} w_j \times P_{ij} \tag{5}$$

where V is the function evaluation value; P_{ij} is the standardized value of the index; w_j is the weight value of index j; $e_j = 1 - H_j$ is the information utility value; H_j is the information entropy of the index j; f_{ij} is the index weight; X_{ij} is the initial value of the index; and max x_j and min x_j are the maximum and minimum of the initial value of the index, respectively.

2.3.4. The Coupling and Coordination Levels

The coupling and coordination level model [38] describes the functional synergy and promotional relationships between PLE functions, which can be calculated as follows:

$$C = \left[\frac{V_1 \times V_2 \times V_3}{(V_1 + V_2)(V_1 + V_3)(V_2 + V_3)}\right]^{\frac{1}{3}} \tag{6}$$

The coupling and coordination levels are mainly used to reflect the interaction among elements and cannot reflect the appropriateness of each element's ratio. Therefore, the coordination level is required to reflect the coordinated development level of the elements [39]. The formula is as follows:

$$T = \alpha V_1 + \beta V_2 + \gamma V_3, \tag{7}$$

$$D = \sqrt{CT} \tag{8}$$

where V_1, V_2, and V_3 are the evaluation values of production function, living function, and ecological function, respectively; and D is the coordination degree, which is the geometric mean of coupling level C and comprehensive development level T. In this research, three types of spaces are considered to be equally important; thus, the three weight coefficients α, β, and γ are equal to one third.

2.3.5. Comparative Advantages

In our study, an index that measured the level of comparative advantages was used to describe the relative advantages [40] that could identify the functions with comparative advantages among the three functions for each area. The formula for calculating the level of comparative advantages is as follows:

$$RCA_{ij} = (X_{ij}/Y_i)/(X_{\omega j}/Y_{\omega}) \tag{9}$$

where X_{ij} represents the jth function value of county i; Y_i represents the sum of all function values of county i; X_{wj} represents the sum of all the jth function values of all of the counties; and Y_w represents the sum of all function values of all counties. Agricultural and industrial production functions essentially differ in their land space use. Thus, they are separated from the comprehensive production functions and are calculated separately in this study. A revealed comparative advantage (*RCA*) value of close to 1 indicates a significant degree of superiority, whereas a *RCA* value of >1 indicates a comparative advantage. The larger the value, the stronger the advantage. An *RCA* value of <1 indicates no comparative advantage. Therefore, the smaller the value, the weaker the advantage.

3. Results

3.1. Pattern Evolution of PLE Functions

3.1.1. Production Function Evolution and Analysis

Considering the temporal changes from 1995 to 2019, the production function of Nanning city showed a continuous upward trend, whereas that of the other prefecture-level cities fluctuated up and then down (Figure 3). From 1995 to 2000, the production functions of the six cities in the study area showed an upward trend with gentle growth. From 2000 to 2009, due to the impact of the external economy, the production function of the cities other than Nanning city declined, whereas that of Nanning city showed gentle growth, which indicated the high stability of its production function. From 2009 to 2019, with the exception of Nanning city, whose production function showed gentle growth, the production function of the other cities tended to be stable. This indicates that, as the requirements for high-quality development increased during the 12th and 13th Five-Year Plan periods, as more energy-consuming industries transformed, their industrial functions were not demonstrated any further. Nanning city's industrial development was mainly based on new industries, such as research technology; thus, its production function continued to grow.

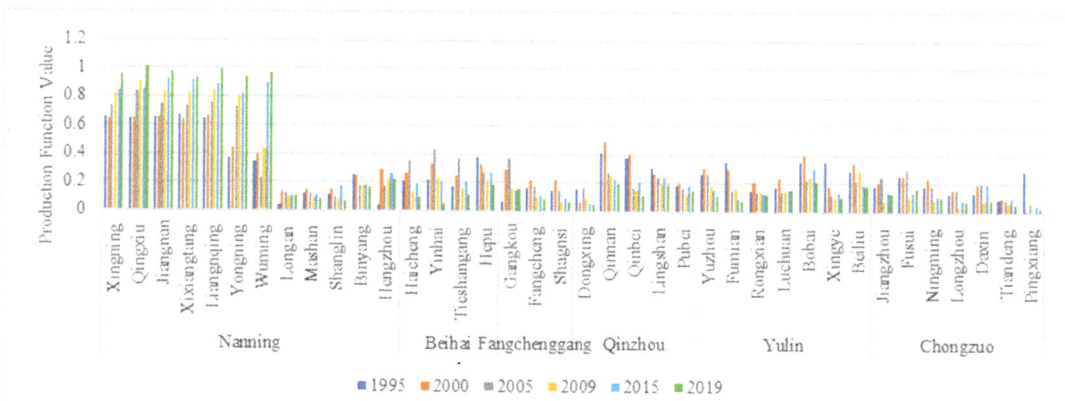

Figure 3. Changing production function trends in the Guangxi Beibu Gulf urban agglomeration from 1995 to 2019.

Considering spatial distribution characteristics, the high-value areas with production functions are mainly distributed in Nanning city, which has gradually become the core area for production functions in the GBG_UA (Figure 4). The high- and low-value areas are distributed with diminishing functions toward their outer edges of the urban agglomeration. That is, the areas with a higher functional value are based around middle-value areas, which then gradually become low-value areas. The production function of the areas adjacent to Nanning city is also relatively high. This indicates that Nanning city, which also has a political function as the capital of Guangxi Zhuang Autonomous Region and boasts advanced tertiary industry, excellent research technologies, and a favorable investment environment, has started to demonstrate its capacity to radiate its functions as a central city in the region. The low-value areas are mainly distributed in the northern area of the region. Due to the harsh terrain of the Shanglin, Mashan, and Shangsi counties, most of these areas are hills and mountains that are not suitable for production functions. Additionally, the Mashan and Shanglin counties suffer from poor traffic conditions and have a weak foundation for agricultural and industrial production, which restrict their production functions. This has led to the formation of low-value area clusters.

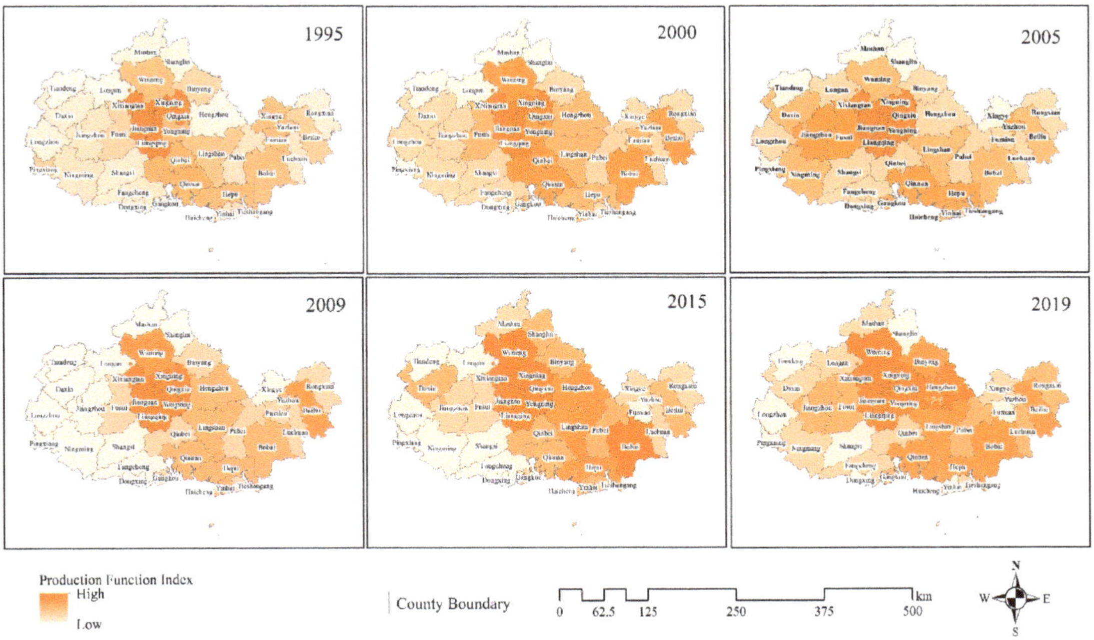

Figure 4. Production function evaluation results in the Guangxi Beibu Gulf urban agglomeration from 1995 to 2019.

3.1.2. Living Function Evolution and Analysis

From 1995 to 2019, the living function of the six prefecture-level cities in the study area showed a cross-fluctuation trend featuring an "up–down–stable" pattern (Figure 5). From 1995 to 2000, the living functions of the cities of Nanning, Beihai, Fangchenggang, and Qinzhou showed an upward trend, whereas those of the cities of Yulin and Chongzuo declined. From 2000 to 2005, the living function of the six cities in the Guangxi Beibu Gulf declined slightly. In contrast to the production function, the living function of the other prefecture-level cities, with the exception of Nanning city, improved from 2005 to 2009, during which the guiding role of the living function among all of the functions was improved as production function declined. From 2009 to 2019, the living function in the study area declined slightly and only the cities of Chongzuo and Nanning witnessed a slight improvement. Through a comparison of the changes in the living functions and the evolution of the living space pattern, the living space area continued to grow, but its layout was not optimized, thus leading to a decline in the living function.

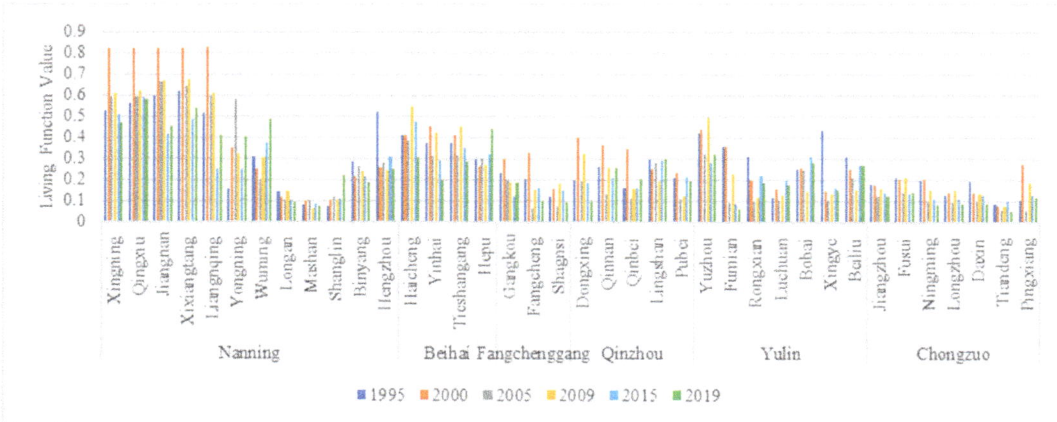

Figure 5. Changing living function trends in the Guangxi Beibu Gulf urban agglomeration from 1995–2019.

Figure 6 shows transitions in several main directions while also considering the evolution of high-value areas with a living function, which were mainly distributed in the urban area of Nanning city in 1995, but that shifted focus along the borders between the cities of Nanning, Qinzhou, and Beihai in 2000, and then extended in a V shape between the cities of Nanning, Beihai, and Yulin in 2005. In 2009, this pattern showed a scattered living space distribution and from 2015 to 2019, there was a slight improvement in the living function. Most of the high-value areas are concentrated in Nanning city and extend toward Beihai city. Overall, the living functions of the Nanning and Beihai cities are much better than they are in the other prefecture-level cities. The counties and districts in the northwestern area of the study area are relatively backward, where their capacity for social and living functions in addition to service levels have yet to be improved, which leads to a low living space quality.

3.1.3. Ecological Function Evolution and Analysis

From 1995 to 2019, the ecological function of the GBG_UA demonstrated small changes and high stability compared with the changes in the production and living functions (Figure 7). From 2009 to 2019, Beihai city, which had a relatively low ecological function, also showed a significant improvement. From 2015 to 2019, the expansion of living and production spaces in Nanning city slightly weakened the city's ecological function. Considering the land use type of the study area from 1995 to 2019, most of the ecological space in the study area was forest land with a high ecological service value, which always accounted for >50% of the entire land space. Therefore, this region largely enjoyed a relatively stable ecological function.

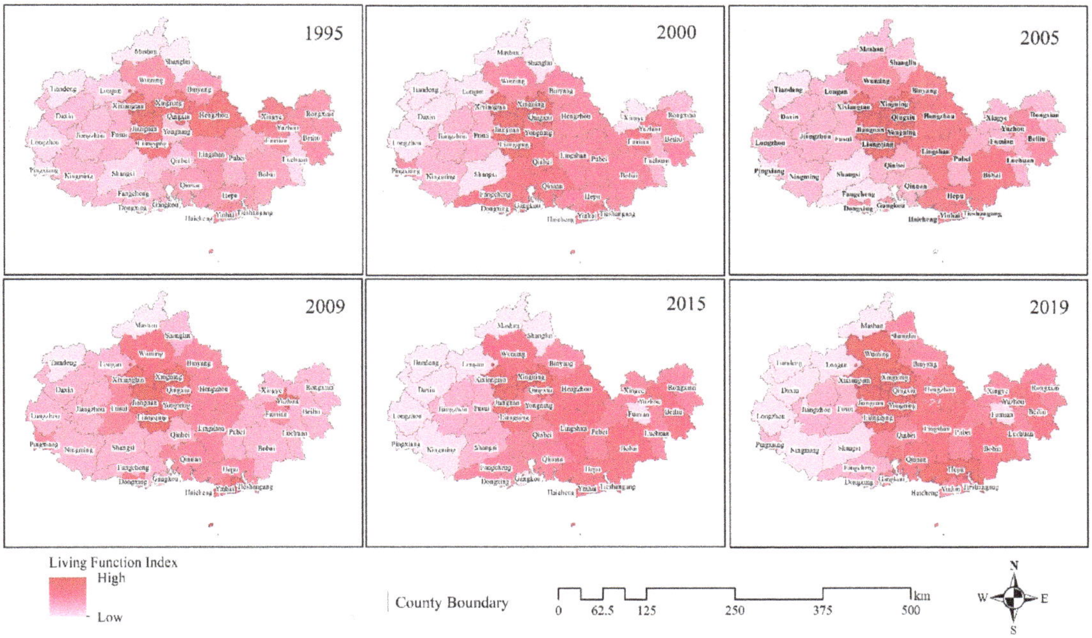

Figure 6. Living function evaluation results in the Guangxi Beibu Gulf urban agglomeration from 1995 to 2019.

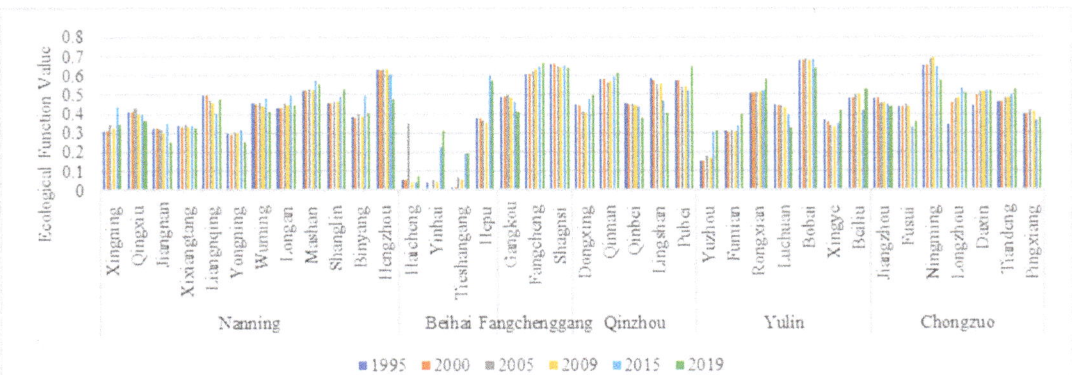

Figure 7. Changing ecological function trends in the Guangxi Beibu Gulf urban agglomeration from 1995 to 2019.

As shown in Figure 8, in terms of the distribution characteristics of the ecological space function, the high-value areas with an ecological function are mainly distributed in hilly and mountainous areas where there are few human activities, especially in those areas comprising the Hundred Thousand Great Mountains. The spatial distribution characteristics for the ecological function in the study area are highly correlated with the ecological landscape distribution. The main landscapes in the areas covered by forests and waters enjoy a strong ecological function. In relative terms, there is a certain gap between the ecological function of Beihai city and that of the other prefecture-level cities. However, in recent years, Beihai city has become a city with a beautiful and comfortable living environment based on its subtropical coastal tourism resources. Both its ecological environment and function have improved.

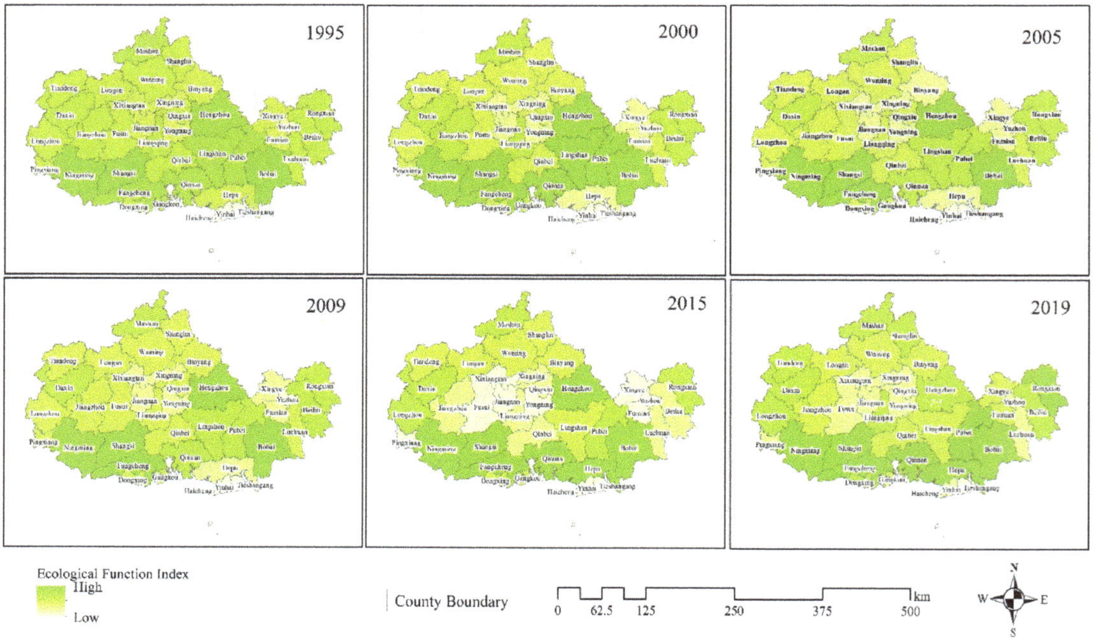

Figure 8. Ecological function evaluation results in the Guangxi Beibu Gulf urban agglomeration from 1995 to 2019.

3.2. Coordination Characteristics of PLE functions

Drawing on the literature [41–45] and the equal interval method, five coordination levels for the study area were identified: imbalance, primary coordination, intermediate coordination, good coordination, and high-quality coordination. The results of the coupling and coordination levels of the study area for the six periods from 1995 to 2019 were obtained (Figure 9).

Considering spatial distribution (Figure 9), the areas with good coordination were mainly the districts and counties of Nanning city, which were stable throughout the study area. This finding indicates that during the research period, the land that was available for PLE functions in Nanning's urban area was arranged reasonably, its urban plan was prepared scientifically, and its land use management policies were implemented well. The imbalanced areas gradually spread from north to west and formed a concentrated contiguous area by 2019. The main counties and cities in the imbalanced areas mostly comprised mountainous terrain, which restricted productive areas, such as in Mashan and Shanglin counties, and in emerging economic areas, such as in the Ningming and Dongxing counties, which indicates the impact of the industrial layout on the coordination level of PLE functions. Meanwhile, the original and traditional agricultural foundation could not satisfy the requirements of high-quality production, leading to an imbalance in the PLE functions. The simultaneous growth of high-quality coordination and primary coordination showed that the coordination of PLE functions in the GBG_UA was developing toward two extremes, the rates of high-quality coordination and imbalance were increasing, and primary coordination maintained the highest proportion, accounting from 55.26% in 1995 to 71.05% in 2019 (Figure 10). Only 18.42% of the study area were in good condition in SDG target 11.a, and the rest were in poor condition.

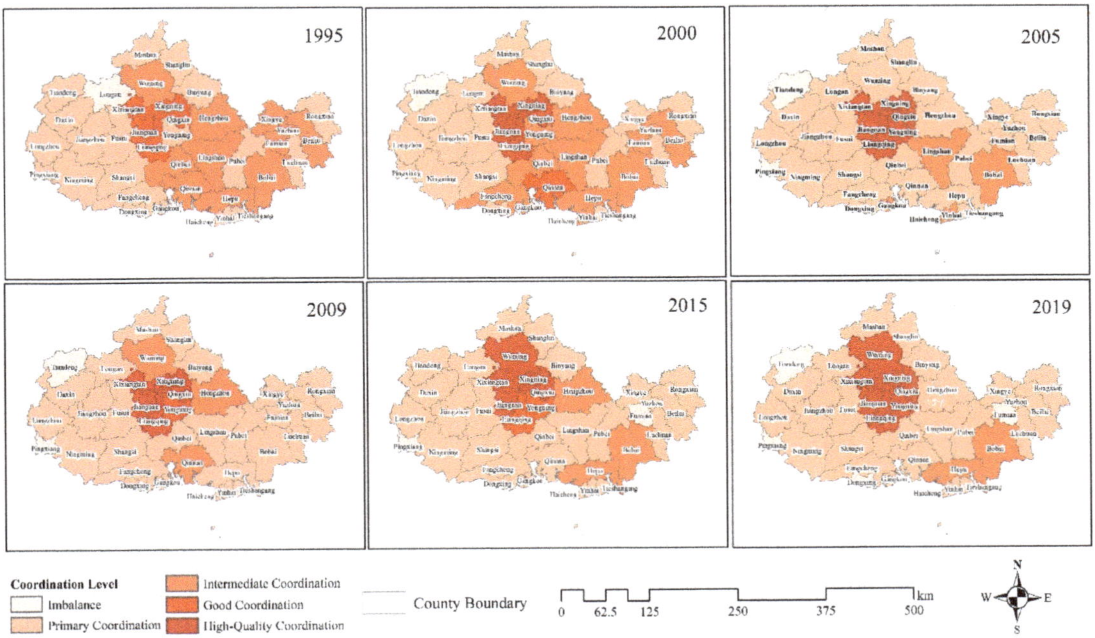

Figure 9. Distribution of land coordination level in the Guangxi Beibu Gulf urban agglomeration from 1995 to 2019.

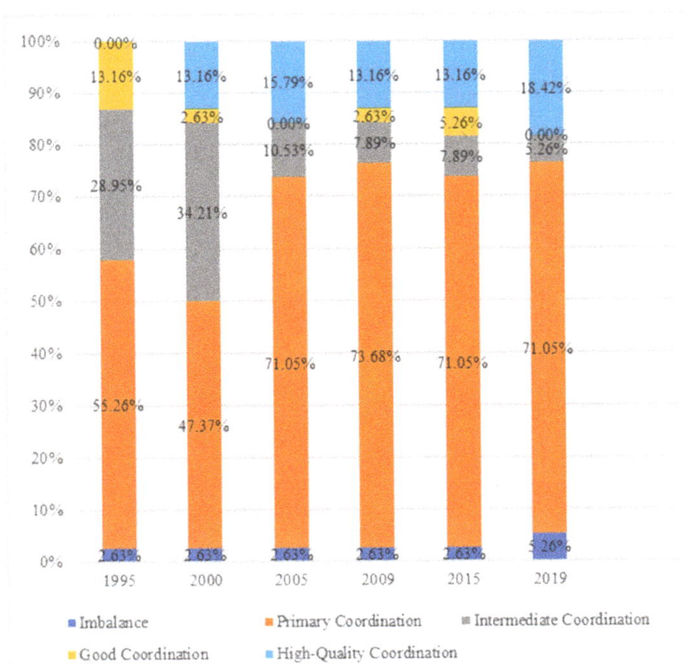

Figure 10. County statistics of land coordination level in the Guangxi Beibu Gulf urban agglomeration from 1995 to 2019.

3.3. Advantageous Areas in PLE Functions

To some extent, the complexity of PLE functions shows the potential of the national land space. An administrative unit is called a single-function advantageous area when it has one advantageous function, a dual-function advantageous area when it has two advantageous functions, and a multifunction advantageous area when it has three advantageous functions. According to the calculation results for the spatial function advantages (Table 3), statistics were obtained for the advantageous areas of the GBG_UA and the results are shown in Table 4.

Table 3. Land function advantage results in the Guangxi Beibu Gulf urban agglomeration.

Administrative District	Agriculture Advantage	Industry Advantage	Living Advantage	Ecological Advantage	Advantage Type
Xingning	1.553	1.473	1.011	0.395	Production–Living
Qingxiu	1.417	1.387	1.139	0.385	Production–Living
Jiangnan	1.591	1.653	0.998	0.297	Production
Xixiangtang	1.493	1.364	1.123	0.363	Production–Living
Liangnqing	1.502	1.632	0.851	0.530	Production
Yongning	1.677	1.623	0.944	0.314	Production
Wuming	1.489	1.447	1.011	0.450	Production–Living
Longan	0.841	0.798	0.638	1.603	Ecological
Mashan	0.835	0.466	0.497	1.921	Ecological
Shanglin	0.563	0.281	1.144	1.475	Living–Ecological
Binyang	1.085	0.755	0.965	1.099	Production–Ecological
Hengzhou	1.143	0.751	1.007	1.016	Production–Living–Ecological
Haicheng	0.233	1.005	2.110	0.278	Production–Living
Yinhai	0.336	0.491	1.434	1.204	Living–Ecological
Tieshangang	0.210	1.020	1.792	0.638	Production–Living
Hepu	0.685	0.615	1.383	0.970	Living
Gangkou	0.088	1.503	1.090	1.265	Production–Living–Ecological
Fangcheng	0.482	0.649	0.551	2.008	Ecological
Shagnsi	0.596	0.280	0.549	2.111	Ecological
Dongxing	0.265	0.487	0.738	2.022	Ecological
Qinnan	0.418	1.361	0.972	1.245	Living–Ecological
Qinbei	0.511	0.890	1.183	1.175	Ecological
Lingshan	1.012	0.657	1.234	0.897	Production–Living
Pubei	0.758	0.641	0.843	1.506	Ecological
Yuzhou	0.414	0.708	1.624	0.839	Living
Fumian	0.847	0.584	0.527	1.823	Ecological
Rongxian	0.653	0.567	0.926	1.519	Ecological
Luchuan	0.922	1.047	1.018	1.010	Production–Living–Ecological
Bobai	1.011	0.590	0.986	1.205	Production–Ecological
Xingye	0.768	0.663	0.932	1.390	Ecological
Beiliu	0.748	0.758	1.111	1.160	Living–Ecological
Jiangzhou	0.925	0.845	0.731	1.421	Ecological
Fusui	0.967	1.199	0.812	1.134	Production–Ecological
Ningming	0.873	0.406	0.497	1.925	Ecological
Longzhou	0.673	0.400	0.607	1.938	Ecological
Daxin	0.815	0.337	0.587	1.896	Ecological
Tiandeng	0.860	0.094	0.408	2.181	Ecological
Pingxiang	0.217	0.338	1.028	1.805	Living–Ecological

Note: Agriculture advantage and industry advantage represent production advantage.

Table 4. Statistical table for the number of land function advantage areas in the Guangxi Beibu Gulf Urban agglomeration.

Advantage Type	Number of Administrative District	Name of Administrative District
Production	3	Jiangnan, Liangqing, Yongning
Living	2	Yuzhou, Hepu
Ecological	15	Longan, Mashan, Fangcheng, Shangsi, Dongxing, Qinbei, Pubei, Fumian, Rongxian, Xingye, Jiangzhou, Ningming, Longzhou, Daxin, Tiandeng
Production–Living	7	Xingning, Qingxiu, Xixiangtang, Wuming, Tieshangang, Lingshan, Haicheng
Production–Ecological	3	Fusui, Binyang, Bobai
Living–Ecological	5	Beiliu, Shanglin, Yinhai, Qinnan, Pingxiang
Production–Living–Ecological	3	Hengzhou, Gangkou, Lunchuan

According to Table 4, the land comprising the GBG_UA is dominated by single-function advantageous areas, which account for 52% of the total area. Of these areas, the advantageous areas for production, living, and ecological functions comprised 7.8%, 5.2%, and 39%, respectively. The proportion of advantageous areas with an ecological function was the highest, which indicates the Guangxi Beibu Gulf's great advantage in terms of its ecological function. There are 15 dual-function advantageous areas, which comprise 39% of the total counties (districts). The multifunction advantageous areas comprised 7.8% of the total counties (districts) and were concentrated in counties and districts dominated by industry and agriculture.

To further explore the spatial distribution of the advantageous functional areas of the GBG_UA, we generated a spatial distribution map (Figure 11). As shown in Figure 10, the advantageous production function areas are significantly adjacent to the advantageous production–living areas and are concentrated in the central part of the study area include the Jiangnan, Liangqing, Yongning, Xingning, Qingxiu, Xixiangtang, Wuming, and Tieshangang districts, and Lingshan county. The administrative units with an advantageous living function include the Yuzhou, Hepu, Yinhai, and Qinnan districts, Beiliu and Pingxiang cities, and Shanglin county, among others. These areas were located in the southeastern part of the study area.

The GBG_UA has many counties and districts with widely distributed advantageous ecological functions. The administrative units with obvious ecological function advantages include the Long'an, Mashan, Shangsi, Pubei, Rongxian, Xingye, Ningming, Longzhou, Daxin, and Tiandeng counties; the Fangcheng, Qinbei, Fumian, and Jiangzhou districts; and the city of Dongxing, with 15 counties being included in total. In addition, a concentrated contiguous ecological area called Shiwan Mountains was formed in the southwestern Guangxi Zhuang Autonomous Region and has obvious ecological function advantages. The advantageous ecological function areas in the GBG_UA comprise a relatively high proportion of the available land space, which indicates the great potential of the study area to provide ecological products and services.

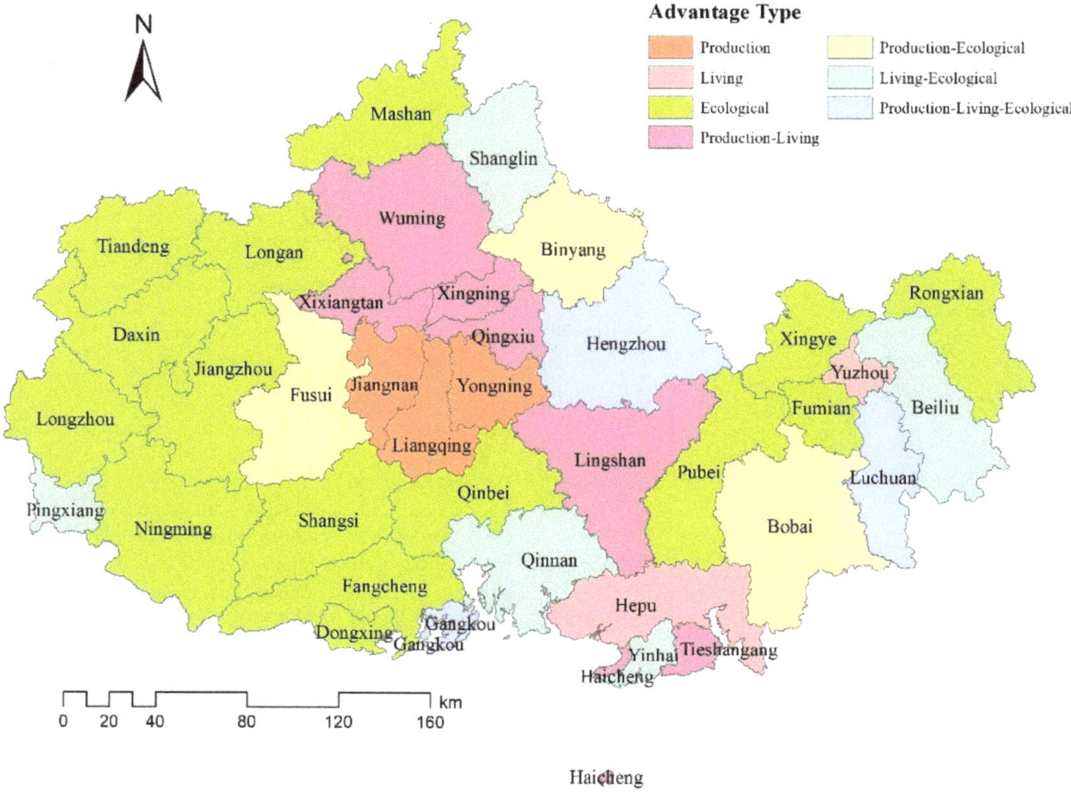

Figure 11. Distribution of land function advantage areas in the Guangxi Beibu Gulf urban agglomeration.

4. Discussion

4.1. The Effectiveness of Our PLE Study

Our study provided a new idea that links the PLE functions coordination with SDG target 11.a. The study of coordinated degree of PLE functions can contribute to the achievement of SDG 1.1.a, which offer useful support for decision makers. In order to achieve SDG target 11.a, on the one hand, decision makers can refer to the coordination degree of PLE functions before planning. On the other hand, decision makers can also use the coordination degree of PLE functions to evaluate the effect of planning implementation afterwards. Although the coordination of PLE functions does not directly calculate the value of SDG target 11.a, their characteristics can reflect a realization of SDG target 11.a (Table 1), helping to reduce the vagueness of SDG target 11.a [1,16]. Our research can help decision makers better understand the coordination of PLE functions and provide them with useful knowledge for SDG target 11.a. Furthermore, research on typical areas of China was carried out by this study according to the international science program of SDGs, the methodology and analytical framework could be easily applied worldwide for supporting the SDG target 11.a and promoting land planning as well as management. This is of great significance for regional development planning and sustainable development in China and is also referential for other countries.

Comparing the evaluation results of PLE functions with the previous research results conducted by Pang et al. and Shen et al. in GBG_UA regions [46,47], it is found that although the evaluation methods or indices are not exactly the same between the studies, the results are highly consistent, such as a relatively higher production function in Nanning and a stable ecological function in the study area. Thus, we believed that the evaluation results of this study are reliable. Compared with previous studies, this study used PLE functions to further evaluate the SDG target 11.a level, so this study is more suitable to consider the implementation of the international plan.

In this research, we have studied production, living and ecological functions, among which ecological functions are related to ecosystem services. Ecosystem services can offer various kinds of benefits for human survival, such as food supply, water conservation, soil conservation, climate regulation and biodiversity protection, and are an essential part of sustainability frameworks [48]. The importance of ecosystem services for SDG has been studied and demonstrated [49,50], and ecosystem services were taken to evaluate the implementation effect of SDG. Although the PLE study did not calculate the value of each type of ecosystem services in detail, it assessed the ecological functions, production functions and living functions in a macro perspective, involving the win–win development of human social system and natural system in a broader perspective.

4.2. Suggestions for the Optimization of PLE Functions

From 1995 to 2019, the number of high-quality coordination counties belonging to urban agglomeration increased (from 0 to 7), and the number of imbalance counties also increased (from 1 to 2), which shows that the coordination of the PLE functions presents a polarized development in the GBG_UA. Primary coordination maintained the highest proportion, accounting from 55.26% in 1995 to 71.05% in 2019, indicating the achievement of SDG target 11.a in the GBG_UA was poor. The transportation, industry, and social services of the GBG_UA should be planned in a unified way, taking advantage of the expansion trend of Nanning city [51], a high-quality coordinated city; emphasizing the driving and radiating role of Nanning city; and affecting the surrounding counties (districts) with good coordination or intermediate coordination to realize high-quality coordination. As for the imbalanced areas, such as Tiandeng county and Fumian district, we can see from the functions value and dominant functions that their production and living functions are low and their ecological functions are high. This indicates that a new ecological industrial structure system should be built, and that can deeply tap into the ecological value and benefits of the area to create green ecological products, and to support the coordinated development of the PLE functions.

The main problem in the utilization of land space with production advantageous functions is the contradiction between environmental protection and economic development. Therefore, while promoting the rational utilization of production space and living space, we should also emphasize the protection of the ecological environment and the development of ecological industry. For example, while developing port logistics and fisheries, the Tieshangang district should also pay attention to the protection of mangroves [52], delimit protection zones, and carry out blue carbon action and a carbon sink economy.

For living function advantageous areas, we should emphasize their geographical advantages and further improve the short board of production and ecological functions. We should pay attention to the joint development of the surrounding high production function cities and collaborate with them with the help of the driving role of the surrounding big cities. For example, Chongzuo city, which is located on the main channel of the "Nanning–Langshan–Hanoi–Guangning" economic corridor, the most convenient channel for China to ASEAN, and Yulin city, which comprises the industrial transfer and processing base in eastern China, should be included in the development of co-urbanization and transportation in the GBG_UA, and to strengthen traffic planning.

On the basis of environmental protection, we should emphasize the value of ecology and realize that ecological benefits can be transformed into economic benefits. Further improving the level of ecological utilization, breaking the state of negative protection and realizing the improvement of the production function should be emphasized. The study area has unique geographical conditions and many characteristic and precious species, such as grapefruit in Rong county, wild camellia produced in Shangsi county (which represents 90% of the world's supply), and chili padi in Tiandeng county, and it is advisable to delimit characteristic functional areas and to provide policy support to create high-quality ecological industries and to realize an ecological back-feeding economy.

4.3. Limitations and Future Work of the Study

There are many indicators affecting PLE functions. This study selects some representative indicators according to previous studies, but there will inevitably be deficiencies. The evaluation index system will continue to be improved in the future. Besides, we will jump out of the restrictions of administrative regions and look for high resolution remote sensing inversion data that can indicate various indicators, to better reflect the details of spatial differences. This paper evaluated PLE functions and SDG target 11.a from 1995 to 2019, but the future trends are unknown. The development scenario simulations can solve this problem [53]. In the future, we will simulate PLE functions and SDG target 11.a under multiple scenarios for the year 2030. Policy makers will be aware of which scenarios can help to realize SDGs, so as to formulate demand-oriented development plans.

5. Conclusions

Based on land use data and socioeconomic statistical data, this study used a comprehensive evaluation model, coupling and coordination degree, and comparative advantage degree to analyze the pattern evolution, coordination characteristics and advantageous areas of PLE functions in the GBG_UA from 1995 to 2019. Our study could offer useful support for the related land management agencies, help policy-makers to assess regional procedures toward achieving SDG target 11.a and inform them the coordinated development economy, society, and environment. The main conclusions are as follows:

(1) When considering the spatiotemporal distribution of PLE functions, the study area has a relatively stable ecological function, a good ecological foundation, and fluctuating production and living functions.

(2) When considering the coordination characteristics of PLE functions, high–high and low–low clustering effects were observed. The coordination level has developed toward two extremes, and primary coordination maintained the highest proportion, accounting from 55.26% in 1995 to 71.05% in 2019, indicating the achievement of SDG target 11.a in the GBG_UA was low.

(3) Considering the advantageous areas for PLE functions, the region mostly comprises single-function advantageous areas and a few the multifunction advantageous areas, including 20 single-function advantage counties (districts), 15 dual-function advantage counties (districts), and three multi-function advantage counties (districts), which indicates the lack of diversified land use structures in this region and that development planning should be formulated in combination with the local functional advantages.

Author Contributions: Conceptualization, Z.L.; methodology, Z.L.; writing—original draft, Z.L.; supervision, W.J.; formal analysis, W.J.; funding acquisition, W.J.; writing—review and editing, W.J. and K.P.; resources, C.L.; investigation, C.L.; data curation, Y.L. (Yanshun Li); software, Y.L. (Yanshun Li); visualization, Y.L. (Yurong Ling); validation, Y.D. All authors have read and agreed to the published version of the manuscript.

Funding: This research was funded by the National Natural Science Foundation of China (Grant No. 42101369, No. U21A2022, and No. 42164001), and the Open Project of Beijing Key Laboratory for Remote Sensing of Environment and Digital Cities.

Institutional Review Board Statement: Not applicable.

Informed Consent Statement: Not applicable.

Data Availability Statement: Data are available from the authors upon reasonable request as the data needs further use.

Conflicts of Interest: The authors declare no conflict of interest.

References

1. United Nations. *Transforming Our World: The 2030 Agenda for Sustainable Development*; United Nations: New York, NY, USA, 2015.
2. Zhang, J.Z.; Wang, S.; Zhao, W.W.; Liu, Y.X.; Fu, B.J. Research progress on the interlinkages between the 17 Sustainable Development Goals and their implication for domestic study. *Acta Ecol. Sin.* **2019**, *39*, 8327–8337.
3. Fang, C.L. Important Progress and Prospects of China's Urbanization and Urban Agglomeration in the Past 40 Years of Reform and Opening-Up. *Econ. Geogr.* **2018**, *38*, 1–9.
4. Wang, Z.Y.; Yang, T.X.; Liu, H.L.; Huang, Y.P.; Liu, F.T. Coordinate Spatial Development Strategy of Densely Urbanized Area: Wuhan Metropolitan Circle. *Planners* **2018**, *34*, 21–27.
5. Li, Z.; Jiang, W.G.; Wang, W.J.; Lv, J.X.; Deng, Y. Study on the wetland leading service function of Jing-Jin-Ji Urban Agglomeration based on the ecosystem service value. *J. Nat. Resour.* **2019**, *34*, 1654–1665. [CrossRef]
6. Jiang, D.; Lin, G.; Fu, J.Y. Discussion on scientific foundation and approach for the overall optimization of "Production-Living-Ecological" space. *J. Nat. Resour.* **2021**, *36*, 1085–1101. [CrossRef]
7. Allen, C.; Metternicht, G.; Wiedmann, T. Initial progress in implementing the Sustainable Development Goals (SDGs): A review of evidence from countries. *Sustain. Sci.* **2018**, *13*, 1453–1467. [CrossRef]
8. Caprotti, F.; Cowley, R.; Datta, A.; Broto, V.C.; Gao, E.; Georgeson, L.; Herrick, C.; Odendaal, N.; Joss, S. The New Urban Agenda: Key Opportunities and Challenges for Policy and Practice. *Urban Res. Pract.* **2017**, *10*, 367–378. [CrossRef]
9. McCarton, L.; O'Hogain, S.; Reid, A. Resilient Cities and Communities. *Worth Water* **2021**, *11*, 173–178.
10. Lawanson, T.; Oyalowo, B.; Nubi, T. Global Goal, Local Context: Pathways to Sustainable Urban Development in Lagos, Nigeria. In *Housing and SDGs in Urban Africa. Advances in 21st Century Human Settlements*; Nubi, T.G., Anderson, I., Lawanson, T., Oyalowo, B., Eds.; Springer: Singapore, 2021; Volume 3, pp. 13–35.
11. Cochran, F.; Daniel, J.; Jackson, L.; Neale, A. Earth Observation-based Ecosystem Services Indicators for National and Subnational Reporting of the Sustainable Development Goals. *Remote Sens. Environ.* **2020**, *244*, 111796. [CrossRef]
12. Ni, P.F.; Marco, K.; Guo, J.; Xu, H.D.; Peng, X.H. Assessment of the Progress of Chinese Cities in Implementing UN Sustainable Development Goal 11. *Urban Environ. Stud.* **2021**, *3*, 36–55.
13. Zarina, P. Local responses to global sustainability agendas: Learning from experimenting with the urban sustainable development goal in Cape Town. *Sustain. Sci.* **2017**, *12*, 785–797.
14. Abubakar, I.R.; Aina, Y.A. The Prospects and Challenges of Developing More Inclusive, Safe, Resilient and Sustainable Cities in Nigeria. *Land Use Policy* **2019**, *87*, 104105. [CrossRef]
15. Erblin, B.; Caterina, C.; Giancarlo, C. Unpacking SDG target 11.a: What is it about and how to measure its progress? *City Environ. Interact.* **2022**, *14*, 100080.
16. Klopp, J.M.; Petretta, D.L. The urban sustainable development goal: Indicators, complexity and the politics of measuring cities. *Cities* **2017**, *63*, 92–97. [CrossRef]
17. Hansson, S.; Arfvidsson, H.; Simon, D. Governance for sustainable urban development: The double function of SDG indicators. *Area Dev. Policy* **2019**, *4*, 217–235. [CrossRef]
18. Duan, Y.M.; Wang, H.; Huang, A.; Xu, Y.Q.; Lu, L.H.; Ji, Z.X. Identification and spatial-temporal evolution of rural "production-living-ecological" space from the perspective of villagers' behavior—A case study of Ertai Town, Zhangjiakou City. *Land Use Policy* **2021**, *106*, 105457. [CrossRef]
19. Zhang, Y.N.; Long, H.L.; Tu, S.S.; Ge, D.Z.; Ma, L.; Wang, L.Z. Spatial identification of land use functions and their trade-offs/synergies in China: Implications for sustainable land management. *Ecol. Indic.* **2019**, *107*, 105550. [CrossRef]
20. Kienast, F.; Bolliger, J.; Potschin, M.; de Groot, R.S.; Verburg, P.H.; Heller, I.; Wascher, D.; Haines-Young, R. Assessing landscape functions with broad- scale environmental data: Insights gained from a prototype development for Europe. *Environ. Manag.* **2009**, *44*, 1099–1120. [CrossRef]
21. Li, G.; Fang, C. Quantitative function identification and analysis of urban ecological-production-living spaces. *Acta Geogr. Sin.* **2016**, *71*, 49–65.
22. Zou, L.L.; Liu, Y.S.; Wang, J.Y.; Yang, Y.Y. An analysis of land use conflict potentials based on ecological-production-living function in the southeast coastal area of China. *Ecol. Indic.* **2021**, *122*, 107297. [CrossRef]
23. Ge, C.H.; Zhang, Z.G. Coordinated Development of Wuhan Metropolitan Circle From Economic Geography Viewpoint. *Planners* **2018**, *34*, 39–45.
24. Zhang, X.S.; Xu, Z.J. Functional Coupling Degree and Human Activity Intensity of Production–Living–Ecological Space in Underdeveloped Regions in China: Case Study of Guizhou Province. *Land* **2021**, *10*, 56. [CrossRef]
25. Fu, J.C.; Zhang, S.L. Functional Assessment and Coordination Characteristics of Production, Living, Ecological Function—A Case Study of Henan Province, China. *Int. J. Environ. Res. Public Health* **2021**, *18*, 8051. [CrossRef] [PubMed]
26. Li, J.S.; Sun, W.; Li, M.Y.; Meng, L.L. Coupling coordination degree of production, living and ecologicalspaces and its influencing factors in the Yellow River Basin. *J. Clean. Prod.* **2021**, *298*, 126803. [CrossRef]
27. Chen, H.J.; Yang, Q.Y.; Su, K.C.; Zhang, H.Z.; Lu, D.; Xiang, H.; Zhou, L.L. Identification and Optimization of Production-Living-Ecological Space in an Ecological Foundation Area in the Upper Reaches of the Yangtze River: A Case Study of Jiangjin Districtof Chongqing, China. *Land* **2021**, *10*, 863. [CrossRef]
28. Fang, C.l. Progress and the future direction of research into urban agglomeration in China. *Acta Geogr. Sin.* **2014**, *69*, 1130–1144.

29. Ling, Z.Y.; Li, Y.S.; Jiang, W.G.; Liao, C.M.; Ling, Y.R. Dynamic Change Characteristics of "Production-living-ecological Spaces"of Urban Agglomeration Interlaced with Mountains, Rivers and Sea: A Case Study of the Beibu Gulf Urban Agglomeration in Guangxi. *Econ. Geogr.* **2022**, *42*, 18–24.
30. Liu, J.; Zhang, Z.; Xu, X.; Kuang, W.; Zhou, W.; Zhang, S.; Li, R.; Yan, C.; Yu, D.; Jiang, N. Spatial patterns and driving forces of land use change in China during the early 21st century. *J. Geogr. Sci.* **2010**, *20*, 483–494. [CrossRef]
31. Liu, J.; Liu, Y.; Li, Y. Classification evaluation and spatial-temporal analysis of "production-living-ecological" spaces in China. *Acta Geogr. Sin.* **2017**, *72*, 1290–1304.
32. Xie, H.L.; Yao, G.; Liu, G.Y. Spatial evaluation of the ecological importance based on GIS for environmental management: A case study in Xingguo county of China. *Ecol. Indic.* **2015**, *51*, 3–12. [CrossRef]
33. Cheng, J.J.; Hu, S.F.; Zhang, Y.G.; Qi, X.Z.; Jin, S.L. Tourism Destination of Rural Residents Based on Rough Set and Analytic Hierarchy Process Method. *Sci. Geogr. Sin.* **2016**, *36*, 1885–1893.
34. Wang, P.; Liu, M.; Liu, H.D. Evaluation of tourism utilization potential of intangible cultural heritage resources in Shanxi Province. *Reg. Res. Dev.* **2017**, *36*, 92–98.
35. Ling, Z.Y.; Liu, R. Risk Assessment on Regional Water Scarcity in Guangdong Province Based on Principal Component Analysis. *Resour. Sci.* **2010**, *32*, 2324–2328.
36. Zhao, L.; Zhu, Y.M.; Fu, M.C.; Zhang, P.T.; Cao, Y.G. Comparative study on intensive use of rural residential land based on principal component analysis and entropy method. *Trans. Chin. Soc. Agric. Eng.* **2012**, *28*, 235–242.
37. Xu, L. *Research on the Optimization of Geographical Spatial Pattern of Urban Agglomeration in the Middle Reaches of the Yangtze River base on the "Production, Living and Ecological" Function*; Huazhong Agricultural University: Wuhan, China, 2017.
38. Bian, Z.X.; Cheng, X.F.; Yu, M.; Cui, W.; Shi, S.L. The Proportionality of the functions of production, life and ecology in connection zone between Shenyang and Fushun. *Chin. J. Agric. Resour. Reg. Plan.* **2016**, *37*, 84–92.
39. Wang, C.; Tang, N. Spatio-temporal characteristics and evolution of rural production-living-ecological space function coupling coordination in Chongqing Municipality. *Geogr. Res.* **2018**, *37*, 1100–1114.
40. Li, Y.; Sun, Z.G.; Liu, E.Y.; Shao, C.X. Spatio-temporal evolution of regional structure about crops and livestock in China based on the theory of comparative advantage. *J. Nat. Resour.* **2021**, *36*, 1149–1162. [CrossRef]
41. Zhang, J.X. Assessment of landscapes utilization quality and its coupling and coordination based on "Production, Living and Ecological": A case study of the Southern Jiangsu Region. *J. Agric. Sci.* **2017**, *38*, 57–63.
42. Ni, W.Q. The Coupling and Coordination Relationship and Their Spatial Pattern of Urban Land Use Economic, Social and Ecological Benefits of Cities in Three Largest Urban Agglomerations in China. *Urban Dev. Stud.* **2016**, *23*, 69–77.
43. Wang, L.J.; Dong, X.F.; Liu, X.G. Spatio-temporal characteristics of coupling coordination between the quality of population urbanization and land urbanization: The case of 12 central cities in Gansu Province, China. *Chin. J. Appl. Ecol.* **2016**, *27*, 3335–3343.
44. Guo, Y.H. *Study on the Comprehensive Evaluation and Its Spatial Differences of the Coupling Coordination Degree of Regional Production Space, Living Space and Ecology Space in Jiangsu Province*; Nanjing Agricultural University: Nanjing, China, 2017.
45. Jin, G.; Deng, X.Z.; Zhang, Q.; Wang, Z.Q.; Li, Z.H. Comprehensive function zoning of national land space for Wuhan metropolitan region. *Geogr. Res.* **2017**, *36*, 541–552.
46. Pang, L.; Zhou, X.; Wei, L.G. Function Evaluation and Coupling Coordination Analysis of Land Space "Productive—Living-Ecological" in Guangxi Zhuang Autonomous Region During 2009–2018. *Bull. Soil Water Conserv.* **2021**, *41*, 237–244.
47. Shen, S.K.; Lu, Y.; Hua, C.; Qin, D. Quantitative evaluation of the "Production living Ecological" Function in the Nanliu River Basin. *Chin. J. Agric. Resour. Reg. Plan.* **2020**, *41*, 147–155.
48. Dangles, O.; Casas, J. Ecosystem services provided by insects for achieving sustainable development goals. *Ecosyst. Serv.* **2019**, *35*, 109–115. [CrossRef]
49. Reyers, B.; Selig, E.R. Global targets that reveal the social–ecological interdependencies of sustainable development. *Nat. Ecol. Evol.* **2020**, *4*, 1011–1019. [CrossRef] [PubMed]
50. Wood, S.L.R.; Jones, S.K.; Johnson, J.A.; Brauman, K.A.; Chaplin-Kramer, R.; Fremier, A.; Girvetz, E.; Gordon, L.J.; Kappel, C.V.; Mandle, L.; et al. Distilling the role of ecosystem services in the sustainable development goals. *Ecosyst. Serv.* **2018**, *29*, 70–82. [CrossRef]
51. National Development and Reform Commission; Ministry of Housing and Urban-Rural Development of the People's Republic of China. *Beibu Gulf Urban Agglomeration Development Plan*; National Development and Reform Commission: Beijing, China, 2017.
52. Curnick, D.J.; Pettorelli, N.; Amir, A.A.; Balke, T.; Barbier, E.B.; Crooks, S.; Dahdouh-Guebas, F.; Duncan, C.; Endsor, C.; Lee, S.Y.; et al. The value of small mangrove patches. *Science* **2019**, *363*, 239. [CrossRef]
53. Peng, K.F.; Jiang, W.G.; Ling, Z.Y.; Hou, P.; Deng, Y.W. Evaluating the potential impacts of land use changes on ecosystem service value under multiple scenarios in support of SDG reporting: A case study of the Wuhan urban agglomeration. *J. Clean. Prod.* **2021**, *307*, 127321. [CrossRef]

diversity

Article

Coupling Relationship between Ecosystem Service Value and Socioeconomic Development in the Qinba Mountains, China

Qi Ma [1,*], Mengquan Zhou [1,2], Jiechao Liu [1], Jing Zhao [1] and Meijun Xi [3,*]

1 School of Tourism & Research Institute of Human Geography, Xi'an International Studies University, Xi'an 710128, China
2 College of Civil Engineering and Architcture, Guilin University of Technology, Guilin 541004, China
3 School of Economics, Xi'an University of Finance and Economics, Xi'an 710100, China
* Correspondence: maqi86@126.com (Q.M.); meijun_xi@126.com (M.X.)

Abstract: The degree of coordination between ecosystem services and the level of socioeconomic development has essential implications for regional sustainability. The coordinated development of ecology and economy is a major theoretical and practical problem for the Qinba Mountains, which is once one of the 14 contiguous destitute areas in China. Based on the land use and cover change (LUCC) data interpreted by medium-resolution remote sensing images, this study adopted the modified equivalent factor method to calculate ecosystem service value (ESV) and its temporal and spatial variation in the Qinba Mountains. A comprehensive index system was constructed to measure the socioeconomic development level and to reveal the coupling relationship between the ESV and socioeconomic development. The results show that: (1) for 2000–2015, the ESV in these areas was increasing, the proportion of forest ESV was the largest, and it increased significantly. (2) The level of socioeconomic development was constantly improving, and the differences within the region were gradually emerging. (3) Except for the Hantai District, which has been in a highly uncoordinated state, the degree of coordination between ESV and socioeconomic development has improved year by year, and most counties were in a state of medium coordination, or above. The results contribute to a scientific basis for decision making regarding ecological environmental protection and green economic development in the Qinba Mountains, and have positive significance for promoting the construction of ecological civilization and sustainable development in the study area.

Keywords: ecosystem service value (ESV); socioeconomic development; coupling coordination degree; entropy weight method; Qinba Mountains

Citation: Ma, Q.; Zhou, M.; Liu, J.; Zhao, J.; Xi, M. Coupling Relationship between Ecosystem Service Value and Socioeconomic Development in the Qinba Mountains, China. *Diversity* **2022**, *14*, 1105. https://doi.org/10.3390/d14121105

Academic Editors: Michael Wink and Giovanni Bacaro

Received: 30 October 2022
Accepted: 6 December 2022
Published: 12 December 2022

Publisher's Note: MDPI stays neutral with regard to jurisdictional claims in published maps and institutional affiliations.

1. Introduction

The ecological environment is the basis of human dependence. With the rapid population growth, food shortages, resource depletion and other global problems, ecosystem function degradation, environmental quality decline, and other phenomena emerge endlessly, and many negative economic development benefits are highlighted [1,2]. Coordinating the balance between economic development and ecological environmental protection, improving ecosystem services, has become the only way to construct ecological civilization and the sustainable development [2–4]. Ecosystem services are the benefits that humans obtain directly or indirectly from an ecosystem [5]. As a bridge between natural and socioeconomic processes, ecosystem services provide a new theoretical basis for studying the coupling of human and natural systems [1,6,7].

There have been several explorations and practices conducted on the impact of socioeconomic activities on ecosystem services at home and abroad. International research on ESV and socioeconomic development stems from the rise in "ecological economics," focusing on the complex interaction between the economy and the environment, as well as the characteristics of spatial relationships [8–10]. Grossman et al. (1991) proposed the environmental Kuznets curve (EKC) [11], which became an important tool to analyze the direct relationship between regional economic development and the environment [12]. With the EKC model, Cristina et al. (2016) [13] analyzed the research status of the economic value of ecosystem services in Spain and clarified the importance of economic assessment in the study of ESV. Constanza et al. (1997) [14] proposed the evaluation method of ESV, which made the principle and method of ESV estimation clearer in the scientific sense. This method, which is more intuitive, easy to use, and requires less data, is widely applied to ecological and economic problems at the global and regional scales [15,16]. In addition, there are other methods to study the relationship between environmental quality and economic development. Destek et al. (2018) [17] used the ecological footprint to characterize the ecological environment and studied the relationship between the ecological environment and economic development in European countries. Medeiros et al. (2020) [18] analyzed the trade-off between economic and environmental factors through multi-objective programming to further optimize land use in the Midwest of Brazil.

Related research in China is relatively late. After Xie et al. (2008, 2015) [19,20] improved the calculation and evaluation methods of ESV based on Constanza's research [14], considerable exploration and empirical work on coupling and coordination between ecology and economics has been conducted by scholars from various disciplines, including geography, ecology, and forest management [21]. The research areas involve administrative regions [22] and river basins [23,24], as well as economic belts and urban agglomerations [25–27], etc. The main concerns are the environment and the economy [28], ESV of forest and tourism economy [29], urbanization and ecosystem services [30,31], etc. For example, Sun et al. (2015) [32] explored the relationship between economic growth and ESV in the Beijing-Tianjin-Hebei region, conducting an in-depth analysis on the degree of coupling and coordination between the two and the economic benefits of the ecosystem. With the help of Xie et al.'s research results (2008, 2015) [19,20], Zhu et al. (2022) [33] revised and obtained the equivalent factor table of ESV in the Guanzhong region, and used the coupling degree model to explore the coupling relationship between ESV and economic development in this region. Chen et al. (2018) [34] measured the construction of ecological civilization by coupling the three subsystem indicators of ecological environment-economy-society. Feng et al. (2022) [35] analyzed the degree of coupling and coordination between different relocated population density and ESV variables. Wang et al. (2022) [26] calculated the ESV and tourism urbanization degree of the Chang-Zhu-Tan urban agglomeration, and analyzed the coordination relationship between them. Combined with multiple perspectives, these studies illustrate the relationship between ecological protection and economic development at different scales [36,37].

In summary, scholars at home and abroad have fundamentally the same cognition direction regarding ESV. Most methods of calculating the ESV on a regional scale use equivalent factor tables or value substitution. The research on the coupling relationship between changes in ESV and socioeconomic development mainly focuses on economically active areas, such as administrative regions, economic belts, and urban agglomerations. Relatively little attention has been paid to the contiguous destitute areas. Socioeconomic indicator systems are mostly confined to economic indicators, and there is a lack of comprehensive indicators that reflect people's living standards, social security, and socioeconomic scale and structure. Therefore, this study took the Qinba Mountains, one of 14 contiguous destitute areas in China, as the research object, and then revised the "Chinese land ecosystem services value equivalent factor table," discussing the temporal and spatial characteristics of its ESV changes, combined with the comprehensive level of socioeconomic development. The objectives of this study are: (1) to evaluate the comprehensive status of ESV and socioeconomic development in contiguous destitute areas, (2) to reveal the spatial and temporal heterogeneity of the coupling relationship between the ESV and socioeconomic development, and (3) to explore the interaction mechanism and the dynamic relationship of coupling and coordination between the ESV and socioeconomic development. This study can provide a reference for the development of ecological compensation policy and the in-depth study of green development in contiguous destitute areas.

2. Materials and Methods

2.1. Overview of the Study Area

The contiguous destitute areas in the Qinba Mountains cover six provinces in Sichuan, Hubei, Shaanxi, Henan, Chongqing, and Gansu. This paper takes the Southern Shaanxi section of the Qinba Mountains, which is composed of the cities of Ankang, Hanzhong, and Shangluo in Shaanxi Province, as the research object. There are 28 counties in total (Figure 1), with a total area of 70,234.93 km^2. In 2020, the population of the study area was 7,746,100, and the GDP was CNY 0.34 trillion (China yuan). The north side of the area is made up of the Qinling Mountains, the south side is the Bashan Mountains, and the middle part is the Hanjiang River, presenting the feature description that, "There is a river in the middle of two mountains." The terrain is dominated by hilly and mountainous areas at a medium altitude; in terms of climate, the north subtropical humid climate and the warm temperate humid climate are dominant, with cold winters and hot summers, and the four seasons are distinct [38]. The average annual temperature is 12–18 °C, and the annual rainfall is 700–1400 mm. The complex and diverse landforms and climatic environments have nurtured the abundant animal and plant resources in the Qinling Mountains, which is a key area for the protection of rare and endangered animals and plants in the country and the world, including the Qinling, Daba, and Minshan-Hengduan biodiversity priority areas. It is also an essential water source conservation area for China's South-to-North Water Diversion Project, playing an important strategic role in this endeavor.

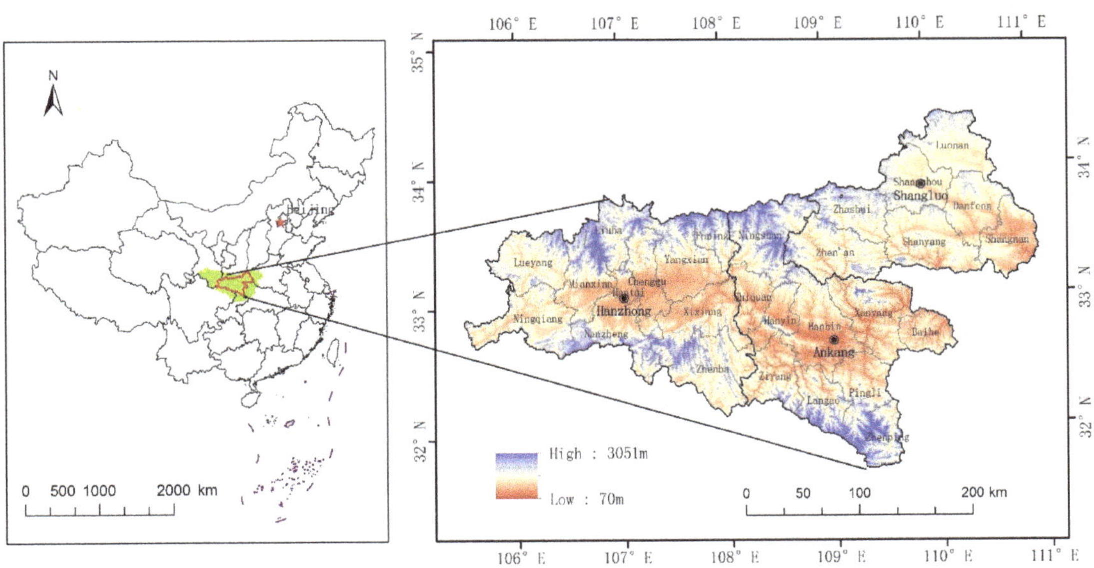

Figure 1. Overview of the study area.

2.2. Methods

This study is mainly divided into three steps. First of all, the ESV of different land use types in the Qinba Mountains from 2000 to 2015 was evaluated. Secondly, the socioeconomic development level of different counties was estimated. Finally, the analysis of their coupling coordination degree was conducted from 2000 to 2015 using the coupling and coordination model. The framework of this study is presented in Figure 2.

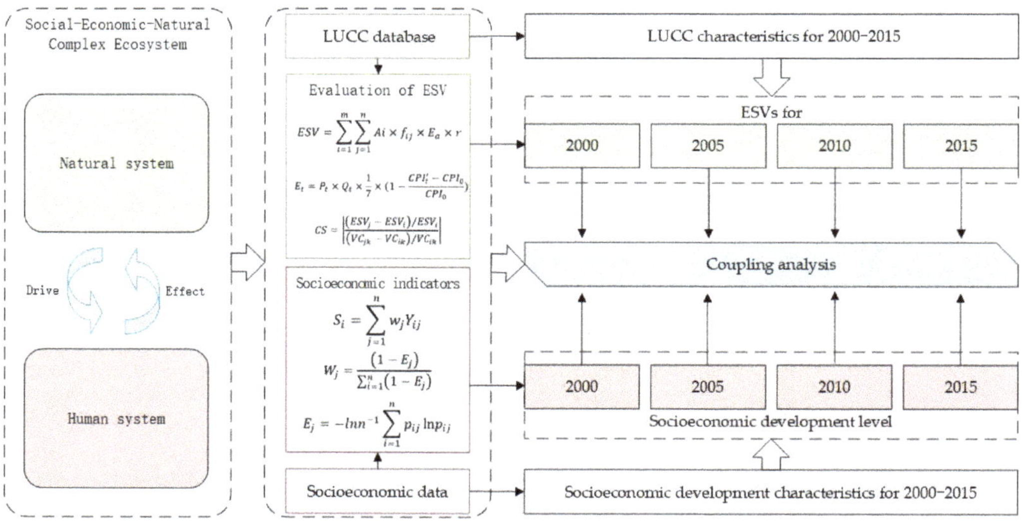

Figure 2. Framework of this study.

2.2.1. Calculation of ESV

(1) Calculation model of ESV

Xie et al. (2008) [19] formulated a table of equivalent factors of ESV per unit area of different terrestrial ecosystems in China (Table 1). Then, this established method was developed and optimized again by Xie et al. (2015) [20]. The formula used to calculate ESV was:

$$ESV = \sum_{i=1}^{m} \sum_{j=1}^{n} Ai \times f_{ij} \times E_a \times r \quad (1)$$

where ESV is the total value of ecosystem services (yuan); Ai is the area of type i land ecosystem; f_{ij} is the equivalent value per unit area of the j type ecosystem service of the i land ecosystem; E_a is the ecological ESV coefficient of the food production per unit area of the regional farmland ecosystem; and r is the equivalent factor correction coefficient.

Table 1. Equivalent value per unit area of ecosystem services in China.

Categories	Sub-Categories	Farmland	Forest	Grassland	Water Area	Unused Land	Construction Land
Supplying services	Food production	1	0.33	0.43	0.53	0.02	0
	Raw material	0.39	2.98	0.36	0.35	0.04	0
Regulating services	Gas regulation	0.72	4.32	1.5	0.51	0.06	0
	Climate regulation	0.97	4.07	1.56	2.06	0.13	0
	Hydrological regulation	0.77	4.09	1.53	18.77	0.07	0
	Waste treatment	1.39	1.72	1.32	14.85	0.26	0
Supporting services	Soil formation and retention	1.47	4.02	2.24	0.41	0.17	0
	Biodiversity protection	1.02	4.51	1.87	3.43	0.40	0
Cultural services	Recreation and culture	0.17	2.08	0.87	4.44	0.24	0

(2) Equivalent correction coefficient r

Due to the differences between ecosystems in different regions, relevant amendments should be made when calculating specific regions. According to the biomass factor table of farmland ecosystems in different provinces of China, developed by Xie et al. (2005) [39], the regional correction of China's ESV was expressed in the table, and the correction coefficient of Shaanxi Province was 0.51. According to the relevant data from the Shaanxi Regional Statistical Yearbook, the average grain yield of the Qinba Mountains in southern Shaanxi from 2000 to 2015 was 3227.33 kg/hm^2, which was 0.936 times that of the average grain yield of 3446.47 kg/hm^2 in Shaanxi Province during the same period. Therefore, the farmland ESV coefficient was revised to 0.477 of the national average.

(3) ESV of a standard equivalent E_t

Xie et al. (2008) [19] put the value of one equivalent factor as 1/7 of the national average grain yield market price. Many studies directly used the average value of food price over a certain year or several years to explore the evolution of ESV. However, there will be corresponding changes in different periods due to different natural, social and cultural conditions. The results obtained by calculating only the average value are quite different from the actual situation. For eliminating the impact of inflation, we introduced the consumer price index (CPI) accumulation coefficient to modify and develop the method for evaluating the value equivalent factor in unit area [40,41]. The formula for E_t is as follows:

$$CPI'_t = CPI_0 \times \frac{CPI_1}{100} \times \frac{CPI_2}{100} \cdots \frac{CPI_t}{100} (t = 0, 1, 2, \ldots, \mathrm{n}) \tag{2}$$

$$E_t = P_t \times Q_t \times \frac{1}{7} \times \left(1 - \frac{CPI'_t - CPI_0}{CPI_0}\right) \tag{3}$$

where CPI'_t is the CPI accumulation coefficient; CPI_t is the CPI in the t year (this study takes the CPI in 2000 as the benchmark, that is $CPI_0 = 100$); E_t is the economic value of a standard unit of ESV equivalent factor in the study area and the unit is yuan·hm^{-2}; P_t is the unit price of grain in the t year, the unit is yuan·kg^{-1}; and Q_t is the grain output per unit area, and the unit is kg·hm^{-2}. The calculation result is shown in the following table (Table 2).

Table 2. Equivalent value table of ecological services per unit area.

Years	National Grain Price/(yuan·kg^{-1})	Unit Grain Yield/(kg·hm^{-2})	Unit Output/(yuan·hm^{-2})	Unit Equivalent Factor/(yuan)	CPI Accumulation Coefficient	Corrected Unit Equivalent Value/(yuan)
2000	0.967	5133	4964	709	100	339
2005	1.347	5897	7943	1135	107	505
2010	2.076	6353	13,189	1884	124	688
2015	2.326	7011	16,307	2329	142	646

(3) The method of Sensitivity test

To test whether the ESV is sensitive to the ESV coefficient and whether the evaluation results are credible, this study introduces the coefficient of sensitivity (CS) in economics [42]. Through the adjustment method of increasing or decreasing the equivalent coefficient, we compare the changes before and after the ESV, and judge the dependence of the total ESV on the change of the equivalent coefficient. The equation is as follows:

$$CS = \left| \frac{(ESV_j - ESV_i)/ESV_i}{\left(VC_{jk} - VC_{ik}\right)/VC_{ik}} \right| \tag{4}$$

where CS is the coefficient sensitivity of a value, and a CS greater than 1 indicates that the ESV is elastic to the ESV coefficient, showing low reliability; a CS less than 1 indicates that the former is not elastic to the equivalence coefficient, and the result can be considered reliable. ESV_j and VC_{jk} represent the adjusted total value and equivalent coefficient of ecosystem services, respectively. The former unit is yuan and the latter unit is yuan/hm^2. In the same way ESV_i and VC_{ik}, respectively, represent the total value and equivalent coefficient of ecosystem services before adjustment. k represents different land use types.

2.2.2. Calculation of Socioeconomic Development Level of the Qinba Mountains

This study refers to the previous findings [27,33], combined with the availability of data, to weigh the level of the indicators of socioeconomic development, and finally chooses GDP per capita and local fiscal revenue per capita to reflect the scale of socioeconomic development; the proportion of the tertiary industry and the industrial added value indicators per capita reflect regional economic structure; the per capita net income of rural residents and the number of beds in health institutions per 1000 people reflect social security; the urbanization rate and total per capita social consumer goods reflect people's living standards. The weight of each index is determined by the entropy weight method [43]; then, the socioeconomic development level score of each district and county in the Qinba Mountains is calculated. The calculation process is as follows:

(1) Data standardization processing. Since the index data obtained from statistical yearbooks or local websites is not uniform in nature, dimension, order of magnitude, etc., we cannot directly use data indicators for in-depth analysis. Therefore, the range method is used to standardize the indicators.

(2) Calculate the entropy weight. The calculation formula of information entropy is as follows:

$$E_j = -lnn^{-1} \sum_{i=1}^{n} p_{ij} \ln p_{ij} \tag{5}$$

where $p_{ij} = \frac{Y_{ij}}{\sum_{i=1}^{n} Y_{ij}}$ and at the same time, the condition of $p_{ij} = 0$ is satisfied, $p_{ij} lnp_{ij} = 0$, and the range of entropy value should be satisfied that $e_{ij} \in [0,\ 1]$.

The weight of each indicator is as follows:

$$W_j = \frac{(1 - E_j)}{\sum_{i=1}^{n} (1 - E_j)} \tag{6}$$

where Y_{ij} is the index value after standardization using the range method, x_{ij} is the original index value, minx_{ij} is the minimum value of each index, maxx_{ij} is the maximum value of each index, n is the number of statistical indicators, E_j the information entropy of the jth index, and W_j is the weight of the jth indicator. The calculation results of the weights of each indicator are as follows (Table 3):

Table 3. Evaluation index system and weight of socioeconomic development level.

Index	Weights
GDP per capita (yuan)	0.105
Local fiscal revenue per capita (yuan)	0.112
Proportion of the tertiary industry (%)	0.058
Industrial added value per capita (yuan)	0.106
Urbanization rate (%)	0.301
Total per capita social consumer goods (yuan)	0.091
Number of beds in health care institutions per 1000 people (sheets)	0.129
Per capita net income of rural residents (yuan)	0.097

(3) Calculate the comprehensive index score. After using the entropy weight method to objectively determine the index weight, the socioeconomic development level of each

district and county in the Qinba mountains are calculated according to the index weight. Calculated as follows:

$$S_i = \sum_{j=1}^{n} w_j Y_{ij} \quad (7)$$

2.2.3. Coupling and Coordination Model

(1) Calculate the coordination index. The coupling degree is used to reflect the degree of interdependence between systems or elements [33]. The coordination index is an indicator to measure the degree of interdependence between the two. In this study, the coupling coordination index is used to measure the coupling coordination between ecology and socioeconomic development level in the Qinba Mountains. The equation is as follows:

$$D = \sqrt{C \cdot T} \quad (8)$$

$$C = 2 \cdot \sqrt{(U_1 \cdot U_2)/(U_1 + U_2)^2} \quad (9)$$

$$T = a \cdot U_1 + b \cdot U_2 \quad (10)$$

where D is the coordination index, $D \in [0, 1]$. The closer D is to 1, the more coordinated the two systems or elements are; and the closer D is to 0, the lower the degree of coordination between them is. C is the coupling degree, $C \in [0, 1]$. The closer C is to 1, the better the coupling state between the two systems or elements, and the closer to 0, the worse the coupling state between the two; T is the comprehensive coordination index of the two systems or elements, U_1 is the standardized ecosystem services, and U_2 is the score of socioeconomic development level; a and b are the undetermined coefficient, with $a + b = 1$. Since the relationship between socioeconomic development and ecological protection is mutually restrictive and interrelated, we use a value of 0.5 for both a and b, in line with previous studies [31].

(2) Divide the types of coupling coordination. In order to more intuitively explain the coupling development status between ESV and socioeconomic development, the coupling coordination degree is divided into five levels by using the equal spacing method, combined with the actual situation of the study area, and referring to the existing research [33] (Table 4).

Table 4. Types and levels of development of coupling and coordination between ESV and socioeconomic development.

Coupling Coordination Degree (D)	Coordination Level
$0 \leq D \leq 0.2$	Highly uncoordinated
$0.2 < D \leq 0.4$	Basically uncoordinated
$0.4 < D \leq 0.6$	Lowly coordinated
$0.6 < D \leq 0.8$	Mediumly coordinated
$0.8 < D \leq 1$	Highly coordinated

2.3. Data Sources

The data used in the study include food production data, land use data, and economic indicators. The land use and cover change (LUCC) data, which can meet user mapping accuracy demands at a scale of 1:100,000, is provided by the Data Center for Resources and Environmental Science, Chinese Academy of Sciences (RESDC) (http://www.resdc.cn, accessed on 18 November 2016). Based on the consistent use of the same methods, this database has been updated every 5 years using the Landsat TM/ETM (1990–2010) and Landsat 8 (2015) remote sensing images, with a spatial resolution of 30 m and the comprehensive evaluation accuracy of more than 93%. The data of Shaanxi Province in 2000, 2005, 2010, and 2015 were downloaded, including data for farmland, forest, grassland, water area, construction land, and unused land (Figure 2) [44,45]. The urbanization rate is replaced by the proportion of construction land in county land use. The data on grain prices and other economic indicators are derived from the China Yearbook of Agricultural Price Survey (2005, 2010, 2016) and the Shaanxi Regional Statistical Yearbook (2000–2016).

3. Results and Analysis

3.1. Evalution Analysis of ESV

3.1.1. Changes in Ecosystem Types

Table 5 shows the structure of land use types in the Qinba Mountains for 2000–2015. The grassland area made up the largest proportion in the study area, accounting for 41.23% in 2015, followed by forest land and farmland, accounting for 31.85% and 25.63%, respectively. For 2000–2015, the change in all kinds of the land area was relatively obvious. The grassland area reduced significantly, decreasing by 395.00 km^2, followed by farmland, decreasing by 58.52 km^2. The forest area increased significantly by 230.09 km^2. The proportion of construction land area increased continuously, from 0.53% in 2000 to 0.77% in 2015, with a total increase of 167.60 km^2. Due to the protection of water conservation areas by the South-to-North Water Transfers Strategy, the water area increased year by year. Overall, the area of all types changed greatly during 2000–2010, but the change was small during 2010–2015.

Table 5. Changes in land use types in the Qinba Mountains from 2000 to 2015.

Land-Use Types	Percentage of Land Use Types (%)				Change in Land Area (km^2)			
	2000	2005	2010	2015	2000–2005	2005–2010	2010–2015	2000–2015
Farmland	25.72	25.39	25.70	25.63	−231.14	219.60	−46.98	−58.52
Forest	31.52	31.60	31.87	31.85	58.63	184.55	−13.09	230.09
Grassland	41.80	41.97	41.23	41.23	120.29	−514.95	−0.34	−395.00
Water area	0.42	0.43	0.48	0.49	5.89	33.77	3.29	42.96
Unused land	0.01	0.01	0.02	0.03	0.02	6.26	6.60	12.88
Construction Land	0.53	0.60	0.70	0.77	46.32	70.77	50.51	167.60

Figure 3 shows the spatial distribution of ecosystems in the Qinba Mountains for 2000–2015. Grassland, as the main component, was mainly distributed in the middle and low mountains, while the forest land was distributed in the middle and high mountains, and the farmland was mainly distributed in the river valley areas. The change in the ecosystem was mainly distributed on both sides of the Hanjiang River and the Danjiang River, mainly in the low mountain and hilly areas. Due to the rapid development of urbanization (e.g., the urbanization rate of Hantai District increased from 6.20% in 2000 to 12.37% in 2015.), the ecosystems types in the flat top basins where the Hantai, Hanbin, and Shangzhou districts are located, were undergoing positive changes. Land for urban construction and rural development expanded by 167.60 km^2 over the past 16 years, occupying a large amount of arable land. Secondly, the implementation of ecological restoration policies, such as the Grain for Green project and the South-to-North Water Transfers from 2002 to 2008, has ensured the stability of the water conservation function in this area. The forest area increased gradually, with an additional area of 230.09 km^2 from 2000 to 2015, while grassland area gradually decreased, with a total decrease of 395.00 km^2 in this period.

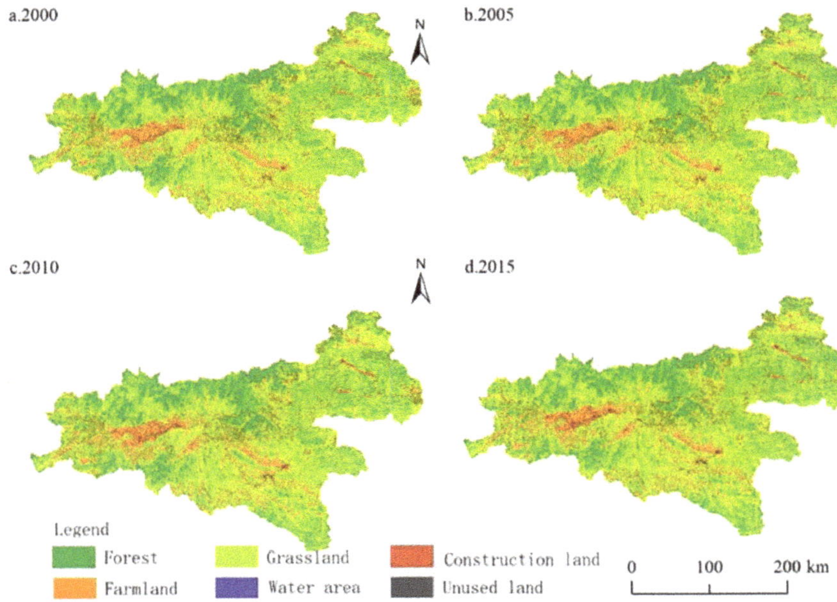

Figure 3. Spatial pattern of ecosystems in the Qinba Mountains in (**a**) 2000, (**b**) 2005, (**c**) 2010, and (**d**) 2015.

3.1.2. Temporal Changes in ESV

The ESV of the Qinba Mountains generally showed an upward trend, with a total increase of CNY 34.623 billion for 2000–2015. In 2000–2005 and 2005–2010, the value of ecosystem services increased evenly. The total ESV decreased slightly by CNY 4.777 billion from 2010 to 2015. During this period, the ESV increased in farmland, forest, grassland, water area, and unused land. Forest ESV increased the most, reaching CNY 19.514 billion, followed by grassland and farmland, with a total increase of CNY 145.67 billion. Among the five land use types, the ESV of forests accounts for the largest share of the total value, reaching 55.77%, followed by grassland and farmland, accounting for 42.92%, and unused land accounts for the smallest share. Except for unused land, water ecosystem service value grows the fastest, followed by forest, farmland, and grassland (Table 6).

Table 6. Changes in the value of ecosystem services in different land types in the Qinba Mountains from 2000 to 2015 (CNY 100 million).

Year	Farmland	Forest	Grassland	Water Area	Unused Land	Construction Land	Total
2000	48.25	210.51	115.95	4.57	0.00	0.00	379.28
2005	70.99	314.58	173.52	6.94	0.00	0.00	566.03
2010	97.97	432.39	232.39	10.52	0.01	0.00	773.28
2015	91.72	405.65	218.15	9.97	0.02	0.00	725.51
ESV increment	43.47	195.14	102.20	5.40	0.02	0.00	346.23
Average weight %	12.64	55.77	30.28	1.31	0.00	0.00	100.00
Growth rate%	90.09	92.70	88.14	118.16	438.20	0.00	/

The proportion and change rate of the value for each ecosystem service in 2000, 2005, 2010, and 2015 were calculated and integrated. Over the 16 years, the proportion of each functional value fluctuated in a small range, and the proportion composition was relatively stable (Table 7). These data indicate that the functions of soil formation and retention and biodiversity protection occupy the largest share of the composition, followed by hydrological regulation, climate regulation, and gas regulation (Table 7). It is worth noting that, although the four functions of soil formation and retention, biodiversity protection, gas regulation, and climate regulation accounted for a large proportion, their proportions showed a decreasing trend year by year for 2000–2015. From the perspective of increment change in value, for 2000–2015, the value of nine ecosystem services all increased, to a certain extent. Among these, the soil formation and retention and the biodiversity protection function value increments were the largest; the value increase for each function changed by nearly 50%, and for 2010–2015, the value increase of each function decreased slightly. The change in the growth rate in each period is characterized by a gradual decline. For 2010–2015, there was a small negative growth. Overall, the value of ecosystem services changed significantly for 2000–2015. The forest is the most important ecosystem in the Qinba Mountains. In 2015, the forest area accounted for 32.10% of all land use types, and its ESV accounted for 55.77%.

Table 7. Changes in the value of different ecosystem services in the Qinba Mountains from 2000 to 2015.

Categories	Proportion (%)				ESV Increment (CNY 100 Million)				Annual Rate of Change (%)			
	2000	2005	2010	2015	2000–2005	2005–2010	2010–2015	2000–2015	2000–2005	2005–2010	2010–2015	2000–2015
Food production	3.40	3.38	3.38	3.38	6.25	7.01	−1.64	11.62	48.42	36.59	−6.25	90.06
Raw material	7.46	7.46	7.49	7.49	13.94	15.66	−3.59	26.02	49.27	37.07	−6.19	91.93
Gas regulation	13.63	13.63	13.62	13.62	25.48	28.16	−6.51	47.13	49.30	36.50	−6.18	91.19
Climate regulation	13.73	13.73	13.72	13.72	25.65	28.39	−6.56	47.48	49.25	36.52	−6.19	91.15
Hydrological regulation	13.82	13.83	13.87	13.87	25.88	28.96	−6.59	48.24	49.38	36.99	−6.15	92.07
Waste treatment	9.48	9.47	9.49	9.49	17.65	19.78	−4.52	32.91	49.07	36.89	−6.16	91.49
Soil formation and retention	16.18	16.17	16.13	16.12	30.17	33.19	−7.72	55.64	49.18	36.26	−6.19	90.68
Biodiversity protection	15.53	15.53	15.52	15.52	29.02	32.08	−7.42	53.68	49.27	36.48	−6.18	91.14
Recreation and culture	6.77	6.78	6.78	6.78	12.71	14.03	−1.64	11.62	49.46	36.54	−6.15	91.51

3.1.3. Spatial Patterns of Change in the ESV

For 2000–2005, the ESV change rate for each county was 48.46–50.50%; for 2005–2010, the ESV change rate for each county was from 33.06% to 35.15%, and for 2010–2015, the ESV change rate was from −7.75% to 6.01%. The change rate of ESV in the counties in the Qinba Mountains over the past 16 years was above 80% for all counties, and the range of change was between 84.29% and 93.29%; the change in the average ecosystem service value of each county was 3606.26–6694.95 yuan/hm^2, with large spatial heterogeneity. Among these, the change rate of ESV in the whole area of Ankang City is relatively obvious, concentrated in Hanbin, Shiquan, Ziyang, and Langao counties and other places in the central district of Ankang City. The variability of Hanzhong and Shangluo cities is relatively small, while Shangzhou and Lueyang counties show prominent changes, in comparison with other counties. On the whole, the change in the counties in the Qinling Mountains is smaller than those in the Bashan Mountains (Figure 4a). According to the average value of land change (Figure 4b), the four counties in Ankang City showed little change, and the average land change value in Hanzhong and Ankang cities, located in the Bashan Mountains, is at a lower level, and the average land change values in Shangluo, Ankang City, and the northern part of Hanzhong City in the Qinling Mountains is relatively large. Thus, the ESV in the Qinling Mountains has a greater contribution to the region and is more stable than that of the Bashan Mountains.

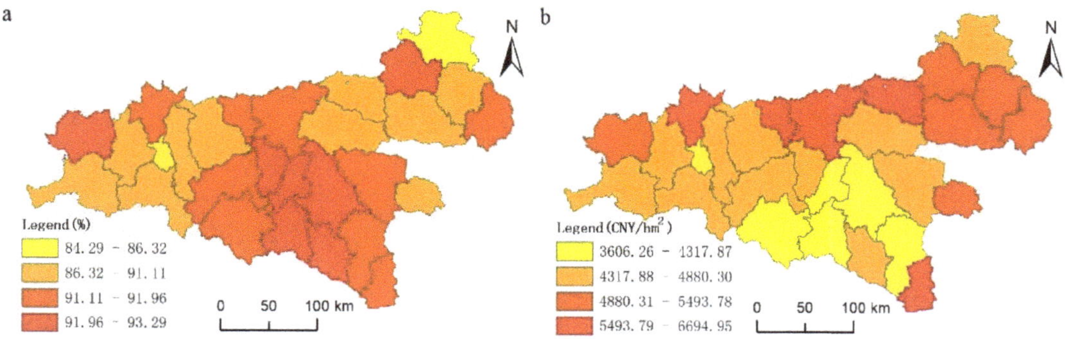

Figure 4. (**a**) Rate of change in ecosystem service value; (**b**) distribution of average ecosystem service value in the Qinba Mountains from 2000 to 2015.

3.1.4. Sensitivity Analysis

By increasing or decreasing the ESV of each land type by 50%, the sensitivity index of the ESV of each land type in four periods was obtained (Table 8). The sensitivity index was less than 1, indicating that the ecosystem services value coefficient of each land type in four periods was reliable. The sensitivity index of each type of land use, from large to small, is forest, grassland, farmland, water area, and unused land. Among these, forest land has the greatest impact and contribution on the total value of regional ecosystem services, and unused land makes the lowest contribution. The sensitivity index of various types of land differed little in the four periods of 2000, 2005, 2010, and 2015. The sensitivity index of forest land and water area increased slightly over time, while that of grassland decreased slightly.

Table 8. Coefficient of sensitivity derived from equivalence coefficient adjustment.

Land Use Types	2000	2005	2010	2015
Farmland	0.127	0.125	0.127	0.126
Forest	0.555	0.556	0.559	0.559
Grassland	0.306	0.307	0.301	0.301
Water area	0.012	0.012	0.014	0.014
Unused land	0.000	0.000	0.000	0.000

3.2. Analysis of Socioeconomic Development Level

3.2.1. Time Change of Socioeconomic Development Level

The score for the socioeconomic development level of each county in the Qinba Mountains for 2000–2015 was obtained using the comprehensive index method (Table 9). For 2000–2015, the socioeconomic development level of the Qinba Mountains increased significantly. The average level of socioeconomic development in 2000 was 0.025, 0.111 in 2005, 0.287 in 2010, and 0.577 in 2015. Over the past 16 years, the level of socioeconomic development has increased by 23 times. On the whole, from 2000 to 2015, the rapid economic growth of the counties in the Qinba Mountains benefited from the introduction of various national economic policies and ecological compensation measures. After 2000, the country began to fully implement the Western Development Strategy, led by the state to increase investment in the Western economy, laying the foundation for the rapid development of the Western region economy. The government of the Qinba mountains seized the opportunity to adjust the economic structure and change the mode of development, adopting the concept of circular economy to promote regional development. Due to this, the socioeconomic development of southern Shaanxi has achieved remarkable improvement.

Table 9. Scores of socioeconomic development levels of counties in the Qinba Mountains.

County	Si			
	2000	2005	2010	2015
Hanbin	0.018	0.098	0.301	0.583
Hanyin	0.022	0.082	0.294	0.602
Shiquan	0.030	0.086	0.315	0.569
Ningshan	0.028	0.114	0.300	0.557
Ziyang	0.010	0.180	0.289	0.521
Langao	0.026	0.112	0.280	0.582
Pingli	0.024	0.112	0.286	0.578
Zhenping	0.026	0.129	0.290	0.555
Xunyang	0.006	0.157	0.284	0.553
Baihe	0.019	0.074	0.309	0.597
shangzhou	0.006	0.085	0.299	0.610
Luonan	0.004	0.080	0.290	0.625
Danfeng	0.003	0.100	0.290	0.607
Shangnan	0.010	0.080	0.303	0.607
Shanyang	0.010	0.101	0.281	0.608
Zhen'an	0.162	0.213	0.154	0.471
Zhashui	0.021	0.074	0.339	0.566
Hantai	0.023	0.064	0.269	0.644
Nanzheng	0.034	0.077	0.248	0.641
Chenggu	0.009	0.104	0.270	0.617
Yangxian	0.001	0.084	0.349	0.567
Xixiang	0.005	0.056	0.328	0.611
Mianxian	0.015	0.072	0.320	0.592
Ningqiang	0.001	0.125	0.302	0.572
Lueyang	0.137	0.236	0.261	0.367
Zhenba	0.008	0.091	0.278	0.623
Liuba	0.006	0.159	0.270	0.566
Foping	0.041	0.158	0.236	0.565

3.2.2. Regional Differences in Socioeconomic Development Level

The socioeconomic development level of the counties in the Qinba Mountains showed a continuous upward trend for 2000–2015. However, the socioeconomic development within the region is unbalanced (Table 9, Figure 5). In 2000, there was little difference in socioeconomic level among counties in southern Shaanxi, except for Zhen'an County of Shangluo City and Lyuyang County of Hanzhong City, which had outstanding socioeconomic development levels, relying on local characteristic industries. In 2005, the socioeconomic development level of Zhen'an and Lueyang county continued to maintain their advantages. In addition, Ziyang and Xunyang counties in Ankang City relied on local tea brands and industrial and mining enterprises to promote economic development. Differences in the level of regional economic development have been highlighted. In 2010, with the national policy of southern Shaanxi boosting the economy of southern Shaanxi, the overall level of socioeconomic development of each county has improved, and Zhen'an and Lueyang counties no longer have a noticeable advantage. Zhashui and Yang counties, as well as other areas with beneficial transportation locations, have shown rapid development. In 2011, the formulation of the circular economy policy in southern Shaanxi coordinated the development of the inter-regional economy. In 2015, the economy developed rapidly in various regions, and the internal differences were obvious. Hantai District was in a dominant position, but Zhen 'an and Lueyang counties were still in a catch-up position. It was different traffic locations and resource advantages in different counties that led to different economic development rates.

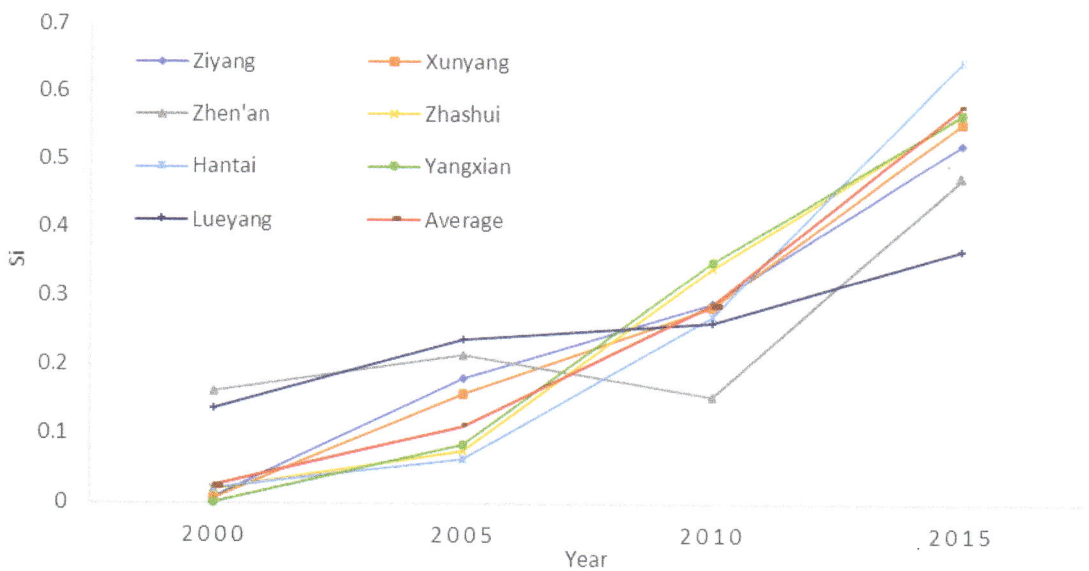

Figure 5. Socioeconomic development level of each county in the Qinba Mountains.

3.3. Coordination Analysis of ESV and Socioeconomic Development

The ecological and socioeconomic coupling coordination status of the study area improved during 2000–2015, and the number of highly incoordinated and basically coordinated counties decreases year by year, from the initial basically uncoordinated status to the final mediumly coordinated status. In 2000, 72.40% of the regions showed a basically uncoordinated status, and Hantai, Ningqiang, Yangxian, and Danfeng counties were in the highly uncoordinated state. By 2005, most of the counties designated as basically uncoordinated in 2000 had changed to a low coordination status, and the low coordination area accounted for about 84.05%. Zhen'an, Hanyin, and Baihe counties were still in a basically uncoordinated state, and only Hantai District was left as a highly uncoordinated area. In 2010, the medium coordination status was dominant, followed by the low coordination status, accounting for 61.15% and 38.07%, respectively. In 2015, medium coordination gradually replaced the low coordination status, and the medium coordination counties accounted for as high as 83.05% of the area, Ningshan and Shanyang counties showed a high degree of coordination, and the low coordination status (e.g., Foping, Shiquan, and Hanyin counties) did not change (Figure 6). It is worth noting that during the four periods, Hantai District has always shown a highly uncoordinated status. Further analysis shows that the socioeconomic development of Hantai District was better from 2000 to 2015, but its ESV showed negative growth. The rapid development of the economy led to the deterioration of the environmental situation, and finally led to the imbalance of economic and ecological development. On the whole, the ecological and socioeconomic coupling coordination level of the study area has increased year by year.

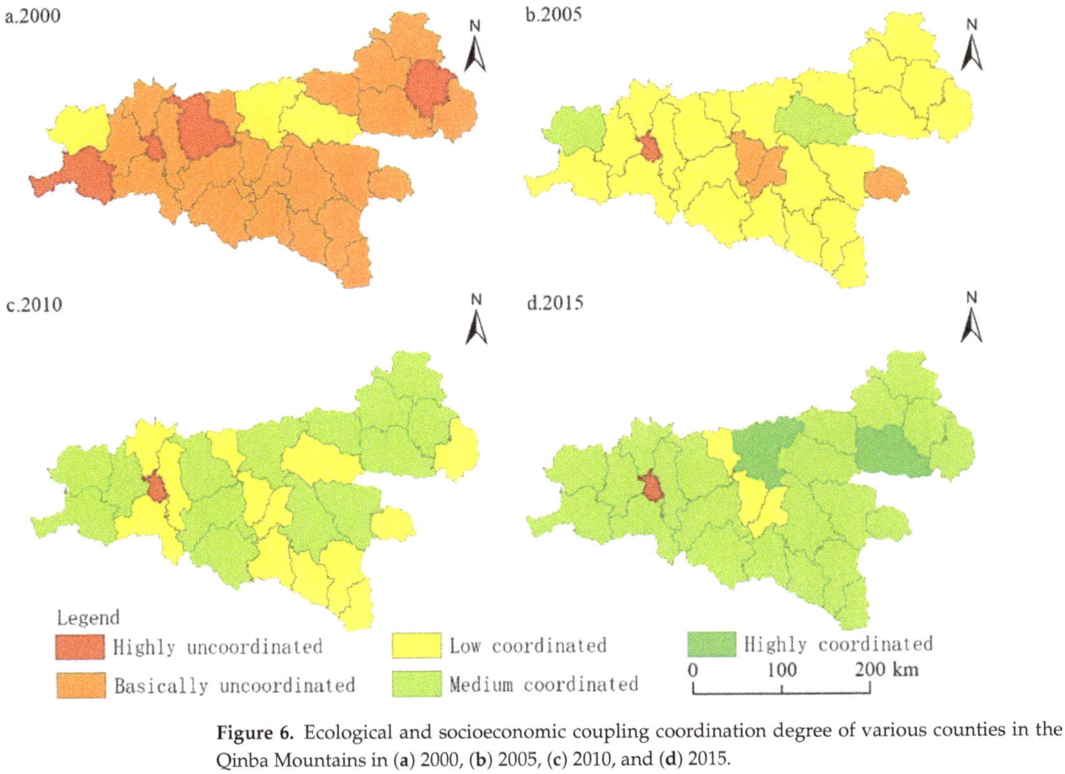

Figure 6. Ecological and socioeconomic coupling coordination degree of various counties in the Qinba Mountains in (**a**) 2000, (**b**) 2005, (**c**) 2010, and (**d**) 2015.

4. Discussion

The natural system is a fundamental condition for the coordinated development of ecological protection and socioeconomic development. How to balance ecological and environmental protection with socioeconomic development is an important issue to be addressed in the construction of ecological civilization. Without a better ecological environment, the cost of blind economic development will lead to the degradation of ecosystem functions, and the economic effect will drop sharply [46,47]. Therefore, to achieve the sustainable development of ecology and economy in the Qinba Mountains, more attention should be paid to continuing to implement the Grain for Green project, optimizing the land use structure, and improving the efficiency of land allocation in future development, according to local conditions. First of all, we should maintain the current coordinated development trend, promoting the development model of a green circular economy, such as in Ningshan and Shanyang counties, and then promote sustainable development throughout the Qinba Mountains. Faced with the problem of uncoordinated ecological and economic development in the Hantai District, the change in regional land use has led to serious ecosystem degradation [25,48]. It is necessary to fully consider the carrying capacity of the regional ecological environment. In the two aspects of strengthening ecological and environmental protection and adjusting the economic structure, we must strike a balance between development and protection.

Compared with the same research content in the existing studies, the current coupling studies focus on the urban agglomerations [26,33], economic belts [27,28], and other areas around the large towns or the economically active areas in the east. Relatively little attention has been paid to the special geographical location of ecological function in contiguous destitute areas. Compared with the ESV assessment study using the same study area [49], the research results of the two in terms of the year-by-year change of ESV and regional differentiation are basically the same, which proves that the results of this study are reliable. This study further discusses the coupling relationship between ecology and economy, in combination with the change in socioeconomic development level in the study area. However, the value evaluation method based on the equivalent value factor also has shortcomings and defects, such as over-reliance on data interpreted by remote sensing satellite images [1]. Ecosystems are complex and diverse, and the food yield data cannot fully explain the problem. For example different vegetation coverage, stand structures, and the estimation of land type value of regional locations are not fully considered in the assessment [50]. Moreover, the urban green space in the construction is neglected in the calculation process. Whether this approach is scientific or not, it still deserves further study and reflection. At the same time, ESV is affected by climate change and cultural and economic factors [47,51], the assessment of ecosystem services is sometimes inaccurate [52,53]. In future studies, it will be necessary to use a more accurate computational model of the ESV and to employ a variety of methods to correct it. We should also start with more dimensions of ecological environment elements and more diverse socioeconomic development data. Further research is needed to explore the mechanisms of coupling and mutual feedback between ecology and economy. The relationship between economy and ecology is both separate and interconnected. First of all, socioeconomic development will coerce or facilitate ecosystem services. Secondly, weak ecosystem services can slow down or inhibit social and economic development. Only by coordinating the relationship between the two above can we achieve sustainable development [8,9,29,54]. The value realization mechanism for ecological products should be established and improved to promote the effective conversion of ESV. The coordinated development of the social-ecological systems is an important condition for the sustainable development of contiguous destitute areas, which have some capacity for self-organization and are affected by both human activities and climate change. The variations in the relationship between the economy and ecology show nonlinear and non-stationary complex dynamic features. The ecosystem is an important support for human well-being and socioeconomic development [6]. Only under the premise of reasonable ESV and sustainable transformation can a regional economy develop with high quality and attract more investment. Blindly pursuing the conversion ESV can easily lead to problems such as the overburdening of the ecosystem's carrying capacity, etc. [55]. In short, we should focus more on development and protection.

5. Conclusions

(1) In terms of changes in ESV, changes in forest area have the greatest impact on ESV. With the increase in forest area, the ecosystem service value of the study area also increased. For 2000–2015, the forest area increased by 230.09 km^2, and the value of ecosystem services in the study area increased by CNY 19.514 billion.

(2) In the aspect of socioeconomic development, with the help of relevant policies, the level of socioeconomic development in the study area has risen steadily, and the vitality of

socioeconomic development has continued to improve. The socioeconomic development level score rose from 0.025 in 2000 to 0.577 in 2015. In the course of development, there is still the problem of unbalanced economic development within the region.

(3) In the coordinated development of natural ecology and socio-economy respect, Hantai District has been highly uncoordinated. The coordinated development of the counties in the Qinba Mountains is in good condition, and the degree of coupling coordination shows a trend of improvement. In 2015, 83.05% of the districts were mediumly coordinated, and Ningshan and Shanyang counties were highly coordinated.

Author Contributions: Conceptualization, Q.M. and M.X.; methodology, M.Z.; formal analysis, M.Z., J.L. and M.X.; data curation, M.Z. and Q.M.; writing—original draft preparation, M.Z.; writing—review and editing, M.Z., J.L., J.Z., M.X. and Q.M.; funding acquisition, Q.M. All authors have read and agreed to the published version of the manuscript.

Funding: This research was funded by the Research Project on Major Theoretical and Practical Problems of Philosophy and Social Sciences in Shaanxi Province (Program No. 2022ND0284), the Natural Science Basic Research Program of Shaanxi (Program No. 2021JQ-769), the Scientific Research Project of Shaanxi Provincial Education Department (Program No. 21JK0153), and the Innovation Capability Support Program of Shaanxi (Program No. 2022KRM038).

Institutional Review Board Statement: Not applicable.

Data Availability Statement: Land use data, with medium-resolution, can be purchased from the funders by request. Other data are available from the authors upon reasonable request, as the data are required for further use.

Conflicts of Interest: The authors declare no conflict of interest.

References

1. Fu, J.; Zhang, Q.; Wang, P.; Zhang, L.; Tian, Y.; Li, X. Spatio-Temporal Changes in Ecosystem Service Value and Its Coordinated Development with Economy: A Case Study in Hainan Province, China. *Remote Sens.* **2022**, *14*, 970. [CrossRef]
2. Rozelle, S.; Huang, J.; Zhang, L. Poverty, population and environmental degradation in China. *Food Policy* **1997**, *22*, 229–251. [CrossRef] [PubMed]
3. Li, S.; Zhang, C.; Liu, J.; Zhu, W.; Ma, C.; Wang, Y. The tradeoffs and synergies of ecosystem services: Research progress, development trend, and themes of geography. *Geogr. Res.* **2013**, *32*, 1379–1390.
4. Yang, T.; Kuang, W.; Liu, W.; Liu, A.; Pan, T. Optimizing the layout of eco-spatial structure in Guanzhong urban agglomeration based on the ecological security pattern. *Geogr. Res.* **2017**, *36*, 441–452.
5. Millennium Ecosystem Assessment. *Ecosystems and Human Well-Being: Synthesis*; Island Press: Washington, DC, USA, 2005; p. 137.
6. Hou, P.; Wang, Q.; Zhai, J.; Liu, H.; Yang, M. Progress of integrated ecosystem assessment: Concept, framework and challenges. *Geogr. Res.* **2015**, *34*, 1809–1823.
7. Braat, L.C.; de Groot, R. The ecosystem services agenda: Bridging the worlds of natural science and economics, conservation and development, and public and private policy. *Ecosyst. Serv.* **2012**, *1*, 4–15. [CrossRef]
8. Stern, D.I.; Common, M.S.; Barbier, E.B. Economic growth and environmental degradation: The environmental Kuznets curve and sustainable development. *Word Dev.* **1996**, *24*, 1151–1160. [CrossRef]
9. Arrow, K.; Bolin, B.; Costanza, R.; Dasgupta, P.; Folke, C.; Holling, C.S.; Jansson, B.O.; Levin, S.; Mäler, K.G.; Perrings, C.; et al. Economic growth, carrying capacity, and the environment. *Ecol. Econ.* **1995**, *15*, 91–95. [CrossRef]
10. Bateman, I.J.; Harwood, A.R.; Mace, G.M.; Watson, R.T.; Abson, D.J.; Andrews, B.; Binner, A.; Crowe, A.; Day, B.H.; Dugdale, S.; et al. Bringing Ecosystem Services into Economic Decision-Making: Land Use in the United Kingdom. *Science* **2013**, *341*, 45–50. [CrossRef]
11. Grossman, G.; Krueger, A.B. Environmental Impacts of a North American Free Trade Agreement. In *NBER Working Paper No. 3914*; National Bureau of Economic Research: Cambridge, MA, USA, 1991; pp. 1–57.
12. Zhao, H.; Wu, J. Measurement of environmental pollution transfer in Beijing-Hebei Region based on the Environmental Kuznets Curve. *China Popul. Resour. Environ.* **2020**, *30*, 90–97.
13. Cristina, Q.-S.; Berta, M.-L.; Fernando, S.-M.; Maria, L.; Carlos, M.; Javier, B.; Marina, G.-L. Ecosystem services values in Spain: A meta-analysis. *Environ. Sci. Policy* **2016**, *55*, 186–195.
14. Costanza, R.; d'Arge, R.; De Groot, R.; Farber, S.; Grasso, M.; Hannon, B.; Limburg, K.; Naeem, S.; ONeill, R.V.; Paruelo, J.; et al. The value of the world's ecosystem services and natural capital. *Nature* **1997**, *387*, 253–260. [CrossRef]
15. Costanza, R.; De Groot, R.; Sutton, P.; van der Ploeg, S.; Anderson, S.J.; Kubiszewski, I.; Farber, S.; Turner, R.K. Changes in the global value of ecosystem services. *Global. Environ. Change* **2014**, *26*, 152–158. [CrossRef]

16. Wang, W.; Guo, H.; Chuai, X.; Dai, C.; Lai, L.; Zhang, M. The impact of land use change on the temporospatial variations of ecosystems services value in China and an optimized land use solution. *Environ. Sci. Policy* **2014**, *44*, 62–72. [CrossRef]
17. Destek, M.A.; Ulucak, R.; Dogan, E. Analyzing the environmental Kuznets curve for the EU countries: The role of ecological footprint. *Environ. Sci. Pollut. Res.* **2018**, *25*, 29387–29396. [CrossRef]
18. Medeiros, G.; Florindo, T.; Talamini, E.; Fett Neto, A.; Ruviaro, C. Optimising Tree Plantation Land Use in Brazil by Analysing Trade-Offs between Economic and Environmental Factors Using Multi-Objective Programming. *Forests* **2020**, *11*, 7. [CrossRef]
19. Xie, G.; Zhen, L.; Lu, C.; Xiao, Y.; Chen, C. Expert Knowledge Based Valuation Method of Ecosystem Services in China. *J. Nat. Resour.* **2008**, *5*, 911–919.
20. Xie, G.; Zhang, C.; Zhang, L.; Chen, W.; Li, S. Improvement of the Evaluation Method for Ecosystem Service Value Based on Per Unit Area. *J. Nat. Resour.* **2015**, *30*, 1243–1254.
21. Li, Z.; Pan, J.; Hu, Y. The Spatio-temporal Variation of Ecological Property Value and Eco-Economic Harmony in Gansu Province. *J. Nat. Resour.* **2017**, *32*, 64–75.
22. Shifaw, E.; Sha, J.; Li, X.; Bao, Z.; Zhou, Z. An insight into land-cover changes and their impacts on ecosystem services before and after the implementation of a comprehensive experimental zone plan in Pingtan island, China. *Land Use Policy* **2019**, *82*, 631–642. [CrossRef]
23. Lei, J.; Chen, Z.; Wu, T.; Li, Y.; Yang, Q.; Chen, X. Spatial autocorrelation pattern analysis of land use and the value of ecosystem services in northeast Hainan island. *Acta Ecol. Sin.* **2019**, *39*, 2366–2377.
24. Qin, M.; Yang, C.; Xu, L.; Gao, Q.; Ding, Y.; Zhang, Y. Analysis of the spatiotemporal evolution of the coordinated development of ecosystem service value and economy in the Yellow River Basin. *J. Earth Environ.* **2022**, *13*, 491–505.
25. Zhang, Y.; Liu, Y.; Zhang, Y.; Liu, Y.; Zhang, G.; Chen, Y. On the spatial relationship between ecosystem services and urbanization: A case study in Wuhan, China. *Sci. Total Environ.* **2018**, *637–638*, 780–790. [CrossRef] [PubMed]
26. Wang, Z.; Tang, G. Coupling of Spatial and Temporal Evolution of Land Ecosystem Service Value and Tourism Urbanization:A case study of Chang-Zhu-Tan Urban Agglomeration. *J. Cent. South. Univ. For. Technol. Soc. Sci.* **2021**, *15*, 88–97.
27. Zhang, Z.; Nie, T.; Gao, Y.; Sun, S.; Gao, J. Study on Temporal and Spatial Characteristics of Coupling Coordination Correlation Between Ecosystem Services and Economic-Social Development in the Yangtze River Economic Belt. *Resour. Environ. Yangtze Basin* **2022**, *31*, 1086–1100.
28. Li, J.; Liang, M.; Zhong, Y. The Spatial-temporal Pattern of the Coordinated Development between Economy and Environment in the Yangtze River Economic Belt and Problem Areas Identification. *Resour. Environ. Yangtze Basin* **2020**, *29*, 2584–2596.
29. Wen, L. Spatiotemporal coupling between forest ecological service function and tourism economic development of western Hunan. *J. Cent. South Univ. For. Technol.* **2021**, *41*, 90–101.
30. Zhao, X.; Du, Y.; Li, H.; Wang, W. Spatio-temporal changes of the coupling relationship between urbanization and ecosystem services in the Middle Yellow River. *J. Nat. Resour.* **2021**, *36*, 131–147. [CrossRef]
31. Maimaiti, B.; Chen, S.; Kasimu, A.; Mamat, A.; Aierken, N.; Chen, Q. Coupling and Coordination Relationships between Urban. Expansion and Ecosystem Service Value in Kashgar City. *Remote Sens.* **2022**, *14*, 2557. [CrossRef]
32. Sun, W.; Miao, Z.; Sun, W. Change in the Value of Ecosystem Services of Beijing-Tianjin-Hebei Area and Its Relationship with Economic Growth. *Ecol. Econ.* **2015**, *31*, 59–62.
33. Zhu, X.; Xue, L. Changes of ecological service value and its coupling with economic development in Guanzhong region. *Chin. J. Ecol.* **2022**, *41*, 769–776.
34. Chen, X.; Zhou, H.; Wang, X. The vulnerability and coordination of the coupling of environment, economy and society in county: A case of Qiqihuar city of Heilongjiang province. *Hum. Geographys.* **2018**, *33*, 94–101.
35. Feng, Q.; Zhou, Z.; Chen, Q.; Zhu, C. Spatial-temporal evolution research of ecosystem service value in ecologically vulnerable karst regions under the perspective of poverty alleiation relocation. *Acta Ecol. Sin.* **2022**, *42*, 2708–2717.
36. Wei, W.; Shi, P.; Wei, X.; Zhou, J.; Xie, B. Evaluation of the coordinated development of economy and eco-environmental systems and spatial evolution in China. *Acta Ecol. Sin.* **2018**, *38*, 2636–2648.
37. Ma, L.; Jin, F.; Song, Z.; Liu, Y. Spatial coupling analysis of regional economic development and environmental polltion in China. *Geogr. Sci.* **2013**, *23*, 525–537. [CrossRef]
38. Shi, Y.; Yang, X.; Wang, T. An evaluation of sustainable livelihood security based on robustness analysis of the Qinba mountain region of Shannxi province, China. *Geogr. Res.* **2016**, *35*, 2309–2321.
39. Xie, G.; Xiao, Y. Study on ecosystem services value of food production in China. *Chin. J. Eco-Agric.* **2005**, *3*, 10–13.
40. Wang, H.; Qin, F.; Zhu, J.; Zhang, C. The effects of land use structure and landscape pattern change on ecosystem service values. *Acta Ecol. Sin.* **2017**, *37*, 1286–1296.
41. Li, Y.; Zhan, J.; Liu, Y.; Zhang, F.; Zhang, M. Response of ecosystem services to land use and cover change: A case study in Chengdu City. *Resour. Conserv. Recycl.* **2018**, *132*, 291–300. [CrossRef]
42. Zhang, Z.; Xia, F.; Yang, D.; Huo, J.; Chen, H. Spatiotemporal characteristics in ecosystem service value and its interaction with human activities in Xinjiang, China. *Ecol. Indic.* **2020**, *110*, 105826. [CrossRef]
43. Zhang, R.; Li, P.; Xu, L. Evaluation and analysis of ecological security based on the improved three-dimensional ecological footprint in Shaanxi Province, China. *Ecol. Indic.* **2022**, *144*, 109483. [CrossRef]
44. Liu, J.; Ning, J.; Kuang, W.; Xu, X.; Zhang, S.; Yan, C.; Li, R.; Wu, S.; Hu, Y.; Du, G.; et al. Spatio-temporal patterns and characteristics of land-use change in China during 2010–2015. *Acta Ecol. Sin.* **2018**, *73*, 789–802.

45. Liu, J.; Kuang, W.; Zhang, Z.; Xu, X.; Qin, Y.; Ning, J.; Zhou, W.; Zhang, S.; Li, R.; Yan, C.; et al. Spatiotemporal characteristics, patterns, and causes of land-use changes in China since the late 1980s. *Geogr. Sci.* **2014**, *24*, 195–210. [CrossRef]
46. Zhou, J.; Gao, F. Analysis on the Coupling and Coordination Relationship between Grassland Eco-Environment and Livestock Husbandry Economy: Taking Inner Mongolia as an Example. *Ecol. Econ.* **2019**, *35*, 170–176.
47. Chen, W.; Zeng, J.; Zhong, M.; Pan, S. Coupling Analysis of Ecosystem Services Value and Economic Development in the Yangtze River Economic Belt: A Case Study in Hunan Province, China. *Remote Sens.* **2021**, *13*, 8. [CrossRef]
48. Luo, Q.; Zhou, J.; Zhang, Y.; Yu, B.; Zhu, Z. What is the spatiotemporal relationship between urbanization and ecosystem services? A case from 110 cities in the Yangtze River Economic Belt, China. *J. Environ. Manag.* **2022**, *321*, 115709. [CrossRef]
49. Wang, H.; Liu, K.; Li, T.; Yuan, J. Spatial- temporal Variation of Ecosystem Service Value in the South Shaanxi Province Based on 3S Technology. *China Popul. Resour. Environ.* **2015**, *25*, 523–525.
50. Xu, L.; Xu, X.; Luo, T.; Zhu, G.; Ma, Z. Services based on land use: A case study of Bohai Rim. *Geogr. Res.* **2012**, *31*, 1775–1784.
51. Wang, H.; Zhou, S.; Li, X.; Liu, H.; Chi, D.; Xu, K. The influence of climate change and human activities on ecosystem service value. *Ecol. Eng.* **2016**, *87*, 224–239. [CrossRef]
52. Lei, J.; Wang, S.; Wang, J.; Wu, S.; You, X.; Wu, J.; Cui, P.; Ding, H. Effects of Land-Use Change on Ecosystem Services Value of Xunwu County. *Acta Ecol. Sin.* **2019**, *39*, 3089–3099.
53. Hasan, S.; Shi, W.; Zhu, X. Impact of land use land cover changes on ecosystem service value–A case study of Guangdong, Hong Kong, and Macao in South China. *PLoS ONE* **2020**, *15*, e0231259. [CrossRef] [PubMed]
54. Grossman, G.M.; Krueger, A.B. Economic growth and the environment. *Q. J. Econ.* **1995**, *110*, 353–377. [CrossRef]
55. Messerli, B.; Grosjean, M.; Hofer, T.; Núez, L.; Pfister, C. From nature-dominated to human-dominated environmental changes. *Quat. Sci. Rev.* **2000**, *19*, 459–479. [CrossRef]

diversity

MDPI

Article

Effects of Different Fertilizers on Soil Microbial Diversity during Long-Term Fertilization of a Corn Field in Shanghai, China

Chenyan Sha †, Jian Wu †, Jianqiang Wu, Chunmei Ye, Cheng Shen, Jinghua Su and Min Wang *

Shanghai Academy of Environmental Sciences, Shanghai 200235, China
* Correspondence: saeswangm78@163.com
† These authors contributed equally to this work.

Abstract: The long-term applications of different fertilizers (chicken manure, swine manure, and organic fertilizer) on the microorganisms of a corn field were investigated. The microbial communities during four periods (seedling, three-leaf, filling and mature periods) were comprehensively studied with molecular biology technology. Results showed that most nutrient contents (organic matter, nitrogen, phosphorus, and potassium) and levels of several heavy metals (As, Pb, and Cr) in the chicken and swine manures were higher than those in the organic fertilizer. The alpha diversity varied during the long-term fertilization, and the chicken manure was the best fertilizer to maintain the abundance of microorganisms. The microbial community of soil changes over time, regardless of the addition of different fertilizers. The correlations between environmental factors and microbial communities revealed that nutrient substances (available nitrogen, available potassium, and NO_3-N) were the most significant characteristics with the chicken and swine manures, while organic matter and nitrogen exhibited similar effects on the microbial structure with the organic fertilizer. The Pearson correlations of environmental factors on genus were significantly different in the organic fertilizer tests compared with the others, and *Pseudomonas*, *Methyloligellaceae*, *Flavobacterium*, and *Bacillus* showed significant correlations with the organic matter. This study will provide a theoretical basis for improving land productivity and sustainable development in corn fields.

Keywords: different fertilizers; long-term fertilization; microbial community; environmental factors; correlation

Citation: Sha, C.; Wu, J.; Wu, J.; Ye, C.; Shen, C.; Su, J.; Wang, M. Effects of Different Fertilizers on Soil Microbial Diversity during Long-Term Fertilization of a Corn Field in Shanghai, China. *Diversity* **2023**, *15*, 78. https://doi.org/10.3390/d15010078

Academic Editors: Maria Teresa Ceccherini Guicciardini and Ipek Kurtboke

Received: 26 October 2022
Revised: 15 December 2022
Accepted: 20 December 2022
Published: 6 January 2023

1. Introduction

The application of manure as fertilizer to agricultural land is a common practice around the world [1]. Livestock and poultry breeding is the pillar industry of China's agriculture. In order to ensure the effective supply of livestock and poultry products, this industry has been developing intensively and on a large scale in recent years. The treatment and disposal of a large amount of livestock and poultry manure produced by large-scale breeding is an effective way to reduce livestock and poultry waste and to recycle resources. Microorganisms, nutrients, heavy metals, and antibiotics change the soil environment after livestock application, while the biomass, community diversity and composition, and functional flora of soil microorganisms are affected as well [2].

Long-term or short-term livestock and poultry manure application or combined application with straws and chemical fertilizers could affect the physical and chemical characteristics of the soil, thereby significantly affecting the community structure, abundance, and diversity of farmland soil microorganisms. Qin et al. [3] found that long-term fertilization has a significant impact on the bacterial community structure of black soil. Giacometti et al. [4] also found that under a condition of long-term fertilization, the bacterial abundance of farmland soil increased significantly with the increase in swine manure application, while its community diversity significantly decreased. Xun et al. [5] showed that the short-term fertilization of swine manure improved soil nutrient content and had a significant impact on soil microbial diversity. Li et al. [6] revealed that the application of chicken and swine manure can effectively increase the diversity and abundance of soil bacterial communities in farmland. The changes in amino-acid carbon sources and carbohydrates in the soil were the main reasons that affected the soil microbial diversity changes. Hence, improving soil microbial diversity helps enhance the stability of farmland ecosystem functions and their resistance and resilience against environmental disturbances.

The changes in the microbial community after livestock manure application have been widely studied. It was reported that an increase in soil nutrient content and the introduction of exogenous bacteria affect the growth and reproduction of soil microorganisms. An increase in soil nutrient content enhances the growth and reproduction of eutrophic microorganisms in farmland soil, while it reduces the abundance of oligotrophic flora [7,8]. Adding organic or inorganic fertilizers can reduce the relative abundance of oligotrophic Acidobacteriain black and farmland soil in Northeast China [9]. However, in one wheat-corn rotation system, the application of swine manure and straw or the combined application of swine manure fertilizers inhibited the growth of soil dominant bacteria, such as Bacteroidetes, Acidobacteria, and Gemmatimonadetes, in a short period [10]. Changes in soil nitrogen content have a significant impact on soil high-abundance bacteria. Short-term nitrogen application may reduce the abundance of Actinobacteriaand Nitrospirae in farmland soil [11], while long-term nitrogen application is beneficial for the deformation of the growth of Proteobacteria and Actinobacteria [12]. Actinomycetes are more sensitive to the addition of exogenous nutrients. Meanwhile, untreated livestock and poultry manure contain a large number of pathogenic microorganisms, such as Salmonella, *E. coli*, Mycobacterium tuberculosis, etc. Most of them can survive in the soil for a long time after entering the farmland, and some pathogenic bacteria such as *E. coli* can continue to survive and reproduce after entering the soil. It is easy for them to enter the human body through the food chain and cause food-borne diseases, which seriously threatens the health of the soil environment [13]. However, comparisons between the microbial communities with livestock manure and those with organic fertilizer in the long-term fertilization of corn were rarely considered and need to be further studied.

In the applications of different fertilizers, the nutrients and pollutants have greater impacts on the functional microbes and functional flora of the soil. Accordingly, the correlations between environmental factors and microorganisms should be studied. When the nutrients in fertilizers enter the soil, they affect the soil nutrient cycle and the decomposition of organic matter [6]. Changes in soil nutrients after the application of manure also have a great impact on the functional plants involved in the turnover of soil organic matter or the utilization of carbon sources. Li et al. [14] studied the effects of adding livestock manure on soil functional microorganisms and found that the application of chicken and swine manure can effectively increase the carbon source utilization capacity of the soil microbial community by changing the soil nutrients in the farmland. Guo et al. [15] evaluated the effect of swine manure on the carbon conversion of soil during long-term application, finding that it increased the diversity of soil microorganisms and improved the soil microbial organic matter turnover capacity and carbon source utilization rate. In addition, after pollutants such as heavy metals and antibiotics in livestock manure enter the soil, they can also reduce the abundance of ammonia-oxidizing bacteria, inhibit the nitrification and denitrification of soil microorganisms, and reduce the utilization of carbon sources by microorganisms, thereby affecting the biogeochemical cycles of nitrogen and carbon in the soil [16]. Therefore, the comparison of nutrients, heavy metals, and their correlations with microbial communities between different fertilizers was necessarily investigated.

In this study, different fertilizers such as chicken manure, swine manure, and organic fertilizer were applied in the long-term fertilization of corn. The physical and chemical characteristics and heavy metals of different fertilizers were analyzed. Moreover, the diversity and richness of the microbial communities as well as the microbial compositions and structures were comprehensively compared. In particular, the correlations between environmental factors and microbial communities with the applications of different fertilizers were further investigated. This study will provide a theoretical basis for improving land productivity and sustainable development in corn fields.

2. Materials and Methods

2.1. Experimental Location

This research experiment was carried out in Hongxing Village, Zhongxing Town, Chongming Island (121°09′–121°54′, 31°27′–31°51′ N). This area is located at the mouth of the Yangtze River and belongs to the northern subtropical region. The climate is mild and humid. The annual average sunshine number is 2094.2 h, the annual average temperature is 15.3 °C, the annual average rainfall is 1025 mm, and the relative air humidity is above 80%. The natural geographical condition of the region is suitable for the development of the planting industry. The experimental base was established in 2010 and grows representative crops in the Chongming District, mainly yellow corn and cauliflower. The plot selected for this experiment covers an area of 300 m^2, mainly corn planting plots.

2.2. Experimental Set-Up

Chicken manure (CM), swine manure (SM), and organic fertilizer (OF) were selected as the experimental manures. The chicken manure was collected from a chicken farm, while the swine manure and organic fertilizer were collected from a swine farm in Chongming. The organic fertilizer was composted under aerobic conditions from swine manure, mushroom, and straw wastes.

As a common crop in Chongming, yellow corn was selected as the planting crop in this experiment. Direct sowing was conducted in all the experiments. The sowing time was early April in 2020. The planting density was 52,500 plants per mu. Other cultivation and management measures were the same as those in the field.

Ten treatments including nine fertilization treatments and one control treatment (CK) were designed. Each treatment had a repetition of 3 times. The concentration gradient of the manure application was set in accordance with the standard procedures in "Technical Specifications for Returning Livestock and Poultry Manure to Field (GB/T 25246-2010)", which were 2 $kg{\cdot}m^{-2}$, 4 $kg{\cdot}m^{-2}$, and 6 $kg{\cdot}m^{-2}$. Different concentrations of the chicken manure, swine manure, and organic fertilizer were applied as base fertilizers to the soil at one time before sowing the corn and then plowed after spreading fertilization.

Thirty experimental plots (3 m × 3 m) were set up, and a 0.5 mm thickness impermeable membrane was used to block the plots. The buried depth of the impermeable membrane was 30 cm to prevent mutual interference between the plots. At the same time, a channel for on-site monitoring and sample collection was isolated between each plot, with a width of about 20 cm. The layout of the experimental plots is shown in Figure 1.

2.3. Soil Sample Collection

The soil samples were collected in late April (seedling period), mid-May (three-leaf period), early July (filling period), and late August (mature period) in 2020.According to the "S" sampling method, five points were randomly selected from each sample plot and rhizosphere soil samples were collected from each of the 5 points by using a sterile stainless-steel soil drill (0–20 cm depth). Finally, the samples of five points were mixed evenly. A PVC soil respiration ring base (soil collar) with a diameter of 20 cm and a height of 12 cm was pre-embedded in each parallel plot on the site. After the corn was sowed, it was embedded in the soil with a depth of 10 cm and the uppercut was 2 cm above the ground. In order to avoid experimental errors, the roots of the corn plants should be avoided when the base is embedded and placed in the gaps between the plants. In order to eliminate the influence of green plant photosynthesis and surface litter on the measurement, the day before each measurement, the plants and litter in the ring should be removed, and the surface soil should not be disturbed. Soil samples of 500 g from each plot were collected. After removing debris, earthworms, and plant residues in each soil sample, each soil sample was divided into two portions: the first portion was stored at 4 °C for the analysis of physical and chemical characteristics, whereas the other was stored at −80 °C for DNA extraction and microbial analysis.

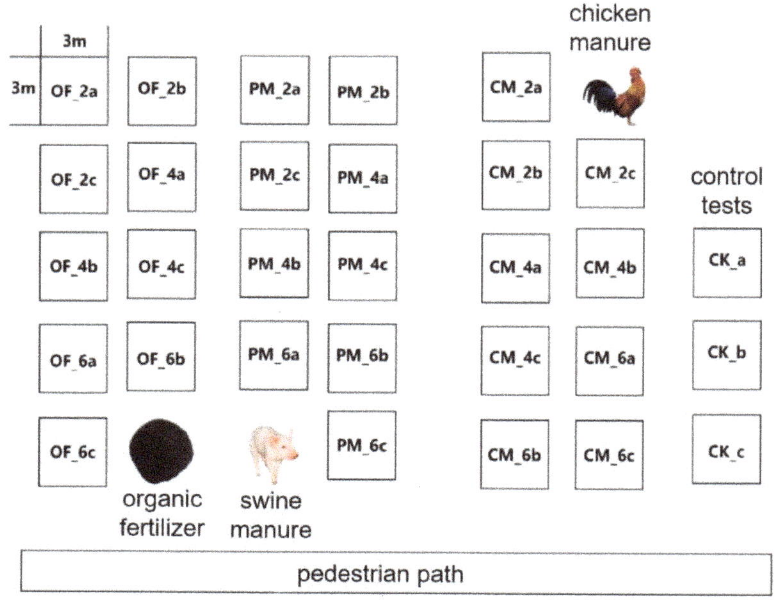

Figure 1. The experimental site layout (CK: control; CM: chicken manure; OF: organic fertilizer; SM: swine manure;PM represents swine manure).

2.4. *Analysis Methods*

2.4.1. Physical and Chemical Characteristics

The samples brought back to the laboratory were placed in the room for natural air drying, and, after grinding, they were passed through 10-mesh and 100-mesh sieves and placed in a refrigerator at 4 °C for further analysis. Organic matter (OM) and organic carbon (OC) were measured using a total organic carbon analyzer (TOC-VCPH, Shimadzu, Japan). Total nitrogen (TN), total phosphorus (TP), total potassium (TK), available nitrogen (AN), available phosphorus (AP), and available potassium (AK) were measured according to the Standard Methods [17]. Metal ion concentrations such as lead (Pb), cadmium (Cd), arsenic (As), mercury (Hg), and Chromium (Cr) were determined using an inductively coupled plasma emission spectrometer (ICP-AES, Agilent, Santa Clara, CA, USA).

2.4.2. Microbial Community Characterization

Microbial DNA was extracted from sludge using the E.Z.N.A.® Soil DNA kit (Omega Bio-Tek, Norcross, GA, USA) according to the manufacturer's protocols and then pooled together. DNA extracts were checked on 1% agarose gel, and DNA concentration and purity were determined with a Nano Drop 2000 UV-vis spectrophotometer (Thermo Scientific, Wilmington, NC, USA). For the bacterial community, the bacterial 16S rRNA genes were amplified using the universal bacterial primers 27F (5′-AGRGTTYGATYMTGGCTCAG-3′) and 1492R (5′-RGYTACCTTGTTACGACTT-3′). The primers were tailed with PacBio barcode sequences to distinguish each sample. The amplification reactions (20 μL volume) consisted of 5 × 4 μL of FastPfu buffer, 2 μL of 2.5 mM dNTPs, 0.8 μL of forward primer (5 μM), 0.8 μL of reverse primer (5 μM), 0.4 μL of FastPfu DNA Polymerase, 10 ng of template DNA, and DNase-free water. The PCR amplification was performed as follows: initial denaturation at 95 °C for 3 min, followed by 27 cycles of denaturing at 95 °C for

30 s, annealing at 60 °C for 30 s and extension at 72 °C for 45 s, single extension at 72 °C for 10 min, and end at 4 °C (ABI Gene Amp® 9700 PCR Thermocyclerm, Santa Clara, CA, USA). The PCR reactions were performed in triplicate. After electrophoresis, the PCR products were purified using the AM Pure® PB beads (Pacifc Biosciences, Menlo Park, CA, USA) and quantified with a Quantus™ Fluorometer (Promega, Fitchburg, WI, USA). The optimized CCS reads were clustered into operational taxonomic units (OTUs) using UPARSE 7.1 with a 97% sequence similarity level. The most abundant sequence for each OTU was selected as a representative sequence. To minimize the effects of sequencing depth on alpha and beta diversity measures, the number of 16S rRNA gene sequences from each sample was rarefied to 6000, which still yielded an average Good's coverage of 99.09%. The alpha diversity indices were calculated in MOTHUR (http://www.mothur.org (accessed on 9 September 2020)). The Pearson correlation coefficient was calculated to evaluate the correlations between environmental factors and individual microorganisms according to Ping et al. [18].

2.5. Statistical Analysis

Excel 2019, SPSS 24, and Origin 2017 were used for the data analysis and graphing. The graphs in Figure 2 were built with Origin. The differences between the different treatment groups were compared using the one-way analysis of variance method, while the multiple comparison was performed using the LSD (Least Significance Difference) method ($p < 0.05$).

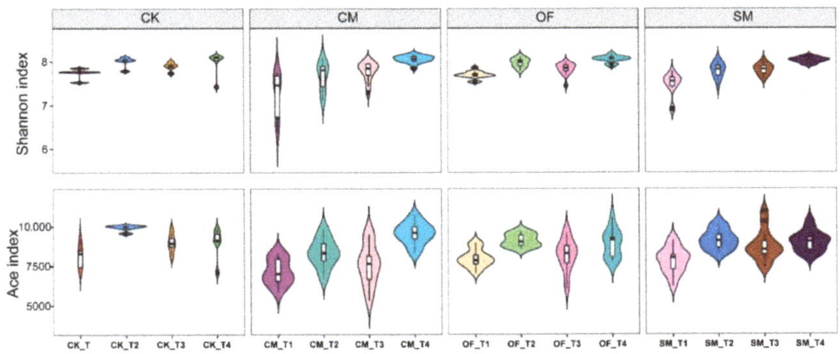

Figure 2. The Shannon and Ace indices of microbial community with different fertilizers (CK: control; CM: chicken manure; OF: organic fertilizer; SM: swine manure).

3. Results and Discussion

3.1. Analysis of Physical and Chemical Characteristics

The physical and chemical characteristics of the different fertilizers are shown in Table 1. The pH value of the organic fertilizer was the highest (7.42), and was significantly higher than that of CM and SM ($p < 0.05$). The contents of OM, TN, AN, TK, and AK in CM were the highest among the three fertilizers. The concentrations of TK and AK were significantly higher than those in SM ($p < 0.05$), measuring 81.95% and 118.53% higher, respectively. The highest contents of TP and AP appeared in the swine manure, and its TP concentration was higher than that of CM ($p < 0.05$). OF had the lowest nutrient content, and its OM, TN, TP, TK, AN, AP, and AK contents were significantly lower than those of CM ($p < 0.05$), which were reduced by 34.71%, 37.05%, 45.04%, 1.86%, 80.99%, and 41.45%, respectively. Similarly, the OM, TN, TP, AN, and AP contents of OF were significantly lower than those of SM ($p < 0.05$), which were reduced by 43.38%, 32.91%, 57.50%, and 82.28%, respectively. It was reported that fresh swine manure usually has more soluble phosphorus [19], which is in accordance with this study in which the concentrations of TP and AP in SM were higher than those in CM and OF. Moreover, the chicken manure had the highest nutrient content amongst the livestock manure, which was similar to the results of Ksheem et al. [20]. The nutrient content of OF is relatively low, which may relate to the high temperature during the composting process [21,22].

Table 1. The characteristics of chicken manure, swine manure, and organic fertilizer.

Characteristics	Chicken Manure	Swine Manure	Organic Fertilizer
pH	6.31 ± 1.23 [b]	6.60 ± 0.78 [b]	7.42 ± 0.93 [a]
Moisture (%)	64.2 ± 9.28 [a]	56.9 ± 6.01 [b]	40.8 ± 4.79 [c]
OM ($g \cdot kg^{-1}$)	844 ± 39.92 [a]	790 ± 18.80 [a]	551 ± 25.61 [b]
TN ($g \cdot kg^{-1}$)	34.3 ± 8.23 [a]	32.2 ± 3.08 [a]	21.6 ± 2.46 [b]
AN ($mg \cdot kg^{-1}$)	3.17 ± 0.47 [a]	2.73 ± 0.22 [a]	1.16 ± 0.23 [b]
TP ($g \cdot kg^{-1}$)	18.2 ± 0.23 [b]	29.10 ± 4.22 [a]	12.60 ± 1.27 [c]
AP($mg \cdot kg^{-1}$)	3.02 ± 0.18 [a]	3.24 ± 0.35 [a]	0.574 ± 0.09 [b]
TK ($g \cdot kg^{-1}$)	3.73 ± 0.55 [a]	2.05 ± 0.39 [b]	2.05 ± 0.11 [b]
AK ($mg \cdot kg^{-1}$)	21.7 ± 5.43 [a]	9.93 ± 1.91 [b]	12.7 ± 2.09 [b]

Note: Different letters (a, b, c) indicate that there were significant differences between different fertilizers with each characteristic ($p < 0.05$).

As can be seen in Table 2, the heavy metal contents of the three fertilizers did not exceed the heavy metal concentration limit (NY525-2012) [23]. The concentrations of Cd and Hg were not significantly different between the three fertilizers. The highest concentrations of As and Pb were observed in CM, measuring at 0.64 $mg \cdot kg^{-1}$ and 19.2 $mg \cdot kg^{-1}$, respectively. The As concentration in CM was 48.83% higher than that in SM ($p < 0.05$), while the Pb and As concentrations were significantly higher than those in OF ($p < 0.05$), measuring 28.00% and 357.14% higher, respectively. The As content in SM was also significantly higher than that in OF, measuring 207.12% higher. The Cr concentration (25 $mg \cdot kg^{-1}$) was the highest in SM, followed by CM. The Cr concentrations in CM and SM were significantly higher than that in OF ($p < 0.05$), measuring 48.38% and 61.29% higher, respectively. The analysis results showed that the heavy metals As, Pb, and Cr in the chicken manure or SM were more abundant than in OF.

Table 2. Heavy metal concentrations in chicken manure, swine manure, and organic fertilizer.

Heavy Metal	Chicken Manure	Swine Manure	Organic Fertilizer
Cd ($mg \cdot kg^{-1}$)	0.24 ± 0.02 [a]	0.27 ± 0.04 [a]	0.25 ± 0.00 [a]
Hg ($mg \cdot kg^{-1}$)	0.156 ± 0.04 [a]	0.173 ± 0.01 [a]	0.157 ± 0.01 [a]
As ($mg \cdot kg^{-1}$)	0.64 ± 0.09 [a]	0.43 ± 0.00 [b]	0.14 ± 0.00 [c]
Pb ($mg \cdot kg^{-1}$)	19.2 ± 0.04 [a]	13.3 ± 1.03 [b]	15.0 ± 1.21 [ab]
Cr ($mg \cdot kg^{-1}$)	23 ± 0.97 [a]	25 ± 2.73 [a]	15.5 ± 3.88 [b]

Note: Different letters (a, b, c) indicate that there were significant differences between different fertilizers with each heavy metal.

3.2. Diversity and Richness of Microbial Communities in Long-Term Fertilization

In the microbial community analysis, the rarefaction curves (at a 97% sequence similarity) of all the samples plateaued, demonstrating that the number of pyrosequencing reads was enough to explain the large fraction of OTUs in the soil samples.

Alpha diversity indices can be used to quantitatively estimate different microbes in a dataset and determine the total number of microbes. The Shannon and Ace indices were selected to reflect the diversity and richness of soil microorganisms with different fertilization treatments (Figure 2). The Shannon and Ace indices in the OF and CK soils demonstrated no significant difference in different periods, indicating that the application of organic fertilizer has little effect on the diversity and abundance of soil microbial communities. For the Ace index, there was no significant difference between the SM and CK soils during the four periods, while the Shannon index of the SM soil was lower than that of the CK soil during the four periods and reached a significant level in the first three periods ($p < 0.05$), indicating that the swine manure reduced the diversity of the soil microbial community. However, it had a small effect on the microbial abundance. The application of swine manure reduces the diversity of the soil microbial community, and the reduction in the microbial community diversity leads to a reduction in the efficiencies of the ecosystem functions [24]. The Shannon and Ace indices of the CM soil both increased with time, although they were significantly lower than CK in the first two periods ($p < 0.05$); however, by the mature stage, there was no significant difference between CM and CK. This reveals that the chicken manure greatly increased the diversity and richness of microorganisms during the four periods. In addition, the richness of CM in the last stage was significantly higher than that of OF and SM ($p < 0.05$), indicating that the chicken manure was the best fertilizer to maintain the abundance of microorganisms. Campbell et al. [25] also found that from a long-term perspective, the application of chicken manure is conducive to improving the diversity of soil microbial communities. An increase in microbial community diversity is conducive to enhancing the stability of a soil ecosystem, making it more resistant and able to recover in the face of environmental disturbances [26].

3.3. Microbial Composition and Structure in Long-Term Fertilization

The bacterial communities determined at the phylum and genus levels were further analyzed by comparing populations among the samples. A total of 42 phyla were detected, and the phyla with a relative abundance of less than 0.1% were merged into others (Figure 3a). It can be seen in Figure 3a that Proteobacteria was the most abundant phylum in all soils with different treatments, accounting for 27.95~39.84% of the total bacterial counts in each sample. This result was consistent with previous studies of soil microbial communities [27]. The main phyla present in these soil samples with different treatments were quite similar, including Actinobacteria, Acidobacteria, Acidobacteria, Chloroflexi, Bacteroidetes, Firmicutes, Gemmatimonadetes, Rokubacteria, Patescibacteria, Planctomycetes, Verrucomicrobia, Nitrospirae, and Latescibacteria, and their total abundance accounts for 97.86~98.68%.

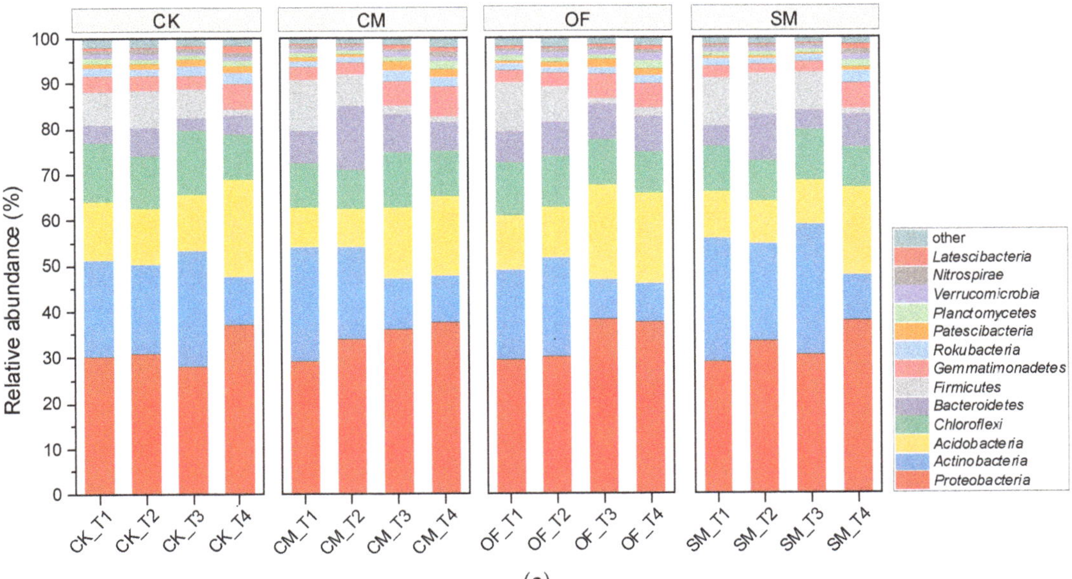

(a)

Figure 3. *Cont.*

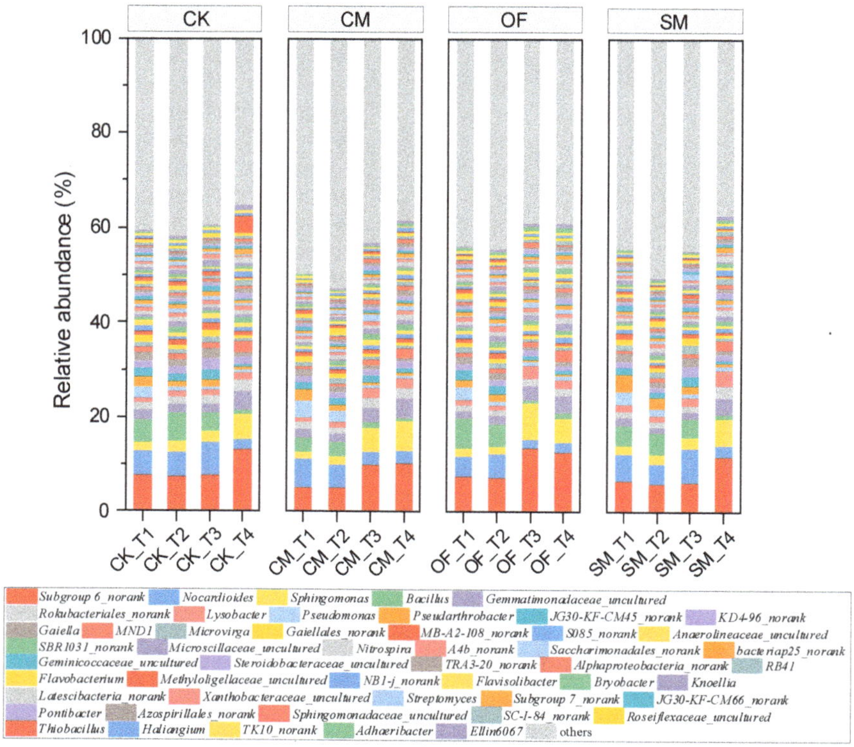

(**b**)

Figure 3. The relative abundances of microorganisms on phylum (**a**) and genus (**b**) levels with different fertilizers (CK: control; CM: chicken manure; OF: organic fertilizer; SM: swine manure; T1: seedling period; T2: three-leaf period; T3: filling period; T4: mature period).

There were certain differences in the second dominant phyla of the soil in each treatment group. At the seedling and three-leaf stage, the second dominant phylum of each treatment group was Actinobacteria, accounting for 19.49–26.91% of the total bacterial abundance. During the filling stage, the second dominant phylum was still Actinobacteriain the CK and SM tests, accounting for 25.26% and 25.84%, respectively. However, the abundance of Acidobacteria was greatly increased in the CM and OF tests, measuring at 14.21% and 19.62%, respectively. It was reported that Acidobacteriais one of the dominant bacteria in soil, and it is easy to accumulate in acid soil [28]. The decrease in soil pH after the application of chicken manure and organic fertilizer may be the main reason for the increased abundance in Acidobacteria. In the mature period, Acidobacteriabecame the second dominant bacteria phylum in the soils of all the treatments, accounting for 16.30–20.48%. From the changes in the dominant microbial phyla in different periods, it can be found that the microbial community of soil changes over time regardless of the addition of different manure, which is consistent with the results of previous studies [29]. The changes in microorganisms on the genus level were similar to those on the phylum level. It can be seen in Figure 3b that the relative abundances of *Subgroup 6_norank*, *Sphingomonas*, *gemmatinonadaceae_uncultured*, and *Lysobacter* were greatly increased in the mature period, while *Nocardioides* and *Bacillus* were significantly inhibited.

Furthermore, the one-way analysis of variance was selected to analyze the differences in the compositions of microorganisms between treatment groups at different periods based on the relative abundances of phylum-level microorganisms. The effects of fertilization on the microbes in the soil with high abundance levels are highly consistent. The application of CM, OF, and SM all increased the relative abundance of Firmicutes and Bacteroidetes. Compared with CK, the relative abundance of Firmicutes in the soil of CM, OF, and SM at the seedling stage was significantly higher ($p < 0.05$), measuring at 68.45%, 70.03%, and 49.77% higher, respectively. It was known that Firmicutes can be enriched at a rapid multiplication rate in a growth environment where soluble organic matter and nutrient concentrations are high [30]. Therefore, it was significantly enhanced with the application of different fertilizers. During the entire growth period, the relative abundance of Bacteroidetes in the fertilized soil was higher than that of CK. Bacteroidetes reached a significant level in the seedling, filling, and mature stages ($p < 0.05$) with the OF fertilization, and they reached a significant level in all periods ($p < 0.05$) in the CM tests. Similarly, the relative abundance was significantly higher in the CK tests after the three-leaf stage ($p < 0.05$). During the entire growth period, the relative abundances of Bacteroidetes with the CM, OF, and SM fertilizations were higher than that in CK by 122.50%, 72.45%, and 55.33%, respectively. Bacteroidetes are eutrophic bacteria, and the increase in soil nutrients after manure application is the main reason for the increase in their abundance [8].

The analysis of the relative abundance of bacteria can only show the affected phyla and genera after fertilization, while the influence of fertilization on the bacterial structure is difficult to judge. Therefore, a principal component analysis (PCA) was used to study the similarities and differences in the microbial community structures between different fertilization treatment tests on the OTU level [31]. It was apparent that the four types of treatments during four periods can be divided into two categories (Figure 4). During the corn filling period, the samples of different treatment groups were clearly distinguished, indicating that the microbial community structures of the different treatment groups were quite different at this time. Among them, the SM and CK sample points are closer, and the OF and CM sample points are closer, indicating that the microbial community structures of the SM and CK samples were quite similar during this period. However, the microorganisms with OF are more similar to those with CM. This difference may be related to the nutrient content of the soil. At the maturity stage, the sample points of the different treatment groups are distinguished to a certain extent, but they are not clearly separated, indicating the microbial community structures between the different groups were relatively similar at this time. That was because the microorganisms continued to change over time, and they became stable during the mature period, indicating that the microbial structures were relatively sensitive to the change in temperature.

3.4. Correlations between Environmental Factors and Microbial Communities

In order to determine the interaction between environmental factors and microbial communities, the RDA redundancy and Pearson correlation coefficient were used to evaluate the effects of external factors on the microorganisms.

The RDA redundancy analysis showed that heavy metals such as, Cd, and Cr were more influential than organic characteristics (OM, OC) in the control tests. However, in the treatments with CM and SM, nutrient substances such as AN, AK, and NO_3-N became the most significant characteristics (Figure 5). That was because the nutrient composition, especially nitrogen, was abundant in the livestock manures [32,33]. An increase in nitrogen provides abundant substrates for microorganisms and stimulates the growth and reproduction of soil bacteria, which has a greater impact on the community structure. In addition, As was also a greatly affected factor in the CM and SM tests since the concentrations of As in CM and SM were significantly higher than that in OF. In the OF tests, organic (OM and OC) and nutrient characteristics (AN and NO_3-N)exhibited similar effects on the microbial structure, and the effect of heavy metals was relatively less than in the other tests, which was in accordance with the characteristics of OF in the previous section.

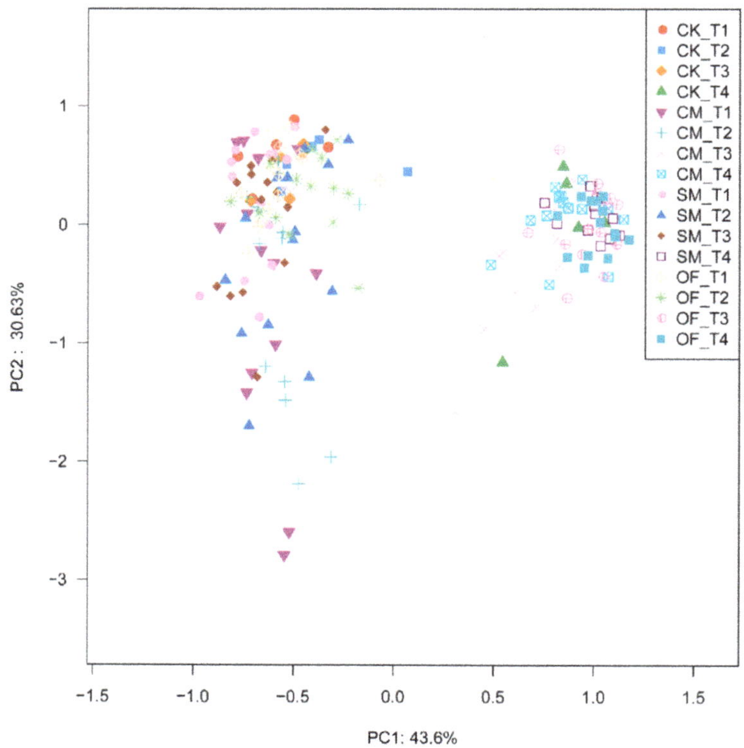

Figure 4. The PCA analysis of all treatment samples (CK: control; CM: chicken manure; OF: organic fertilizer; SM: swine manure; T1: seedling period; T2: three-leaf period; T3: filling period; T4: mature period).

A Pearson correlation analysis can further verify the correlations between soil environmental factors and dominant bacteria on the genus level. More significant correlation coefficients were observed with the fertilizer treatments than with the control tests (Figure 6), indicating that the addition of the fertilizers changed the environmental conditions of the soil and produced a great variety of microorganisms. It was observed that NO_3-N, AN, TP, and AK significantly correlated with most genera in the CM tests (Figure 6a). *Sphingomonadaceae, Saccharimonadales, Microscillaceae, Lysobacter, Chitinophagaceae,* and *Adhaeribacter* all exhibited positive correlations with TP, AN, and AK, while *Roseiflexaceae, Pseudarthrobacter, Nocardioides, Methyloligellaceae, Knoellia, Ilumatobacteracear, Geminicoccaceae, Gaiellales, Gaiella,* and *Bacillus* exhibited negative correlations with them. A similar result was observed in the SM tests, such that NO_3-N, AN, and AK were the most significant factors on genus (Figure 6d). The Pearson correlations of environmental factors on genus in the OF tests were significantly different compared with the other tests. It can be seen in Figure 6b that OM and OC exhibited the maximum number of significant co-efficiencies, revealing that organic compounds were the most significant factor in the OF tests. *Pseudomonas, Methyloligellaceae, Flavobacterium,* and *Bacillus* exhibited significant correlations with the organic matter; therefore, it can be deduced that they could rapidly grow and reproduce with high levels of organic matter in the soil. It was indicated that members of the genus *Pseudomonas* are a ubiquitous component of soil and rhizospheric ecosystems, where they play multifarious roles, such as in the recycling of organic matter [34]. *Methylotrophic* bacteria play a significant role in the biogeochemical cycle, and they are especially involved in phosphorous, nitrogen, and carbon cyclingin soil ecosystems [35,36]. Pijanowsk et al. [37]

revealed that *Pseudomonas* and *Bacillus* had the highest hydrocarbon biodegradation rates in short-term experiments.

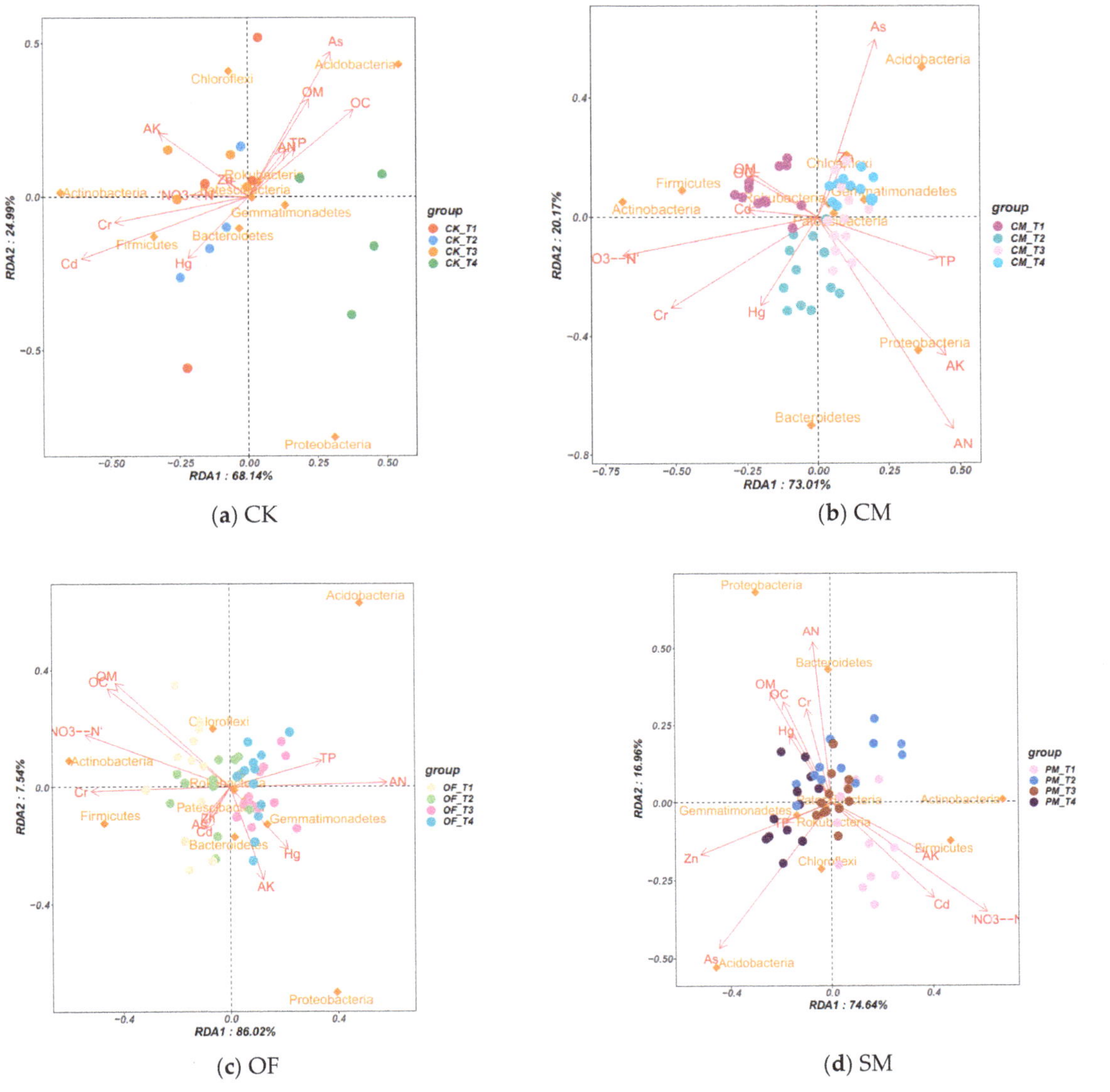

Figure 5. The relationship of environmental factors and microorganisms with different fertilizers (CK: control; CM: chicken manure; OF: organic fertilizer; PM: swine manure; T1: seedling period; T2: three-leaf period; T3: filling period; T4: mature period; OM: organic matter; OC: organic carbon; AN: available nitrogen; AK: available potassium; TP: total phosphorus).

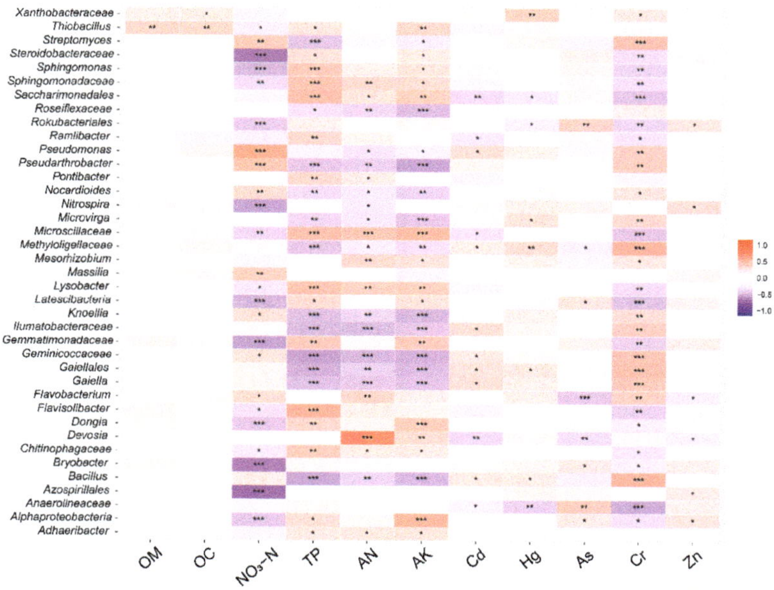

(**a**) CM

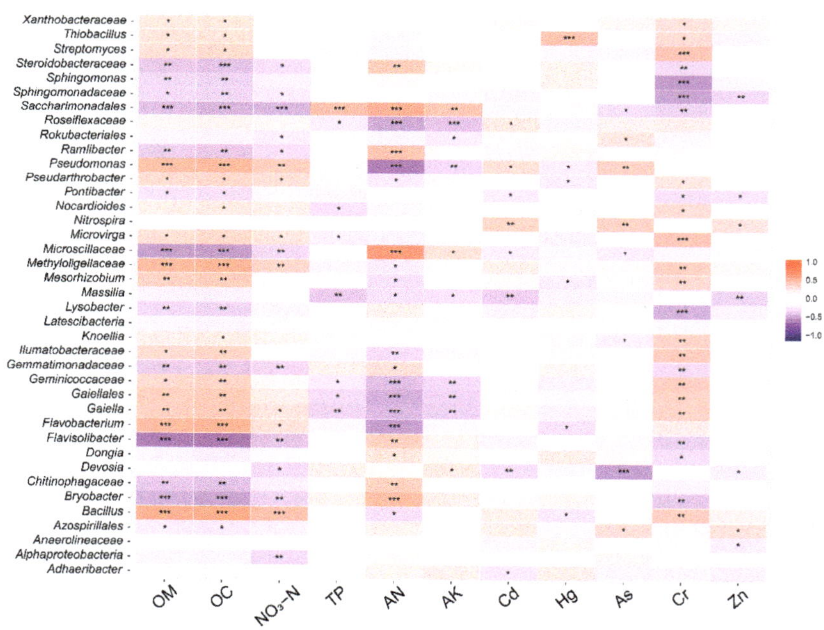

(**b**) OF

Figure 6. *Cont.*

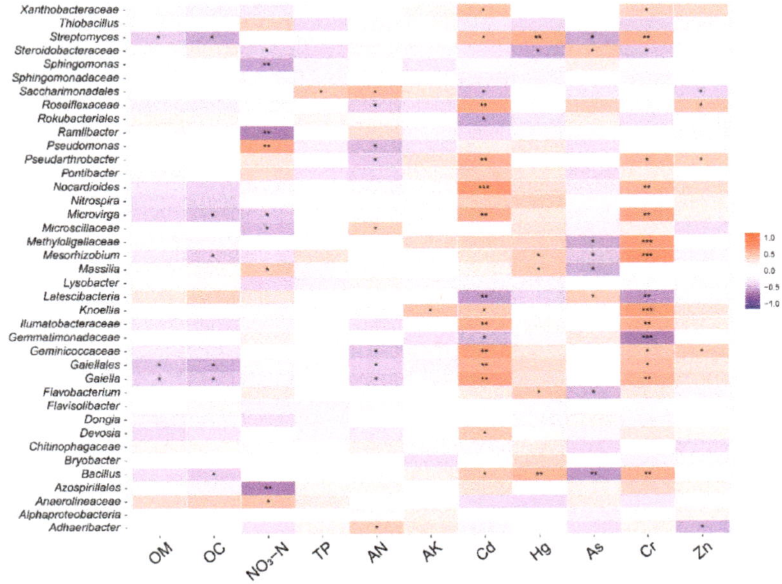

(**c**) CK

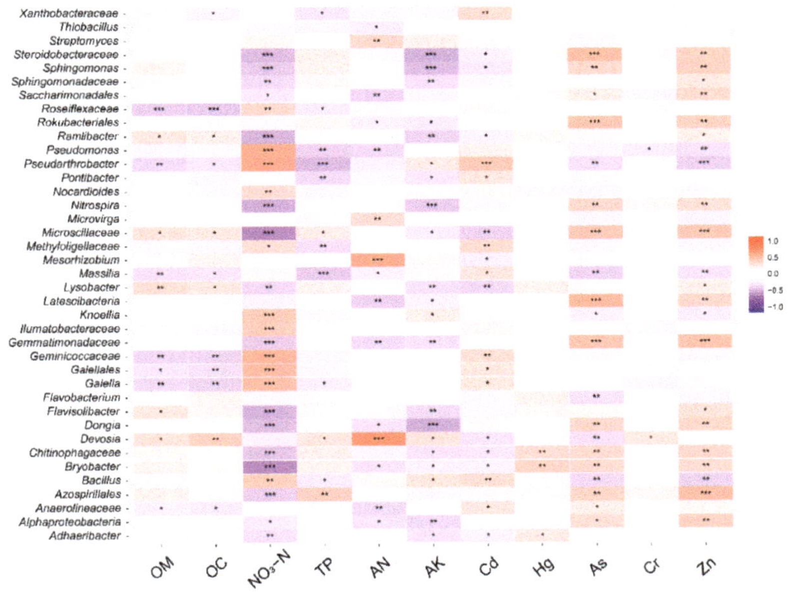

(**d**) SM

Figure 6. The Pearson correlation analysis of environmental factors and main genus of the CM (**a**), OF (**b**), control (**c**), and SM (**d**) tests (CM: chicken manure; OF: organic fertilizer; OM: organic matter; OC: organic carbon; TP: total phosphorus; AN: available nitrogen; AK: available potassium.* indicateds ($p < 0.1$), ** indicateds ($p < 0.05$),*** indicateds ($p < 0.01$)).

4. Conclusions

This study demonstrated that the microbial diversity varied during long-term fertilization, and the chicken manure was the best fertilizer to maintain the abundance of microorganisms. The microbial community of soil changes over time regardless of the addition of different fertilizers. Proteobacteria was the most abundant phylum in all the soils with different treatments. The second dominant phylum of each treatment group was Actinobacteria at the seedling and three-leaf stage, while the abundance of Acidobacteria was greatly increased with the chicken manure and organic fertilizer, becoming the second dominant phylum in the filling and mature periods. The applications of different fertilizers all increased the relative abundances of Firmicutes and Bacteroidetes. The correlations between environmental factors and microbial communities reflected that nutrient substances such as available nitrogen, available potassium, and nitrate became the most significant characteristics in the chicken and swine manure tests, while organic matter and ammonia exhibited similar effects on the microbial structure as with the organic fertilizer. The Pearson correlations of environmental factors on genus in the organic fertilizer tests were significantly different compared with the other tests. *Pseudomonas*, *Methyloligellaceae*, *Flavobacterium*, and *Bacillus* exhibited significant correlations with the organic matter. These results indicated that manure and organic fertilization directly affected soil bacterial diversity and community composition. However, the response mechanism of soil bacteria to organic fertilizer application is complex and needs further study to explain it satisfactorily.

Author Contributions: Conceptualization, C.S. (Chenyan Sha) and M.W.; methodology, C.S. (Chenyan Sha), J.W. (Jian Wu) and C.Y.; software, C.Y., J.W. (Jian Wu) and C.S. (Cheng Shen); validation, C.S. (Chenyan Sha), J.W. (Jian Wu) and J.S.; formal analysis, C.Y. and C.S. (Cheng Shen); investigation, J.W. (Jian Wu), C.Y. and C.S. (Cheng Shen); resources, M.W.; data curation, C.S. (Chenyan Sha) and J.W. (Jianqiang Wu); writing—original draft preparation, C.S. (Chenyan Sha) and J.W. (Jian Wu); writing—review and editing, C.S. (Chenyan Sha) and J.W (Jian Wu); visualization, C.S. (Chenyan Sha), J.W. (Jian Wu) and J.S.; supervision, C.S. (Chenyan Sha) and J.W. (Jian Wu); project administration, M.W.; funding acquisition, C.S. (Chenyan Sha). All authors have read and agreed to the published version of the manuscript.

Funding: This research was funded by the National Natural Science Foundation of China (grant numbers 31100404).

Institutional Review Board Statement: Not applicable.

Informed Consent Statement: Not applicable.

Data Availability Statement: Data can be found within the article.

Acknowledgments: This work was supported by the National Natural Science Foundation of China (grant numbers 31100404).

Conflicts of Interest: The authors declare no conflict of interest.

References

1. Kumar, U.; Kumar Nayak, A.; Shahid, M.; Gupta, V.V.S.R.; Panneerselvam, P.; Mohanty, S.; Kaviraj, M.; Kumar, A.; Chatterjee, D.; Lal, B.; et al. Continuous application of inorganic and organic fertilizers over 47 years in paddy soil alters the bacterial community structure and its influence on rice production. *Agric. Ecosyst. Environ.* **2018**, *262*, 65–75. [CrossRef]
2. Monard, C.; Vandenkoornhuyse, P.; Le Bot, B.; Binet, F. Relationship between bacterial diversity and function under biotic control: The soil pesticide degraders as a case study. *ISME J.* **2011**, *5*, 1048–1056. [CrossRef] [PubMed]
3. Qin, J.; Jiang, X.; Zhou, J.; Ma, M.; Guan, D.; Zhou, B.; Zhao, B.; Du, B.; Li, J. Characteristics and driving factors of soil bacterial and archaeal communities under long-term fertilization regimes in black soil. *J. Plant Nutr. Ferti.* **2015**, *21*, 1590–1598.
4. Giacometti, C.; Cavani, L.; Baldoni, G.; Ciavatta, C.; Marzadori, C.; Kandeler, E. Microplate-scale fluorometric soil enzyme assays as tools to assess soil quality in a long-term agricultural field experiment. *App. Soil. Ecol.* **2014**, *75*, 80–85. [CrossRef]
5. Xun, W.; Xiong, W.; Huang, T.; Ran, W.; Li, D.; Shen, Q.; Li, Q.; Zhang, R. Swine manure and quicklime have different impacts on chemical properties and composition of bacterial communities of an acidic soil. *App. Soil. Ecol.* **2016**, *100*, 38–44. [CrossRef]
6. Li, F.; Chen, L.; Zhang, J.; Yin, J.; Huang, S. Bacterial Community Structure after Long-term Organic and Inorganic Fertilization Reveals Important Associations between Soil Nutrients and Specific Taxa Involved in Nutrient Transformations. *Front. Microbiol.* **2017**, *8*, 187. [CrossRef]

7. Basilio, A.; González, I.; Vicente, M.F.; Gorrochategui, J.; Cabello, A.; González, A.; Genilloud, O. Patterns of antimicrobial activities from soil actinomycetes isolated under different conditions of pH and salinity. *J. Appl. Microbiol.* **2003**, *95*, 814–823. [CrossRef]
8. Francioli, D.; Schulz, E.; Lentendu, G.; Wubet, T.; Buscot, F.; Reitz, T. Mineral vs. Organic Amendments: Microbial Community Structure, Activity and Abundance of Agriculturally Relevant Microbes Are Driven by Long-Term Fertilization Strategies. *Front. Microbiol.* **2016**, *7*, 1446. [CrossRef]
9. Hu, X. Response of soil microbial community to long-term fertilization regimes in different black soil locations of northease China. *Univ. Chin. Acad. Sci.* **2018**, *28*, 751–763.
10. Wu, X. Effects of different fertilization systems on soil microbial properties in wheat-corn rotation system. *Agric. Univ. HEBEI* **2014**. Available online: https://kns.cnki.net/KCMS/detail/detail.aspx?dbname=CMFD201501&filename=1015515560.nh (accessed on 25 October 2022).
11. Wang, C.; Liu, D.; Bai, E. Decreasing soil microbial diversity is associated with decreasing microbial biomass under nitrogen addition. *Soil Biol. Biochem.* **2018**, *7*, 1473–1483. [CrossRef]
12. Dai, Z.; Su, W.; Chen, H.; Barberán, A.; Xu, J. Long-term nitrogen fertilization decreases bacterial diversity and favors the growth of Actinobacteria and Proteobacteria in agro-ecosystems across the globe. *Glob. Chang. Biol.* **2018**, *24*, 3452–3461. [CrossRef] [PubMed]
13. Monaghan, J.M.; Hutchison, M.L. Distribution and decline of human pathogenic bacteria in soil after application in irrigation water and the potential for soil-splash-mediated dispersal onto fresh produce. *J. Appl. Microbiol.* **2012**, *112*, 1007–1019. [CrossRef] [PubMed]
14. Li, J.; Luo, T.; Du, M.; Ni, J.; Gong, J. Effects of livestock and poultry manure amendment and wetting-drying cycle on functional diversity of soil microial community. *Guangdong Agri. Sci.* **2018**, *45*, 68–77.
15. Guo, Y.; Wang, Y.; Wu, P.; Peng, S. Influence of long-term manure application in paddy soil on the functional diversity of microbial community. *Chin. J. App. Environ. Biol.* **2019**, *25*, 0593–0602.
16. Girvan, M.S.; Campbell, C.D.; Killham, K.; Prosser, J.I.; Glover, L.A. Bacterial diversity promotes community stability and functional resilience after perturbation. *Environ. Microbiol.* **2005**, *7*, 301–313. [CrossRef] [PubMed]
17. APHA. *Standard Methods for the Examination of Water and Waste Water*; American Public Health Association, American Water Works Association, Water Environment Federation: Washington, DC, USA, 2012.
18. Ping, Q.; Lu, X.; Zheng, M.; Li, Y. Effect of CaO_2 addition on anaerobic digestion of waste activated sludge at different temperatures and the promotion of valuable carbon source production under ambient condition. *Bioresour. Technol.* **2018**, *265*, 247–256. [CrossRef]
19. Rong, X.; Zhang, Y.; Guo, C.; Xiang, X.; Han, Y.; Si, T.; Yang, L. Effects of pig manure type of organic fertilizr on soil fertility of spring maize planting land and yield of spring maize. *Hunan Agri. Sci.* **2011**, *9*, 45–48.
20. Ksheem, A.M.; Bennett, J.M.; Antille, D.L.; Raine, S.R. Towards a method for optimized extraction of soluble nutrients from fresh and composted chicken manures. *Waste Manage.* **2015**, *45*, 76–90. [CrossRef]
21. Lalander, C.; Nordberg, Å.; Vinnerås, B. A comparison in product-value potential in four treatment strategies for food waste and faeces—Assessing composting, fly larvae composting and anaerobic digestion. *GCB Bioenergy* **2018**, *10*, 84–91. [CrossRef]
22. Mitchell, S.M.; Ullman, J.L.; Bary, A.; Cogger, C.G.; Teel, A.L.; Watts, R.J. Antibiotic Degradation During Thermophilic Composting. *Water Air Soil Pollut.* **2015**, *226*, 13. [CrossRef]
23. *NY 525-2012*; Organic Fertilizer. Ministry of Agriculture of the People's Republic of China: Beijing, China, 2012.
24. Cardinale, B.J. Biodiversity loss and its impact on humanity (vol 486, pg 59, 2012). *Nature* **2012**, *489*, 326. [CrossRef]
25. Campbell, V.; Murphy, G.; Romanuk, T.N. Experimental design and the outcome and interpretation of diversity–stability relations. *Oikos* **2011**, *120*, 399–408. [CrossRef]
26. Van Ruijven, J.; Berendse, F. Diversity enhances community recovery, but not resistance, after drought. *J. Ecol.* **2010**, *98*, 81–86. [CrossRef]
27. Jangid, K.; Williams, M.A.; Franzluebbers, A.J.; Sanderlin, J.S.; Reeves, J.H.; Jenkins, M.B.; Endale, D.M.; Coleman, D.C.; Whitman, W.B. Relative impacts of land-use, management intensity and fertilization upon soil microbial community structure in agricultural systems. *Soil Biol. Biochem.* **2008**, *40*, 2843–2853. [CrossRef]
28. Jones, R.T.; Robeson, M.S.; Lauber, C.L.; Hamady, M.; Knight, R.; Fierer, N. A comprehensive survey of soil acidobacterial diversity using pyrosequencing and clone library analyses. *ISME J.* **2009**, *3*, 442–453. [CrossRef] [PubMed]
29. Lipson, D.A. Relationships between temperature responses and bacterial community structure along seasonal and altitudinal gradients. *FEMS Microbiol. Ecol.* **2010**, *59*, 418–427. [CrossRef]
30. Kabisch, A.; Otto, A.; König, S.; Becher, D.; Albrecht, D.; Schüler, M.; Teeling, H.; Amann, R.I.; Schweder, T. Functional characterization of polysaccharide utilization loci in the marine Bacteroidetes 'Gramella forsetii' KT0803. *ISME J.* **2014**, *8*, 1492–1502. [CrossRef]
31. Cuartero, J.; Özbolat, O.; Sánchez-Navarro, V.; Egea-Cortines, M.; Zornoza, R.; Canfora, L.; Orrù, L.; Pascual, J.A.; Vivo, J.-M.; Ros, M. Changes in Bacterial and Fungal Soil Communities in Long-Term Organic Cropping Systems. *Agriculture* **2021**, *11*, 445. [CrossRef]

32. Wongkiew, S.; Koottatep, T.; Polprasert, C.; Prombutara, P.; Jinsart, W.; Khanal, S.K. Bioponic system for nitrogen and phosphorus recovery from chicken manure: Evaluation of manure loading and microbial communities. *Waste Manage* **2021**, *125*, 67–76. [CrossRef]
33. Montalvo, S.; Huiliñir, C.; Castillo, A.; Pagés-Díaz, J.; Guerrero, L. Carbon, nitrogen and phosphorus recovery from liquid swine wastes: A review. *J. Chem. Technol. Biot.* **2020**, *95*, 2335–2347. [CrossRef]
34. Selvakumar, G.; Panneerselvam, P.; Bindu, G.H.; Ganeshamurthy, A.N. Pseudomonads: Plant growth promotion and beyond. In *Plant Microbes Symbiosis: Applied Facets*; Springer: New Delhi, India, 2015; pp. 193–208.
35. Kumar, M.; Tomar, R.S.; Lade, H.; Paul, D. Methylotrophic bacteria in sustainable agriculture. *World J. Microbiol. Biotechnol.* **2016**, *32*, 120. [CrossRef] [PubMed]
36. Tani, A.; Sahin, N.; Fujitani, Y.; Kato, A.; Sato, K.; Kimbara, K. Methylobacterium species promoting rice and barley growth and interaction specificity revealed with whole-cell matrix-assisted laser desorption/ionization-time-of-flight mass spectrometry (MALDI-TOF/MS) analysis. *PLoS ONE* **2015**, *10*, e0129509. [CrossRef] [PubMed]
37. Pijanowska, A.; Kaczorek, E.; Chrzanowski, Ł.; Olszanowski, A. Cell hydrophobicity of Pseudomonas spp. and Bacillus spp. bacteria and hydrocarbon biodegradation in the presence of Quillaya saponin. *World J. Microbiol. Biotechnol.* **2007**, *23*, 677–682. [CrossRef]

diversity

MDPI

Article

Assessment of Grassland Ecosystem Service Value in Response to Climate Change in China

Yuqing Xu *, Fengjin Xiao and Yaoming Liao

National Climate Center, China Meteorological Administration, Beijing 100081, China; xiaofj@cma.gov.cn (F.X.); lymzxr@cma.gov.cn (Y.L.)

* Correspondence: xuyq@cma.gov.cn

Abstract: The assessment of ecosystem services provides an intuitive source of information on the benefits humans derive from ecosystems. The equivalent factor method was applied to calculate the ecosystem service value (ESV) in combination with net primary productivity (NPP) calculated by the process-based Carnegie–Ames–Stanford approach (CASA) model. This study evaluated grassland ESV and its spatial evolution characteristics in China from 2001 to 2020 and revealed the impact of climate factors. For 2001–2020, the annual grassland ESV ranged from 1.17×10^{12} to 1.51×10^{12} yuan (renminbi, China yuan—the same below; \$$0.15 \times 10^{12}$–\$$0.20 \times 10^{12}$, US dollar), with an average of 1.37×10^{12} yuan (\$$0.18 \times 10^{12}$). The spatial pattern of ESV per unit area of grassland was notably characterized by an increase from northwest to southeast. However, the value of grassland ecosystem services was relatively large (exceeding 10×10^6 yuan; \$$1.30 \times 10^6$) in northern and western provinces and was the lowest (less than 0.2×10^6 yuan; \$$0.03 \times 10^6$) in eastern and southern provinces. In the last 20 years, grassland ESV has increased in most areas of China and has decreased only in some western and northern areas. Compared with the first 10 years, the average ESV of grassland in most areas increased in the last 10 years, usually by less than 20%. However, it decreased in the western and northern parts of China, mainly concentrated in the alpine meadow and alpine grassland of the Qinghai–Tibet Plateau and the grassland around the Yili region of Xinjiang. Precipitation was the main regulating factor of grassland ESV and had a positive impact in 79% grassland areas, especially in northern China. Evapotranspiration and sunshine hours exhibited a marginal impact on ESV, but temperature and relative humidity had no significant effect. Overall, this study contributes to exploring the spatiotemporal patterns of grassland ecosystem service value and the impact of climate factors in China, thereby providing reliable guidance for grassland ecosystem management.

Keywords: ecosystem service value (ESV); grassland; regulator; CASA model; equivalent factor

Citation: Xu, Y.; Xiao, F.; Liao, Y. Assessment of Grassland Ecosystem Service Value in Response to Climate Change in China. *Diversity* **2022**, *14*, 160. https://doi.org/10.3390/d14030160

Academic Editors: Luc Legal, Peng Hou, Weiguo Jiang, Wei Li and Li Zhang

Received: 7 January 2022
Accepted: 21 February 2022
Published: 24 February 2022

1. Introduction

Ecosystem services are the benefits human populations derive, directly or indirectly, from ecosystem functions [1], including provision, regulating, supporting, and cultural services [2], or supply, regulation and cultural services [3,4], or other categories from different classification systems. These services can support human survival and development and play an irreplaceable role in maintaining the dynamic balance of the earth's life support system and environment. However, in recent decades, due to the influence of human activities and natural factors, nearly 60% of global ecosystem services have degraded to varying degrees [2,5–8], which seriously threatens human security and health and has become one of the main problems affecting human sustainable development.

As the world's population increases, the demand for ecosystem supply services and other services is increasing [9]. At the same time, governments and managers always expect the maximization of ecosystem service value [10]. The ecological environment has deteriorated seriously. The interweaving of these factors means that the evaluation of ecosystem services ought to be paid increasing attention and become a research priority [11].

In particular, since the launch of the Millennium Ecosystem Assessment Project in 2001, many countries have assessed ecosystem services at various scales. At the end of the 20th century, some Chinese ecology and eco-economics scholars made a preliminary exploration on the theory, method, and practical application of ecosystem service evaluation. Especially since entering the 21st century, a large number of relative studies have emerged and some valuable research results have been obtained [12]. This has greatly promoted the correct understanding of ecological assets and the active implementation of ecological protection measures. However, these studies have many deficiencies. For example, although many studies focused on the effects of land use (e.g., Han et al., 2021 [13]) or the combined effects of land use and climate change on ecosystem services (e.g., Schirpke et al., 2017 [14]), few studies have addressed the specific effects of climate change or climate factors. In addition, almost all existing studies were based on separate years and rarely involved continuous time series.

Grassland is the production base of animal husbandry and the basis of herdsmen's life and cultural inheritance. On the one hand, grasslands provide multiple ecological services, such as climate regulation, soil and water conservation, wind prevention and sand fixation, soil improvement, and biodiversity maintenance [1,15]. On the other hand, grasslands also have the ability to purify air, water, and soil pollutants through various physical, chemical, and biological processes, which is conducive to ecosystem quality and human health [16]. However, the long-term interference of human activities, coupled with the influence of natural factors, such as climate change, have resulted in a decline in the carrying capacity (e.g., the ability to support grazing) and the ability to resist natural risks (e.g., drought) [17] of grassland ecosystems. As reported, about 50% of global grasslands have been degraded [18–21], and temperate grasslands had lost more than 70% of their natural cover by 1950 [2]. About 22% of China's grassland was degraded before 2000 [22], especially in the grassland of North China [23], while the degraded areas are still increasing year by year [24]. As a result, the sustainable development of regional ecology, economy, and society are directly affected. To some extent, this reflects the lack of understanding of grassland ecosystem service status and function and potential economic value. Under such a severe situation, it is of great practical significance to evaluate the economic value of grassland ecosystem services in China for protecting and restoring the effectiveness of grassland resources and making reasonable decisions for regional ecological protection and economic development.

Over the past century, the earth's climate has undergone significant changes characterized by global warming, which has had a significant impact on global ecosystems and their important services [9]. Studies have shown that climate change has a negative impact on 59% of ecosystem services [25] and this impact is expected to increase rapidly around the world in the future [26–29]. Climate change may profoundly affect the behavior patterns and sensitivities of biotic/abiotic organisms, thereby promoting regulation, support, and cultural services [29] or modifying the relationships and benefits related to ecosystem services [30]. For grassland ecosystems, climate-induced changes had a significant impact on the regulation and cultural services of grasslands in the French Alps, which even exceeded the impact of grassland management decisions [31]. Drought and warming combined with overgrazing led to desertification in some grassland areas in China [32]. The projected future climate scenarios will also have a significant impact on grassland ecosystem services, such as carbon stocks, in northern China [33]. For forest and other ecosystems, climate change has also exhibited a profound impact on ecosystem services there (e.g., Gong et al., 2017; Cui et al., 2021 [34,35]). Although it is well known that climate change is an important cause of ecosystem service change [2,34,36], the exact influencing factors and driving effects remain largely unclear. Therefore, studying the relationship between ecosystem services and climate variation will help to better understand the driving mechanism of ecosystem services change and lay a foundation for adaptation to climate change.

In view of this, this study constructed the index system and method of ESV evaluation and comprehensively assessed the grassland ESV in China for 20 consecutive years. The main objectives of this research were to find out (1) how the value of grassland ecosystem services has evolved over time and space; (2) the long-term economic benefits and ecological conditions of grassland in China; and (3) whether the climatic variation is a potential regulator of ESV change and which variable is the dominant factor involved.

2. Materials and Methods

2.1. Study Area

China is located in the east of Eurasia and on the west coast of the Pacific Ocean. The latitude is 3°51′–53°33′ N, longitude is 73°33′–135°05′ E, and the elevation is −100–8000 m. It crosses five climatic and thermal zones of tropical, subtropical, warm temperate, middle temperate, and cold temperate zones. The annual precipitation ranges from 50–2000 m.

China's grassland area is approximately 4 million km^2, ranking second in the world, and accounting for more than 40% of the national territory [37]. Grassland in China is mainly distributed in the northeast, northwest, and the Qinghai–Tibet Plateau. The grassland is divided into high, medium, and low coverage grassland (Figure 1; [38]). High coverage grassland refers to natural grassland, improved grassland, and mowed grassland covering >50%. This kind of grassland generally has good water conditions and dense grass cover. Medium coverage grassland refers to natural grassland and improved grassland with coverage of 20–50%. Generally, this kind of grassland has insufficient water and sparse grass cover. Low coverage grassland refers to the natural grassland with coverage of 5–20%. This kind of grassland is characterized by a lack of water, sparse grass cover, and poor animal husbandry utilization conditions.

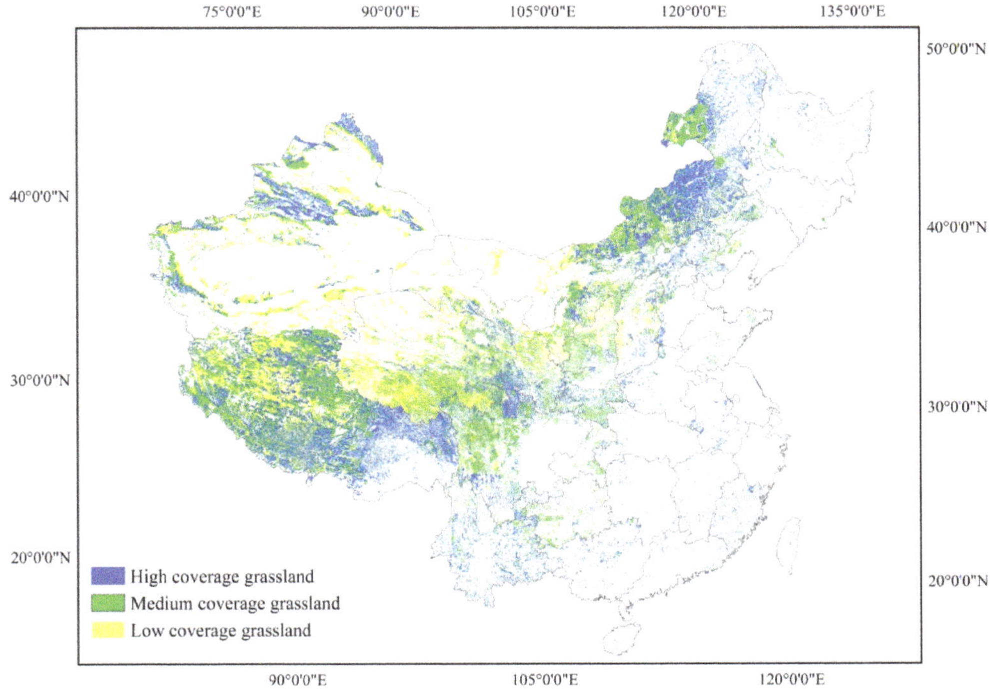

Figure 1. The distribution of the study area.

2.2. ESV Evaluation Method

Xie et al. (2008) [39] showed that, since ecosystem services give different subjective satisfaction to people in different social geographical environments, it will lead to different marginal utility unit values. Therefore, based on Costanza et al. (1997) [1], Xie et al.

(2003, 2008) [39,40] modified the value of each type of ecosystem service by extracting the equivalent weight factors of ecosystem services in China according to a survey of 700 of Chinese ecological experts. Meanwhile, after comparing a large number of results, Xie et al. (2008) [39] confirmed that the unit prices of ecosystem services determined by him were close to those estimated based on material quality, and the two were well comparable. Subsequently, Xie's method was widely used in China. Here, the equivalent factor method modified by Xie et al. (2003) [40] based on Costanza et al. (1997) [1] was applied to calculate the grassland ecosystem service value from 2001 to 2020 in China. The unit price of ecosystem services determined by Xie et al. (2003) [40] is shown in Table 1.

Table 1. Ecosystem service value per unit area of grassland ecosystem type in China (yuan/hm^2).

Categories	Services Item (Sub-Categories)	Unit Price/yuan (p_i)
Regulating service	Gas regulation	707.9
	Climate regulation	796.4
Supporting service	Water conservation	707.9
	Soil formation and protection	1725.5
	Waste treatment	1159.2
	Biodiversity protection	964.5
Provision service	Food production	265.5
	Raw material production	44.3
Cultural service	Recreation and culture	35.4

2.2.1. Model and Index System

The ESV evaluation equation [39,40], relevant indicators, and parameters are as follows:

$$\text{ESV} = \Sigma P_i \times A \tag{1}$$

ESV is the total value of grassland ecosystem services in China. P_i is the revised unit price of ecosystem services i of the grassland ecosystem. $i = 1, 2, \ldots, 9$, respectively, representing nine ecosystem services: gas regulation, climate regulation, water conservation, soil formation and protection, waste treatment, biodiversity protection, food production, raw material production, recreation and culture; A is the area of grassland ecosystem in China.

The revised unit price of ecosystem services is adjusted by using biomass factors as follows:

$$P_i = (b/\text{B})p_i \tag{2}$$

where P_i is the revised unit price of ecosystem services i of the grassland ecosystem, the same as above; B is the biomass per unit area of grassland in China, B = 1322 × 85% kg/(hm^2·a); p_i is the unit price of ecosystem service i under the national average state in Table 1 put forward by Xie et al. (2003) [40]; b is the biomass of the grassland ecosystem, b = NPP/(6 × 0.45).

The parameter setting and calculation process are as follows.

Biomass generally refers to the dry weight of all living biological organisms per unit area, which is the dry matter accumulated by net productivity. At present, there is no report on the relationship between grassland vegetation biomass and NPP at the national scale of long time series in China. Grassland NPP is actually the biomass of vegetation in a year. Therefore, the dry matter weight of grassland NPP was determined as the annual biomass. According to Fang et al. (2010) [41], the average total biomass of grassland in China was 479.56–773 g/m^2. Piao et al. (2001) [42] obtained the average aboveground and underground biomass of 98.0 and 602.5 g/m^2, respectively (underground/aboveground biomass was 6.14) using the model established by *China Grassland Resources Data* in the 1990s. Yang et al. (2010) [43] calculated that the average aboveground and underground biomass of grassland in China were 104.8 and 570.2 g/m^2, respectively (underground/aboveground biomass was 5.44). By integrating all of these multiple research data, we calculated that the average aboveground and underground biomass were

127.9 and 639.3 g/m^2 (underground/aboveground biomass was 5.00). Therefore, the ratio of underground/aboveground biomass is set as 5, that is, aboveground biomass accounts for 1/6 of the total biomass.

The aboveground biomass of grassland vegetation is equal to the grass yield (air-dried weight) minus the water content in air-dried grass. In this study, the moisture content of the air-dried grass is 15% [44]. The yield of air-dried grass per unit area of grassland in China is 1322 $kg/(hm^2 \cdot a)$ [44].

In addition, when the plant biomass (dry matter weight, unit: g) is converted to carbon weight (unit: gC), the conversion coefficient is usually 0.45 [44]. The NPP unit in this study is gC/m^2. Therefore, when NPP is converted to dry matter mass, 0.45 is taken as the conversion coefficient (divided by 0.45).

2.2.2. Assumptions

(1) Grassland area remained unchanged. Due to human disturbance and destruction, the area of major grassland pastoral areas has decreased by 0.4% in recent years in China [45]. Meanwhile, due to the implementation of a large-scale ecological restoration project—the Grain to Green Program (GTGP)—in the past 20 years, the area of artificial grassland in China has been increasing. However, all of these changes in grassland area caused by human activities constituted less than 1%. Therefore, in order to reveal the possible impact of climate change and eliminate the impact of land use type change induced by human activities as much as possible, this study assumed that grassland area did not change in different years, and therefore only 1-year land use type data were used.

(2) ESV based on biomass. Costanza et al. (1997) [1] showed that biomass not only reflects the raw material production capacity of an ecosystem but also has an important impact on other services of the ecosystem during the formation and accumulation of biomass. Xie et al. (2003, 2008, 2015) [39,40,46] assumed that biomass can largely reflect the differences in service capacity of different types of ecosystems. Therefore, it is assumed that the intensity of an ecosystem service is linearly correlated with biomass, that is, the greater the biomass, the stronger the ecosystem service capacity.

2.3. NPP Evaluation

The process-based Carnegie–Ames–Stanford approach (CASA) model, jointly developed by Stanford University, the Nature Conservation Society (TNC), and the World Wide Fund for nature (WWF), is used to calculate vegetation NPP over 2001–2020. Compared with the in situ NPP and MODIS-NPP, the performance of NPP according to the CASA model was verified to be reliable. The CASA model is a satellite-based light use efficiency model. The model expression and parameter setting were detailed in the work of Potter et al. (1993), Luo et al. (2020), and Zhang et al. (2021) [47–49]. In brief, absorbed photosynthetically active radiation (APAR) and actual light use efficiency (ε) are used to estimate NPP; the equation is as follows:

$$\mathrm{NPP}(x, t) = \mathrm{APAR}(x, t) \times \varepsilon(x, t) \tag{3}$$

where NPP is the net primary productivity (gC/m^2); APAR(x, t) represents the absorbed photosynthetically effective radiation (MJ/m) absorbed by pixel x in the month t, which is calculated from the normalized difference vegetation index (NDVI); and $\varepsilon(x, t)$ represents the actual light use efficiency of pixel x in the month t (gC/MJ).

Estimation of APAR

The effective solar radiation absorbed by vegetation and the absorption ratio of the vegetation layer to the incident photosynthetically active radiation are used to estimate APAR, using the following relation:

$$\mathrm{APAR}(x, t) = \mathrm{SOL}(x, t) \times \mathrm{FPAR}(x, t) \times 0.5 \tag{4}$$

where SOL(x, t) represents the total solar radiation (Gc/m^2·month) at pixel x in month t and FPAR(x, t) represents the absorption ratio of the vegetation layer to the incident photosynthetic effective radiation. The constant 0.5 represents the effective solar radiation that the vegetation can use as a (wavelength is 0.4–0.7 μm) proportion of total solar radiation. FPAR is derived based on NDVI; the calculations for FPAR were detailed in the work of (Potter et al., 1993; Luo et al., 2020; Zhang et al., 2021) [47–49].

Estimation of light use efficiency

Light use efficiency refers to the ratio of the chemical potential of dry matter formed per unit area in a given time to the photosynthetic effective radiant energy projected onto the same area during the same period. Environmental factors affecting the photosynthetic capacity of plants, viz., air temperature, soil moisture status, and the difference in atmospheric water and vapor pressures, etc., can regulate the NPP of vegetation.

$$\varepsilon(x, t) \times T_{\varepsilon 1}(x, t) \times T_{\varepsilon 2}(x, t) \times W_{\varepsilon}(x, t) \times \varepsilon_{\max} \tag{5}$$

where $T_{\varepsilon 1}(x, t)$ and $T_{\varepsilon 2}(x, t)$ indicate the stress effect of low temperatures and high temperatures on light use efficiency; $W_{\varepsilon}(x, t)$ is the influence coefficient of water stress, reflecting the influence of water conditions; and $\varepsilon_{\max}$ is the maximum light use efficiency under ideal conditions.

$$T_{\varepsilon 1}(x, t) = 0.8 + 0.02 \times T_{opt}(x) - 0.0005 \times [T_{opt}(x)]^2 \tag{6}$$

where $T_{opt}(x)$ is the optimum temperature for plant growth, representing the average monthly temperature of a region measured in degrees Celsius when the NDVI values reach the maximum in a given year.

$$T_{\varepsilon 2}(x, t) = 1.184/\{1 + \exp[0.2 \times (T_{opt}(x) - 10 - T(x))]\} \times 1/\{1 + \exp[0.3 \times (-T_{opt}(x) - 10 + T(x))]\} \tag{7}$$

When the average monthly temperature $T(x, t)$ is 10 °C higher or 13 °C lower than the optimum temperature $T_{opt}(x)$, then the monthly average temperature $T(x, t)$ is equal to 2(x, t) and the optimum temperature $T_{opt}(x)$ is equal to half of 2(x, t).

$$W(x, t) = 0.5 + 0.5 \times \mathrm{EET}(x, t)/\mathrm{EPT}(x, t) \tag{8}$$

where EET is the actual regional evapotranspiration (mm) and EPT is the potential regional evapotranspiration (mm).

2.4. Data Sources and Processing

2.4.1. Used Data

The data used for the NPP estimation by the CASA model include vegetation index, land use/land cover, and climate (Table 2). Remote sensing data of mod17A3 annual NPP and the monthly mod13A2 normalized vegetation index (NDVI) of global land vegetation have been accessed for 20 years from 2001 to 2020 via the EOS/MODIS portal of NASA. The land use type data for 2015 that were used were from the Data Center for Resources and Environmental Sciences of the Chinese Academy of Sciences (RESDC) [38]. All the climate data were from the high-resolution meteorological dataset downloaded from the Chinese Meteorological Information Center (Table 2). GDP data came from the National Bureau of Statistics. The data used for ESV calculations are shown in Table 2.

At present, both RMB yuan and US dollars were used in the research; RMB yuan was usually used in domestic studies. In order to facilitate the comparison of domestic and international research results, both the units of RMB yuan and US dollars were used in this study. The 2007 benchmark price of 7.68 yuan/US $ was used according to Xie et al. (2008) [39].

Table 2. Input and output data of the model and data source.

Model/Output	Input Data	Data Source
ESV	Unit price	Xie et al. (2003, 2008) [39,40]
	Area	RESDC (Xu et al., 2018) [39]
	b	Fang et al. (1996) [44]
	B	Fang et al. (1996) [44]
	NPP	CASA (as follows)
CASA/NPP	NDVI	The EOS/MODIS portal of NASA (https://ladsweb.modaps.eosdis.nasa.gov, accessed on 1 May 2021)
	Climate data	Chinese Meteorological Information Center (http://cdc.cma.gov.cn, accessed on 1 July 2021)
	Land use type	RESDC (Xu et al., 2018) [38]

2.4.2. Data Analysis

All these spatial data were interpolated or resampled to a 1 km × 1 km resolution using ArcGIS 10.0 (Esri, Redlands, CA, USA) before they were inputted into the models. The spatial resolution of NPP and ESV values output by the model was 1 km × 1 km. According to the test, the annual ESV and climate factor data conform to the normal distribution. Therefore, Pearson's rank correlation analysis was conducted to explore the relationship between grassland ESV and climate factors across an annual range at the national scale. $p < 0.05$ (i.e., 95% confidence level) was defined as the statistical significance level. All statistical analyses were performed using the SPSS version 11.5 software package. All grid data analysis and processing were carried out by ArcGIS10.0.

3. Results

3.1. *The Temporal Distribution of Grassland ESV*

The annual grassland ecosystem service values (ESVs) for the national total ranged from 1.17×10^{12} to 1.51×10^{12} yuan (\$$0.15 \times 10^{12}$–\$$0.20 \times 10^{12}$), with an average of 1.37×10^{12} yuan (\$$0.18 \times 10^{12}$) from 2001 to 2020 (Figure 2), accounting for 4.42% of GDP (48.7×10^{12} yuan; \$$6.34 \times 10^{12}$) (Figure 3). The grassland ESV has increased significantly ($p < 0.001$, $R = 0.892$) over the last 20 years at the rate of 122×10^8 yuan (\$$15.89 \times 10^8$) per year, with a peak occurring in 2020. In general, the mean annual grassland ESV was substantially higher in the last 10 years (2011–2020, 1.44×10^{12} yuan; \$$0.19 \times 10^{12}$) than in the first 10 years (2001–2010, 1.30×10^{12} yuan; \$$0.17 \times 10^{12}$), indicating that the grassland ecological environment and service capacity have been generally improved over recent years.

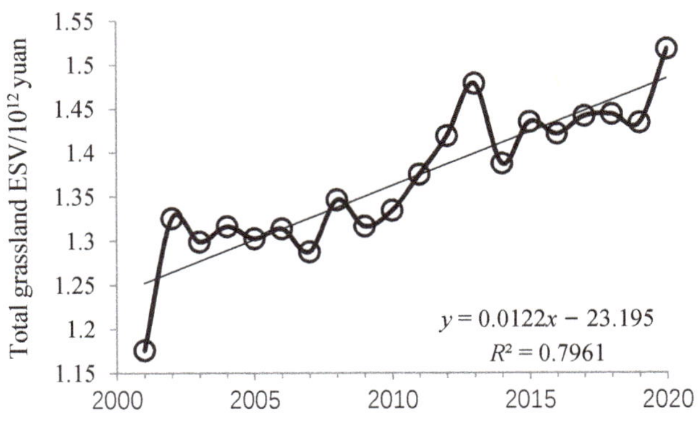

Figure 2. Interannual variation of the grassland ecosystem service value (ESV) for the national total from 2001 to 2020.

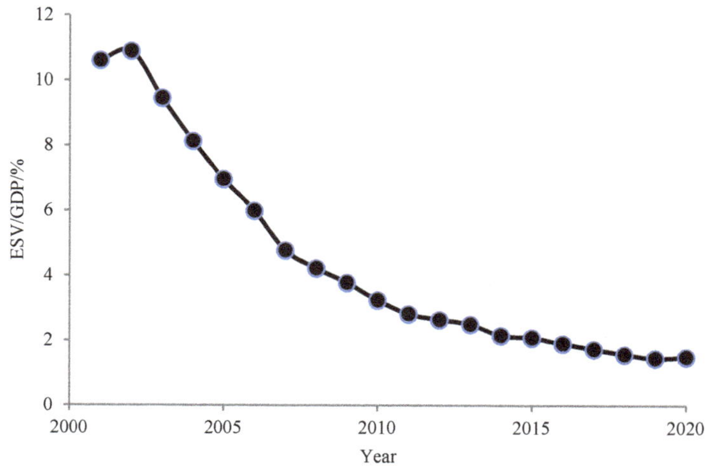

Figure 3. Ratio of grassland ecosystem service value to GDP in China from 2001 to 2020.

In terms of the four major ecosystem services, regulating service accounted for the largest proportion of 52.6% (0.721×10^{12} yuan; \$938.80 $\times 10^8$), followed by supporting service (42%, 0.575×10^{12} yuan; \$748.70 $\times 10^8$), provision service (4.8%, 0.066×10^{12} yuan; \$85.94 $\times 10^8$), and cultural service (0.6%, 0.008×10^{12} yuan; \$10.42 $\times 10^8$) (Figure 4).

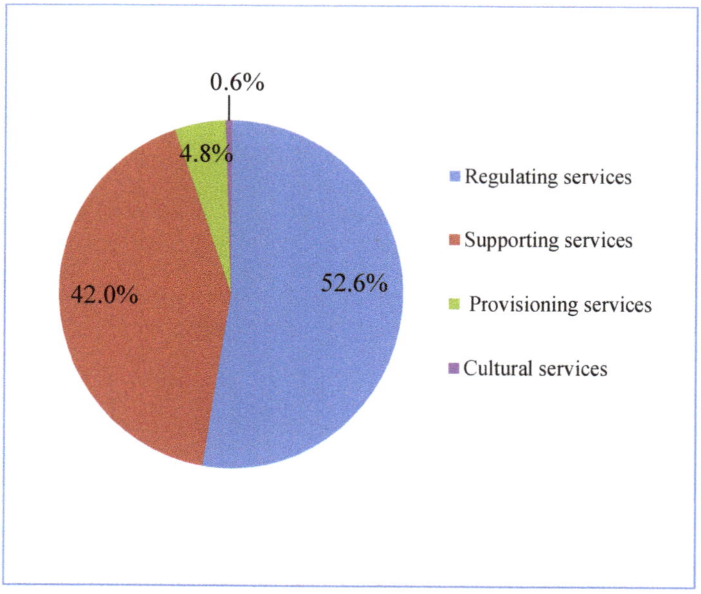

Figure 4. Contribution rate of the four grassland ecosystem services in China.

3.2. *The Spatial Distribution of Grassland ESV*

Spatially, the value of grassland ecosystem service per unit area in China increased from northwest to southeast. The lowest values were less than 0.6×10^6 yuan (\$7.81 $\times 10^4$) per square kilometer, mainly distributed in the desert steppe, alpine meadow, and alpine grassland of western and northern China. The relatively high values of more than 1.5×10^6 yuan (\$19.53 $\times 10^4$) per square kilometer were mostly observed in the savannas of southern China, mainly distributed in Yunnan, Guangxi, Guangdong, and Fujian provinces (Figure 5).

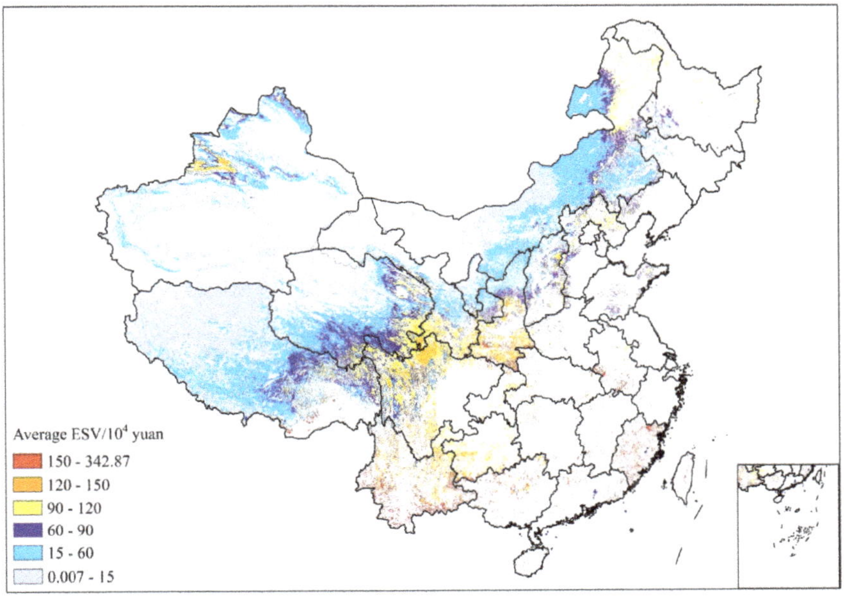

Figure 5. Spatial distribution of mean annual grassland ecosystem service value (ESV) from 2001 to 2020 in China.

However, the value of grassland ecosystem services in the northern and western provinces of China was relatively greater. Inner Mongolia Province in North China, Qinghai Province in Northwest China, and Tibet and Sichuan provinces in Southwest China had the highest values of more than 10×10^6 yuan (\$$130 \times 10^4$). Eastern and southern provinces, such as Zhejiang, Jiangsu, and Hainan, had the lowest ecosystem service values of less than 0.2×10^6 yuan (\$$2.60 \times 10^4$) (Table 3).

Table 3. The mean annual grassland ecosystem service value (ESV) during 2001–2020 in provinces and cities of China.

Province/City	ESV/10^4 yuan	Province/City	ESV/10^4 yuan
Hong Kong	1	Ningxia	44
Shanghai	0	Qinghai	1369
Hainan	12	Shaanxi	561
Guangxi	203	Henan	60
Guangdong	74	Shandong	77
Yunnan	824	Gansu	555
Guizhou	259	Shanxi	247
Fujian	194	Beijing	10
Hunan	68	Tianjin	1
Jiangxi	63	Hebei	231
Zhejiang	19	Liaoning	62
Chongqing	65	Jilin	48
Sichuan	1284	Inner Mongolia	1847
Hubei	68	Heilongjiang	271
Anhui	81	Xinjiang	857
Jiangsu	6	Taiwan	13
Tibet	1303		

Over the past 20 years, the service values of the majority of grassland ecosystems in China have shown an increasing trend ($p < 0.05$, or $R > 0.445$), especially in the warm shrub grass and alpine meadows to the south of Inner Mongolia. The values have decreased only in very few grassland areas, mainly in the alpine meadow and alpine grassland of Tibet Province and the alpine meadow and desert grassland around Ili in Xinjiang Province (Figure 6).

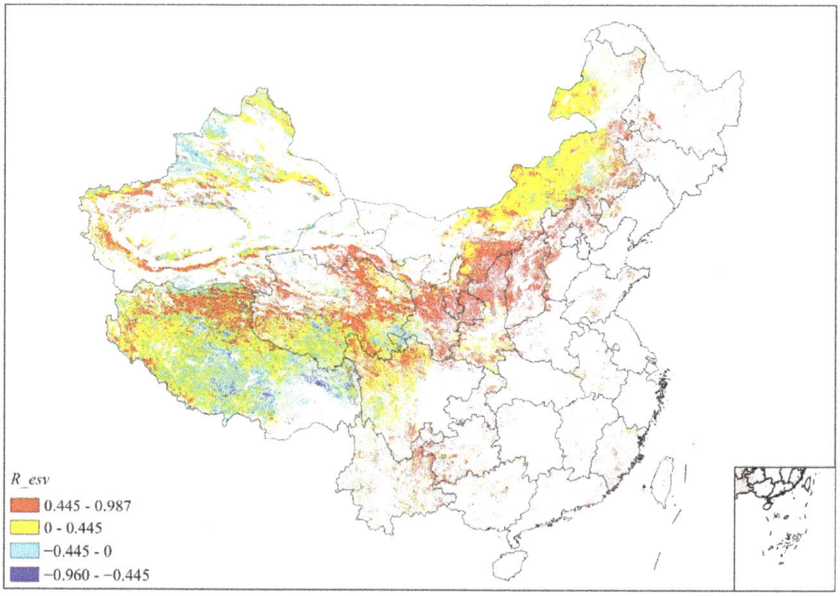

Figure 6. Significance test of regression coefficient of annual grassland ecosystem service value during 2001–2020 in China ($R = \pm 0.445$ represents the significance level $p = 0.05$).

The mean annual grassland ESV was mostly higher in the last 10 years than in the first 10 years of the 21st century, and the growth rate was usually less than 20%. The growth rate of more than 80% was mainly distributed in such mountain areas as Kunlun Mountain, Qilian Mountain, and Tianshan Mountain. However, compared with the first 10 years of the 21st century, the mean annual grassland ESV in the last 10 years decreased in some western and northern areas, mainly in the alpine meadow and alpine grassland of the Tibetan Plateau and the surrounding grasslands of Ili areas in Xinjiang Province (Figure 7).

3.3. The Relationship of Ecosystem Service Value and Climatic Factors

As shown in Figure 8, over the past 20 years, precipitation ($p = 0.030$) and evapotranspiration ($p = 0.004$) in the grassland of China have shown significant increasing trends; the temperature has exhibited a marginal increase ($p = 0.062$), confirming the background of climate warming in recent decades; and the sunshine hours ($p = 0.002$) and relative humidity ($p = 0.014$) have shown a significant decreasing trend.

In order to examine whether the grassland ESV is related to climatic factors, we plotted variations in grassland ESV and temperature, precipitation, evapotranspiration, sunshine hours, and relative humidity over 20 years. The results showed that grassland ESV was significantly positively correlated with precipitation ($p = 0.021$), marginally positively correlated with evapotranspiration ($p = 0.082$), and marginally negatively correlated with sunshine hours ($p = 0.091$), but not significantly correlated with temperature ($p = 0.189$) and relative humidity ($p = 0.166$).

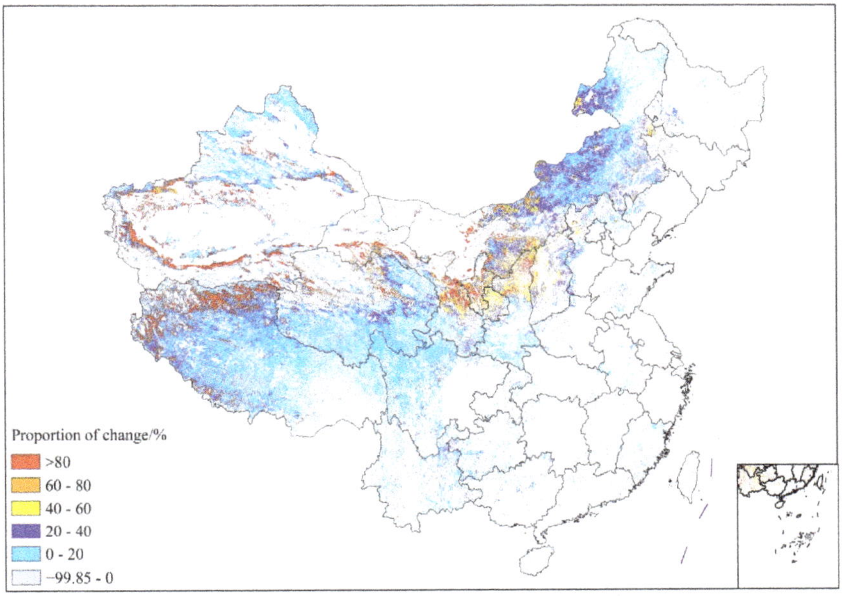

Figure 7. Percentage change of mean annual grassland ecosystem service value (ESV) over 2011–2020 relative to 2001–2011 in China.

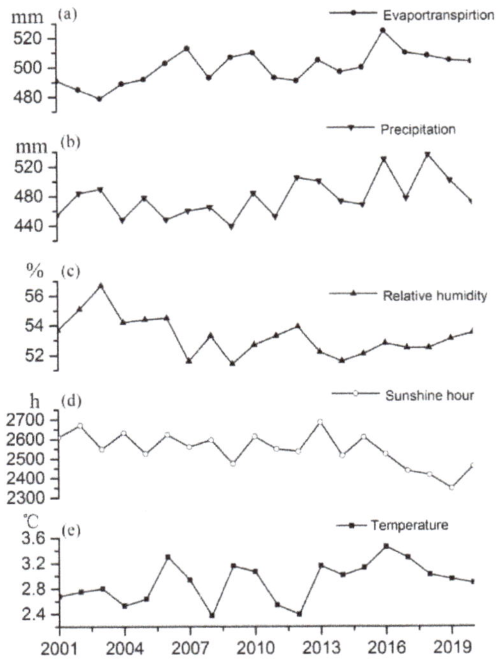

Figure 8. The change trends of evapotranspiration (**a**), precipitation (**b**), relative humidity (**c**), sunshine hours (**d**), and temperature (**e**) in the grassland of China from 2001 to 2020.

As shown in Figure 9, the area with a significant correlation between grassland ESV and precipitation ($p < 0.05$) was the largest (31%, 30% positive and 1% negative), followed by sunshine hours (17%, 7% positive and 10% negative), relative humidity (15%, 13% positive and 2% negative), evapotranspiration (12%, 5% positive and 7% negative), and temperature (10%, 7% positive and 3% negative). Precipitation exhibited a positive impact on ESV in 79% of the grassland areas, especially in the semi-arid grassland areas of central and eastern Inner Mongolia Province, central and northern Qinhai Province, southern Gansu Province, and western Xinjiang Province. Sunshine hours had a negative impact on ESV in 57% of the grassland areas, especially in the desert areas of the Qinghai–Tibet Plateau, the semi-arid area of Loess Plateau, and the mountainous grassland area of Xinjiang Province.

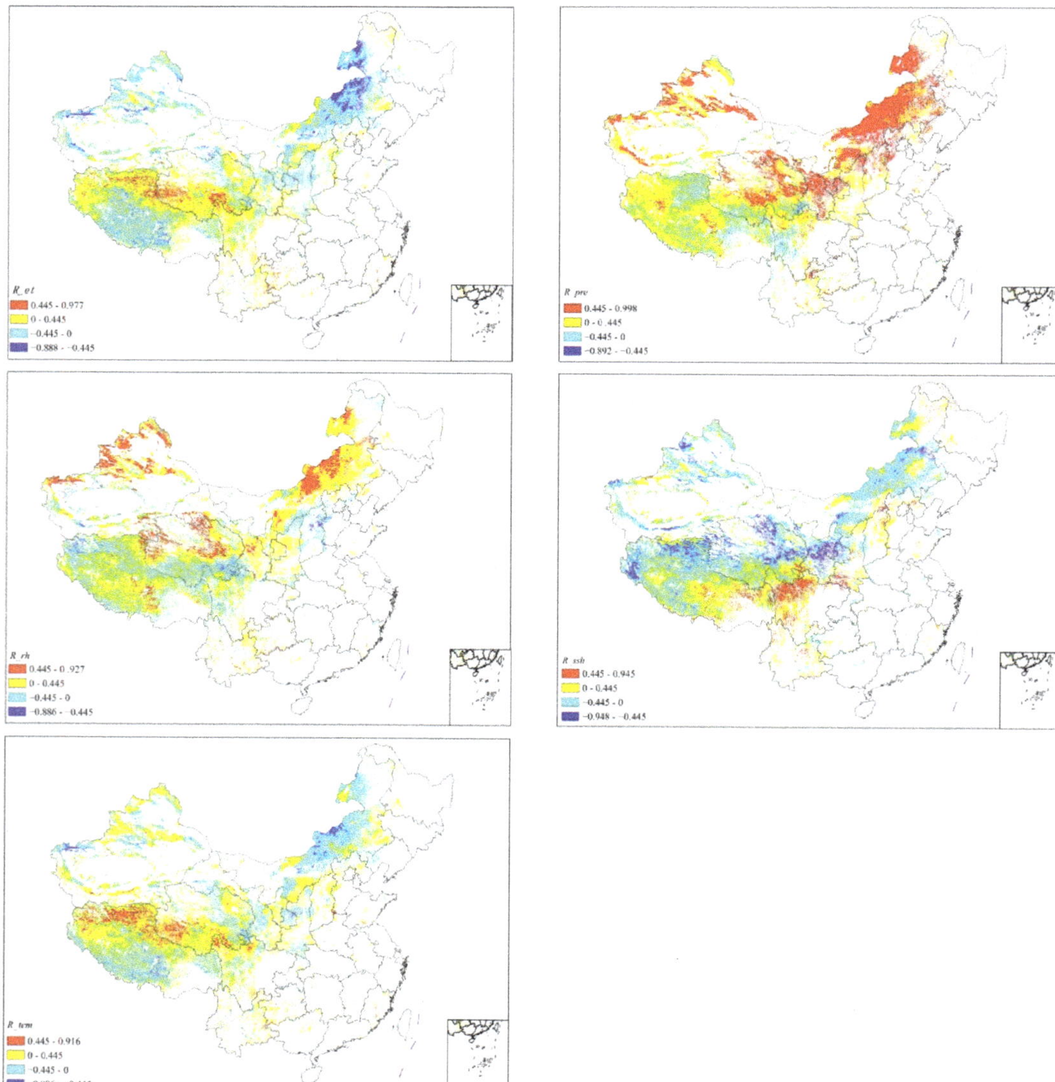

Figure 9. Correlation between grassland ecosystem service value and five climate factors from 2001 to 2020 in China ($R = \pm0.445$ represents the significance level $p = 0.05$).

4. Discussion

4.1. Quantity and Evolution Characteristics of ESVs

Grassland ecosystems can provide abundant benefits for human beings. In our study, the mean annual grassland ESV of 1.37×10^{12} yuan (\$$0.18 \times 10^{12}$) accounted for 4.42% of GDP from 2001 to 2020, which is a huge contribution of spiritual and material wealth. This result is somewhat inconsistent with some previous findings (see Table 4). However, our result does fall within the surveyed and estimated range of 0.87×10^{12}–7.5×10^{12} yuan that is considered more reliable for accounting [50]. The difference in results is mainly attributed to the inconsistency of methods, the evaluation index system, and the research period [51]. Different evaluation objects may also be another reason for inconsistent results (see Table 4).

Table 4. Comparison of grassland ecosystem service value in different studies (i = 1, 2, . . . , n, respectively, representing n ecosystem services).

Studies	Currency Used for ESV		Evaluation Object		Calculation Method/Cited
	yuan/10^{12}	US \$/10^{12}	Value	Grassland	
Our study	1.37	0.18	Total	Total	Σ(ESV_i per area × area)/Xie et al., 2008 [39]
Zhao et al., 2004 [52]	0.88		Indirect	Total	Σ(ES_i × $price_i$)/a method cited for Each ES
Xie et al., 2001 [53]		0.15	Total	Natural	Σ(ESV_i per area × area)/Xie et al., 2001 [53]
Chen et al., 2000 [54]	0.87		Total	Total	Σ(ESV_i per area × area)/Costanza et al., 1997 [1]
Wang et al., 2007 [55]		0.03	Total	Total	Σ(ESV_i per area × area)/Revised from Xie et al., 2001 [53]
Jiang et al., 2007 [56]	1.7		Total	Total	Σ(ES_i × $price_i$)/a method cited for Each ES
Xie et al., 2010 [50]	(0.87–7.5)		Total	Total	Survey and estimation
Liu et al., 2021 [51]	1.38		Total	Total	Σ(ES_i × $price_i$)/a method cited for Each ES

In our study, the linear regression coefficient R^2 of the annual average ESV is about 0.8, indicating that the simulation results of the model are reliable, that is, the increasing trend of ESV in the last 20 years is credible. However, the grassland ESV significantly decreased in some areas. The decline may be directly related to the incomplete matching and uneven distribution of hydrothermal conditions in these areas. All these results indicate that the grassland ecological environment and its quality in China have generally been improved and stabilized in recent years, but the regional development is unbalanced. Grassland ecosystems have stronger potential restoration capacities than forests and other ecosystems, whereas it is difficult to recover if damaged to some extent [57]. Therefore, we should try our best to avoid the destruction of grassland and take measures of protection and reconstruction for some key areas where the value of ecosystem services has declined.

4.2. Key Climatic Controlling Factors

In our study, precipitation significantly affected ESV and exhibited a positive effect in 79% of grassland areas. Meanwhile, in 57% of grassland areas, the ESV was negatively correlated with sunshine hours, especially in desert, semi-arid, and mountain grasslands. These results indicate that precipitation is definitely the most important regulating and stress factor of grassland ecosystems in China. This is mainly due to the fact that about 78% of grassland in China is located in arid and semi-arid areas [58].

The temperature change in grasslands in China confirms that the climate has been warming in recent decades. With the background of climate warming, grassland ESV in China has been increasing significantly over the last 20 years. Even though the significant correlation between grassland ESV and temperature only existed in northwestern alpine grassland and southern tropical–subtropical grassland, and did not exist in some northern grassland, this does not mean that climate change has no effect on northern grassland ESVs. The northern grasslands are relatively drier and water is a more important limiting factor, which may override the influence of temperature, since the effects of temperature and water on vegetation growth in nature may be interactive or fluctuating [59]. Meanwhile, given that the data used were only 20 years old, climate change may influence grassland ESV in the future, as projected by some studies [12,26–29,33].

4.3. Limitations

Our study has some limitations in the quantification of ESV. In terms of the equivalence factor, on the one hand, it was mainly based on biomass [12,60–63], but biomass was not always positively correlated with ecosystem services [64]; on the other hand, it mainly depended on the cognitive level of ecological experts, without considering the spatial heterogeneity of ecosystem services; therefore, the equivalent factor method has certain subjective limitations. Moreover, the unit price of different types of ecosystem services was constant, which does not reflect real life. Furthermore, some anthropogenic activity factors affecting ESV other than climate factors were not considered, and the impact of extreme weather events has not yet been reflected separately in this study. These limitations have introduced greater or lesser amounts of uncertainty into the research results.

Land use change caused by human activities has a significant impact on ecosystem service value [65,66]. Not accounting for the influence of human factors is indeed a limitation of our study. However, changes in grassland area caused by human activities were usually less than 1% in the last 20 years [45], which will not compromise the accuracy of this study. At the same time, the reliability of NPP as determined by the CASA model and the price of ESV in our study were verified, which ensures the correctness of our results.

In addition, this study has some advantages in quantifying the impact of climate change and can achieve the purpose of determining the impact of key climate factors on ESV change. Furthermore, our study can quantitatively demonstrate the long-term evolutionary trends and regional differences in grassland ecosystem service value. These can provide a good decision-making basis for the sustainable utilization of grassland resources and climate change adaptation strategies.

5. Conclusions

(1) From 2001 to 2020, the average annual grassland ESV in China was 1.37×10^{12} yuan ($\$0.18 \times 10^{12}$). The ESV per unit area of grassland increased from northwest to southeast. However, the grassland ESV in northern and western provinces of China was relatively higher and the highest values were more than 10×10^6 yuan ($\$1.30 \times 10^6$).

(2) In most grassland areas, ESV has shown an increasing trend in the past 20 years. At the same time, the mean annual ESV was higher (usually less than 20%) in the last 10 years than in the first 10 years. By contrast, ESV has decreased in some grassland areas of Tibet Province and Xinjiang Province.

(3) With regard to the four major ecosystem services, regulation service accounted for the largest proportion of 52.6% (0.721×10^{12} yuan; $\$938.80 \times 10^8$), followed by supporting (42%, 0.575×10^{12} yuan; $\$748.70 \times 10^8$), provision (4.8%, 0.066×10^{12} yuan; $\$85.94 \times 10^8$), and cultural (0.6%, 0.008×10^{12} yuan; $\$10.42 \times 10^8$) services.

(4) Precipitation was the main regulator of grassland ESV across the 20-year period. It had a positive effect on ESV in 79% of grassland areas. The ESV of grassland was affected by evapotranspiration and sunshine hours to a certain extent, but not by temperature and relative humidity.

All these results indicate that China's grassland ecosystem service has provided a huge amount of spiritual and material wealth for human beings. In general, the grassland ecosystem has been improved on the whole and is in a stable state, while regional development is unbalanced. Particular attention should be given to the areas with declining ESVs. Climatic factors should be considered in the decisions about adaptation plans. These results can not only provide an important reference for eco-environmental protection and sustainable development policies, but also provide an inclusive and in-depth perspective on the complex socio-ecological relationship between ecosystem management decisions and human development.

Author Contributions: Conceptualization, Y.X.; methodology, Y.X.; formal analysis, Y.X., F.X., Y.L.; investigation, F.X., Y.X.; resources, Y.L., F.X.; data curation, F.X., Y.X.; writing—original draft preparation, Y.X.; writing—review and editing, Y.X.; visualization, Y.L.; supervision, Y.L.; project administration, Y.L.; funding acquisition, Y.L. All authors have read and agreed to the published version of the manuscript.

Funding: This research was funded by the National Key Research and Development Program of China (grant no. 2018YFA0606300).

Institutional Review Board Statement: Not applicable.

Informed Consent Statement: Not applicable.

Data Availability Statement: Data are available from the authors upon reasonable request as the data needs further use.

Acknowledgments: We would like to thank the staff of the National Climate Center of the China Meterological Administration (CMA) for providing some climate data and Liu Qiufeng for processing some data. We appreciate the critical and constructive comments from three anonymous reviewers.

Conflicts of Interest: The authors declare that they have no known competing financial interest or personal relationships that could have appeared to influence the work reported in this paper.

References

1. Costanza, R.; d'Arge, R.; de Groot, R.; Farber, S.; Grasso, M.; Hannon, B.; Limburg, K.; Naeem, S.; O'Neill, R.V.; Paruelo, J.; et al. The value of the world's ecosystem services and natural capital. *World Environ.* **1997**, *387*, 3–15. [CrossRef]
2. Millennium Ecosystem Assessment. *Ecosystems and Human Well-Being: Synthesis*; Island Press: Washington, DC, USA, 2005; p. 137.
3. Haines, Y.R.; Potschin, M. *Common International Classification of Ecosystem Services (CICES): 2011 Update*; University of Nottingham: Nottingham, UK, 2011; pp. 1–13.
4. Haines, Y.R.; Potschin, M. *Common International Classification of Ecosystem Services (CICES, Version 4.1)*; European Environment Agency: Copenhagen, Denmark, 2012; pp. 1–17.
5. Costanza, R.; de Groot, R.; Sutton, P.; van der Ploeg, S.; Anderson, S.J.; Kubiszewski, I.; Farber, S.; Turner, R.K. Changes in the global value of ecosystem services. *Glob. Environ. Chang.* **2014**, *26*, 152–158. [CrossRef]
6. Cowie, A.L.; Orr, B.J.; Sanchez, V.M.C.; Chasek, P.; Crossman, N.D.; Erlewein, A.; Louwagie, G.; Cumming, G.S.; Buerkert, A.; Hoffmann, E.M.; et al. Implications of agricultural transitions and urbanization for ecosystem services. *Nature* **2014**, *515*, 50–57. [CrossRef]
7. Schaller, L.; Targetti, S.; Villanueva, A.J.; Zasada, I.; Viaggi, D. Agricultural landscapes, ecosystem services, and regional competitiveness-assessing drivers and mechanisms in nine European case study areas. *Land Use Policy* **2018**, *76*, 735–745. [CrossRef]
8. Chen, J.; Jiang, B.; Bai, Y.; Xu, X.; Alatalo, J.M. Quantifying ecosystem services supply and demand shortfalls and mismatches for management optimisation. *Sci. Total Environ.* **2019**, *650*, 1426–1439. [CrossRef]
9. IPCC. *Climate Change 2014: Impacts, Adaptation, and Vulnerability. Part A: Global and Sectoral Aspects*; Contribution of Working Group II to the Fifth Assessment Report of the Intergovernmental Panel on Climate Change; Field, C.B., Barros, V., Mastrandrea, M.D., Mach, K.J., Abdrabo, M., Adger, W.N., Anokhin, Y.A., Anisimov, O.A., Arent, D., Barnett, J., et al., Eds.; Cambridge University Press: Cambridge, UK, 2014; p. 1132.
10. Lyu, R.; Clarke, K.C.; Zhang, J.; Feng, J.; Jia, X.; Li, J. Spatial correlations among ecosystem services and their socio-ecological driving factors: A case study in the city belt along the Yellow River in Ningxia, China. *Appl. Geogr.* **2019**, *108*, 64–73. [CrossRef]
11. Li, G.; Ren, H. Biomass and net primary productivity of the forests in different climatic zones of china. *Trop. Geogr.* **2004**, *24*, 306–310. (In Chinese) [CrossRef]
12. Xu, Y.; Yu, L.; Zhou, B.; Shi, Y.; Xu, Y. Temporal-spatial dynamic pattern of grassland ecosystem service value under the background of climate change in the future in China. *Ecol. Environ. Sci.* **2017**, *26*, 1649–1658. (In Chinese)
13. Han, X.; Yu, J.; Shi, L.; Zhao, X.; Wang, J. Spatiotemporal evolution of ecosystem service values in an area dominated by vegetation restoration: Quantification and mechanisms. *Ecol. Indic.* **2021**, *131*, 108191. [CrossRef]
14. Schirpke, U.; Kohler, M.; Leitinger, G.; Fontana, V.; Tasser, E.; Tappeiner, U. Future impacts of changing land-use and climate on ecosystem services of mountain grassland and their resilience. *Ecosyst. Serv.* **2017**, *26*, 79–94. [CrossRef]
15. Zhou, H.; Zhao, X.; Tang, Y.; Gu, S.; Zhou, L. Alpine grassland degradation and its control in the source region of the Yangtze and Yellow Rivers, China. *Grassl. Sci.* **2005**, *51*, 191–203. [CrossRef]
16. Yang, Q.; Liu, G.; Giannetti, B.G.; Agostinho, F.; Cecília, M.V.B.; Casazzad, M.C. Emergy-based ecosystem services valuation and classification management applied to China's grasslands. *Ecosyst. Serv.* **2020**, *42*, 101073. [CrossRef]
17. Zheng, X.; Zhang, J. Net value of grassland ecosystem services in mainland China. *Land Use Policy* **2018**, *79*, 94–101. [CrossRef]
18. Harris, R.B. Rangeland degradation on the Qinghai-Tibetan plateau: A review of the evidence of its magnitude and causes. *J. Arid Environ.* **2010**, *74*, 1–12. [CrossRef]

19. Seto, K.C.; Michail, F.; Burak, G.; Reilly, M.K. A meta-analysis of global urban land expansion. *PLoS ONE* **2011**, *6*, e23777. [CrossRef]
20. Gang, C.; Zhou, W.; Chen, Y.; Wang, Z.; Sun, Z.; Li, J.; Qi, J.; Odeh, I. Quantitative assessment of the contribution of climate change and human activities on global grassland degradation. *Environ. Earth Sci.* **2014**, *72*, 4273–4282. [CrossRef]
21. Wang, Z.; Deng, X.; Song, W.; Li, Z.; Chen, J. What is the main cause of grassland degradation? A case study of grassland ecosystem service in the middle-south Inner Mongolia. *Catena* **2017**, *150*, 100–107. [CrossRef]
22. Bao, W.; Shan, W.; Yang, X.; Song, H.; Lan, Y. Ecological crises facing the grassland resources in northern China and their solutions. *Grassl. China* **1998**, *20*, 68–71.
23. Nan, Z. The grassland farming system and sustainable agricultural development in China. *Grassl. Sci.* **2010**, *51*, 15–19. [CrossRef]
24. Feng, C.; Ye, S.; Diao, Z.; Lü, S. Ecological pressures on grassland ecosystems and their conservation strategies in Northern China AU-Zheng, Zhirong. *Chin. J. Popul. Resour. Environ.* **2015**, *13*, 87–91.
25. Runting, R.K.; Bryan, B.A.; Dee, L.E.; Maseyk, F.J.; Mandle, L.; Hamel, P.; Wilson, K.A.; Yetka, K.; Possingham, H.P.; Rhodes, J.R. Incorporating climate change into ecosystem service assessments and decisions: A review. *Glob. Chang. Biol.* **2017**, *23*, 28–41. [CrossRef]
26. Staudinger, M.D.; Grimm, N.B.; Staudt, A.; Carter, S.L.; Chapin, F.S. *Impacts of Climate Change on Biodiversity, Ecosystems, and EcoSystem Services*; United States Global Change Research Progra: Washington, DC, USA, 2012; p. 296.
27. Dow, K.; Berkhout, F.; Preston, B.L.; Klein, R.J.T.; Midgley, G.; Shaw, M.R. Limits to adaptation. *Nat. Clim. Chang.* **2013**, *3*, 305–307. [CrossRef]
28. Nelson, E.J.; Kareiva, P.; Ruckelshaus, M.; Arkema, K.; Geller, G.; Girvetz, E.; Tallis, H. Climate change's impact on key ecosystem services and the human well-being they support in the US. *Front. Ecol. Environ.* **2013**, *11*, 483–493. [CrossRef]
29. Schäfer, L.; Warner, K.; Kreft, S. *Exploring and Managing Adaptation Frontiers with Climate Risk Insurance*; Mechler, R., Bouwer, L.M., Schinko, T., Surminski, S., Linnerooth-Bayer, J., Eds.; Springer Open: Berlin/Heidelberg, Germany, 2018; pp. 317–341. [CrossRef]
30. Augustynczik, A.L.D.; Yousefpour, R. Assessing the synergistic value of ecosystem services in European beech forests. *Ecosyst. Serv.* **2021**, *49*, 101264. [CrossRef]
31. Lamarque, P.; Lavorel, S.; Mouchet, M.; Quetier, F. Plant trait-based models identify direct and indirect effects of climate change on bundles of grassland ecosystem services. *Proc. Natl. Acad. Sci. USA* **2014**, *111*, 13751–13756. [CrossRef]
32. Wang, T.; Wu, W.; Xue, X.; Sun, Q.; Chen, G. Study of spatial distribution of sandy desertification in North China in recent 10 years. *Sci. China J. Earth Sci.* **2004**, *47*, 78–88. [CrossRef]
33. Qiu, L.; Hao, M.; Wu, Y. Potential impacts of climate change on carbon dynamics in a rain-fed agro-ecosystem on the Loess Plateau of China. *Sci. Total Environ.* **2016**, *577*, 267–278. [CrossRef]
34. Gong, S.; Xiao, Y.; Xiao, Y.; Zhang, L.; Ouyang, Z. Driving forces and their effects on water conservation services in forest ecosystems in China. *Chin. Geogr. Sci.* **2017**, *27*, 216–228. [CrossRef]
35. Cui, F.; Wang, B.; Zhang, Q.; Tang, H.; de Maeyer, P.; Hamdi, R.; Dai, L. Climate change versus land-use change—What affects the ecosystem services more in the forest-steppe ecotone? *Sci. Total Environ.* **2021**, *759*, 143525. [CrossRef]
36. Schultner, J.; Dorresteijn, I.; Manlosa, A.O.; von Wehrden, H.; Hylander, K.; Senbeta, F.; Fischer, J. Ecosystem services from forest and farmland: Present and past access separates beneficiaries in rural Ethiopia. *Ecosyst. Serv.* **2021**, *48*, 101263. [CrossRef]
37. National Animal Husbandry and Veterinary Station; Department of Veterinary Medicine; Ministry of Agriculture People's Republic of China. *Grassland Resources in China*; Science and Technology of China Press: Beijing, China, 1996; p. 608. (In Chinese)
38. Xu, X.; Liu, J.; Zhang, S.; Li, R.; Yan, C.; Wu, S. China Many Periods of Land Use Land Cover Remote Sensing Monitoring Data Set (CNLUCC), Chinese Academy of Sciences, Resources and Environment Science Data Center Data Registration and Publication System. 2018. (In Chinese). Available online: http://www.resdc.cn/DOI (accessed on 1 May 2021). [CrossRef]
39. Xie, G.; Zhen, L.; Lu, C. Expert knowledge based valuation method of ecosystem services in China. *J. Nat. Resour.* **2008**, *23*, 911–919. (In Chinese)
40. Xie, G.; Lu, C.; Zheng, D.; Li, S. Ecological assets valuation of the Tibetan Plateau. *J. Nat. Resour.* **2003**, *18*, 189–196. (In Chinese)
41. Fang, J.; Yang, Y.; Ma, W.; Anivar, M.; Shen, H. Ecosystem carbon stocks and their changes in China's grasslands. *Sci. China Life Sci.* **2010**, *53*, 757–765. [CrossRef]
42. Piao, S.; Fang, J. Dynamic vegetation cover change over the last 18 years in China. *Quat. Sci.* **2001**, *21*, 94–302. (In Chinese)
43. Yang, Y.; Fang, J.; Ma, W.H.; Guo, D.; Mohammat, A. Large-scale pattern of biomass partitioning across China's grasslands. *Glob. Ecol. Biogeogr.* **2010**, *19*, 268–277. [CrossRef]
44. Fang, J.Y.; Liu, G.H.; Xu, S.L. Carbon storage in terrestrial ecosystem of China. In *The Measurement of Greenhouse Gas and Their Release and Related Processes*; Wang, G.C., Wen, Y.P., Eds.; China Environmental Science Press: Beijing, China, 1996; pp. 391–397. (In Chinese)
45. Huang, L.; Zhai, J.; Zhu, P.; Zheng, Y. Spatiotemporal evolution characteristics of livestock-carrying pressure in China and its implications for grassland ecosystem conservation pattern. *Acta Geogr. Sin.* **2020**, *75*, 2396–2407.
46. Xie, G.; Zhang, C.; Zhen, L.; Chen, W.; Li, S.; Zhang, L. Improvement of the evaluation method for ecosystem service value based on per unit area. *J. Nat. Resour.* **2015**, *30*, 1243–1254. (In Chinese)
47. Potter, C.S.; Randerson, J.T.; Field, C.B.; Matson, P.A.; Vitousek, P.M.; Mooney, H.A.; Klooster, S.A. Terrestrial ecosystem production: A process model based on global satellite and surface data. *Glob. Biogeochem. Cycles* **1993**, *7*, 811–841. [CrossRef]

48. Luo, Q.; Zhou, J.; Li, Z.; Yu, B. Spatial differences of ecosystem services and their driving factors: A comparation analysis among three urban agglomerations in China's Yangtze River Economic Belt. *Sci. Total Environ.* **2020**, *725*, 138452. [CrossRef]
49. Zhang, Y.; Lu, X.; Liu, B.; Wue, D.; Fu, G.; Zhao, Y.; Sun, P. Spatial relationships between ecosystem services and socioecological drivers across a large-scale region: A case study in the Yellow River Basin. *Sci. Total Environ.* **2021**, *766*, 142480. [CrossRef]
50. Xie, G.; Zhen, L.; Lu, C.; Xiao, Y.; Li, W. Applying value transfer method for eco-service valuation in China. *J. Resour. Ecol.* **2010**, *1*, 51–59.
51. Liu, Y.; Ren, H.; Zhou, R.; Basang, C.; Zhang, W.; Zhang, Z.; Wen, Z. Estimation and dynamic analysis of the service value of grassland ecosystem in China. *Acta Agrestia Sin.* **2021**, *29*, 1522–1532.
52. Zhao, T.; Ouyang, Z.; Jia, L. Ecosystem services and their valuation of China grassland. *Acta Ecol. Sin.* **2004**, *24*, 1101–1110. (In Chinese)
53. Xie, G.; Zhang, Y.; Lu, C.; Zheng, D.; Cheng, S. Ecosystem service value of natural grassland in China. *J. Nat. Resour.* **2001**, *16*, 47–53. (In Chinese)
54. Chen, Z.; Zhang, X. The value of ecosystem benefits in China. *Chin. Sci. Bull.* **2000**, *45*, 17–22. (In Chinese)
55. Wang, R.; Qin, Z. Valuation of Chinese grassland ecosystem services using MODIS data. *Chin. J. Grassl.* **2007**, *9*, 50–53. (In Chinese)
56. Jiang, L.; Qin, Z.; Xie, W. Estimation of Grassland ecosystem services value of china using remote sensing data. *J. Nat. Resour.* **2007**, *22*, 161–169. (In Chinese)
57. Roch, L.; Jaeger, J.A.G. Monitoring an ecosystem at risk: What is the degree of grassland fragmentation in the Canadian Prairies? *Environ. Monit. Assess.* **2014**, *186*, 2505–2534. [CrossRef]
58. Kang, L.; Han, X.G.; Zhang, Z.B.; Sun, O.J. Grassland ecosystems in China: Review of current knowledge and research advancement. *Philos. Trans. R. Soc. B Biol. Sci.* **2007**, *362*, 997–1008. [CrossRef]
59. Xu, Y.; Xiao, F.; Yu, L. Review of spatio—Temporal distribution of net primary productity in forest ecosystem and its responses to climate change in China. *Acta Ecol. Sin.* **2020**, *40*, 4710–4723. (In Chinese)
60. Xie, G.; Zhang, C.; Zhen, L.; Zhang, L. Dynamic changes in the value of China's ecosystem services. *Ecosyst. Serv.* **2017**, *26*, 146–154. [CrossRef]
61. Wang, Y.; Dai, E.; Yin, L.; Ma, L. Land use/land cover change and the effects on ecosystem services in the Hengduan Mountain region China. *Ecosyst. Serv.* **2018**, *34*, 55–67. [CrossRef]
62. Xu, D.; Ding, X. Assessing the impact of desertification dynamics on regional ecosystem service value in North China from 1981 to 2010. *Ecosyst. Serv.* **2018**, *30*, 172–180. [CrossRef]
63. Li, S.; Bing, Z.; Jin, G. Spatially explicit mapping of soil conservation service in monetary units due to land use/cover change for the Three Gorges Reservoir Area, China. *Remote Sens.* **2019**, *11*, 468. [CrossRef]
64. Chen, T.; Feng, Z.; Zhao, H.; Wu, K. Identification of ecosystem service bundles and driving factors in Beijing and its surrounding areas. *Sci. Total Environ.* **2020**, *711*, 134687. [CrossRef]
65. Kim, J.H.; Jobbagy, E.G.; Jackson, R.B. Trade-offs in water and carbon ecosystem services with land-use changes in grasslands. *Ecol. Appl.* **2016**, *26*, 2767. [CrossRef]
66. Liu, J.; Zhang, Z.; Xu, X.; Kuang, W.; Zhou, W.; Zhang, S.; Li, R.; Yan, C.; Yu, D.; Wu, S. Spatial patterns and driving forces of land use change in China in the early 21st century. *Acta Geograph. Sin.* **2009**, *64*, 1411–1420. [CrossRef]

diversity

MDPI

Article

Projections of the Net Primary Production of Terrestrial Ecosystem and Spatiotemporal Responses to Climate Change in the Yangtze River Economic Belt

Li Yu [1], Botao Zhou [2], Yuqing Xu [1], Yongxiang Zhang [1] and Fengxue Gu [3,*]

[1] National Climate Center, China Meteorological Administration, Beijing 100081, China; yuli@cma.cn (L.Y.); xuyq@cma.cn (Y.X.); zhangyx@cma.cn (Y.Z.)
[2] Collaborative Innovation Center on Forecast and Evaluation of Meteorological Disasters, Key Laboratory of Meteorological Disaster, Ministry of Education, Joint International Research of Climate and Environment Change, Nanjing University of Information Science and Technology, Nanjing 210044, China; zhoubt@nuist.edu.cn
[3] Key Laboratory of Dryland Agriculture, Institute of Environment and Sustainable Development in Agriculture, Chinese Academy of Agriculture Sciences, Beijing 100081, China
* Correspondence: gufengxue@caas.cn

Abstract: Evaluating the responses of net primary productivity (NPP) to climate change is essential for regional ecosystem management and adaptations to climate change. The Yangtze River Economic Belt (YREB) is a key ecological functional area and hotspot of carbon sequestration in China due to the high degree of forest coverage. We used a process-based ecosystem model to project terrestrial NPP and analyzed the response to climate change over the 21st century in the YREB under two representative concentration pathway (RCP) scenarios using the regional climate model. The results show that the projection of NPP generally increased by 13.5% under RCP4.5 and 16.4% under RCP8.5 in the middle of the century, by 23% under RCP4.5, and by 35% under RCP8.5 in the late term of the century compared with that from a reference period of current climate conditions (1985–2006). The rate of NPP change under the RCP8.5 scenario is higher than that under the RCP4.5 scenario. Similarly, the NPP is also projected to increase both with 1.5 and 2 °C global warming targets in the YREB. The magnitudes of NPP increment are approximately 14.7% with 1.5 °C and 21% with 2 °C warming targets compared with the current climate, which are higher than the average increments of China. Although NPP is projected to increase under the two scenarios, the tendency of NPP increasingly exhibits a slowdown after the 2060 s under the RCP4.5 scenario, and the growth rate of NPP is projected to drop in more than 31% of regional areas with the additional 0.5 °C warming. In contrast, under the RCP8.5 scenario, the trend in NPP keeps rising substantially, even above 2 °C global warming. However, the NPP in some provinces, including Jiangxi and Hunan, is projected to reduce at the end of the 21st century, probably because of temperature rises, precipitation decreases, and water demand increases. Generally, the NPP is projected to increase due to climate change, particularly temperature increase. However, temperature rising does not always show a positive effect on NPP increasing; the growth rate of NPP will slow down under the RCP4.5 scenario in the mid-late 21st century, and NPP will also reduce by the end of this century under the RCP8.5 scenario in some places, probably presenting some risks to terrestrial ecosystems in these areas, in terms of reduced functions and service decline, a weakened capacity of carbon sequestration, and reduced agricultural production.

Keywords: vegetation NPP; projection; process-based ecosystem model; climate change; global warming target; YREB

Citation: Yu, L.; Zhou, B.; Xu, Y.; Zhang, Y.; Gu, F. Projections of the Net Primary Production of Terrestrial Ecosystem and Spatiotemporal Responses to Climate Change in the Yangtze River Economic Belt. *Diversity* **2022**, *14*, 327. https://doi.org/10.3390/d14050327

Academic Editors: Peng Hou, Weiguo Jiang, Wei Li, Li Zhang and Wenzhi Liu

Received: 25 March 2022
Accepted: 20 April 2022
Published: 22 April 2022

Publisher's Note: MDPI stays neutral with regard to jurisdictional claims in published maps and institutional affiliations.

1. Introduction

Terrestrial ecosystems play a key role in the global carbon cycle, which absorbs approximately 20% and 30% of anthropogenic CO_2 emissions worldwide [1]. The net primary productivity (NPP) of the terrestrial ecosystems is a critical indicator of the terrestrial carbon cycle, as well as being an essential component for the survival of ecosystems that contribute to ensuring the welfare of human beings [2,3]. Exploring NPP responses to climate change and its feedback is crucial for understanding the terrestrial ecosystem dynamics and sustainable development. Terrestrial ecosystem models represent a common framework and a useful method for the ecological research of climate processes, which provide simulations not just for determining the outcomes of terrestrial ecosystems, but also assessing the responses to future stresses [4]. Previous studies have confirmed that the terrestrial NPP is highly sensitive to climate change and other environmental factors [5,6]. Continuous warming has profound effects on the NPP of terrestrial ecosystems, as shown by observational evidence and modeling findings, demonstrating extensive influences on ecosystem functions and services [1,7]. On the global scale, the gross primary productivity (GPP) and NPP have provided evidence of changes, and are projected to increase or remain unchanged, especially in mid-to-high-latitude areas [8,9]. However, on regional scales, the estimations of NPP have varied widely among different studies, and the trends in NPP have exhibited more divergence in different climatic zones [10,11]. Some studies based on ecosystem models and remote sensing data found that the NPP increased on a national scale or within certain typical ecosystem types in recent decades [12,13]; other studies found that the trends of NPP deceased over time in some regions [14]. However, estimations of average NPP ranged considerably on regional scales [15–17]. Furthermore, the projections of NPP showed more inconsistencies in terms of magnitude, trends, and spatial distribution, and were sometimes even controversial when projecting the responses of NPP to future climate change [14,18]. Exploring the spatiotemporal patterns of NPP and dominant factors concretely can develop an understanding about terrestrial carbon sequestration and ecological risks to climate change on different scales; this is also required for facilitating climate change adaptation and ecosystem management.

The Yangtze River is the third largest river in the world, and its development is central to one of three ongoing national strategies of the One Belt and One Road initiative in China, which a project of high international significance. However, it is also one of the regions most sensitive to climate change, being deeply affected by the East Asian Monsoon [19]. The annual average temperature of the Yangtze River Basin has increased more than the global average. Furthermore, the average regional temperature is projected to rise remarkably in the future according to the projections of climate models, as will the frequency of the extreme climatic events occurrence [20,21]. On the other hand, this region has the largest subtropical forest in the world, and the carbon sequestration of artificial forests is considered to represent huge potential for achieving China's carbon neutral target, due to the sufficient water and thermal resources [22,23]. However, a range of ecological and environmental problems, such as the shrinking area of natural forests, ecosystem degradation, soil erosion, reductions in biodiversity, and rapid urbanization, have emerged in the Yangtze River Basin since the 1950s [24,25]. To protect the fundamental ecological barriers and a major grain-producing region of China, a series of guidelines have been issued successively, e.g., "Grain for Green" (since 1990s), the "Outline for the Development Plan of the Yangtze River Economic Belt (YREB)" (2016), and the "Environmental Protection Plan for the YREB" (2017). The future of the YREB is proposed to orient towards ecological priority and green development [26].

With the green development of the YREB clearly confirmed, more and more studies have focused on changes in the ecosystems and its responses to environmental factors. Ke et al. used the Carnegie–Ames–Stanford Approach (CASA) model to simulate the spatial and temporal patterns of vegetation NPP in the Yangtze River Basin from 1982 to 1999, and found that it exhibited a clear increasing tendency of NPP [27]. Zhang et al. estimated that the forest NPP decreased from southeast to northwest by the Lund–Potsdam–Jena (LPJ) model in 1981–2000 in the Yangtze River Basin [28]. However, the estimates of NPP varied in magnitude and pattern, ranging from 262 to 687 g C/m^2 a^1, because of differences in approaches, data, time frames, etc. [29,30]. As for the NPP changes in the future, research by

Miao et al. suggested that terrestrial NPP will decrease in most of the Yangtze River Basin under SRES B2 scenarios using the Atmosphere–Vegetation Interaction Model (AVIM) [30]. Other studies in China also suggested that the Yangtze River Basin is at high ecological risk due to the decrease in ecosystem productivity caused by climate change [31]. However, some studies on the region had projected an increasing tendency of terrestrial NPP [11,18]. In summary, the results of existing studies indicated that there are considerable variations and high uncertainties, whether estimating NPP in recent decades or in projections of NPP in the future in the YREB. According to the implementation of a new regional development philosophy that began in the 1980s, it is also necessary to track the effects of ecosystem programs and understand the dynamics of carbon sequestration in this region, exploring the ecosystem responses and risks associated with climate change in the future.

In this study, terrestrial ecosystem NPP was estimated and projected by a process-based ecosystem model with high-resolution data of vegetation types and Regional Climate Model (RCM) results in 1971–2099, to explore the responses of the ecosystem to climate change in the YREB, which is a significant river basin that covers diverse terrestrial ecosystems in a typical subtropical monsoon region. The major objectives of this study are: (1) to project the spatiotemporal patterns of terrestrial ecosystem NPP in the 21st century, (2) to investigate the changing trends in NPP over different periods and with different global warming targets; and (3) to explore the responses of NPP and potential risks to climate change in the YREB.

2. Materials and Methods

2.1. Study Area

The YRBE is located in the middle of China, and covers nine provinces and two municipalities, over an area of 2.05 billion square kilometers, accounting for 21% of China's landmass (Figure 1). The YREB has a very important ecological role in China, as well as for the rest of East Asia. Approximately 40% of China's available fresh-water resources and more than one-fifth of the total wetland area are here [32]. Grain production, water conservation, and carbon sequestration in the YREB account for 40.2%, 39.2%, and 37.7% of China's total proportions for terrestrial ecosystem services, respectively [33]. Subtropical forests constitute approximately 45% of this region, which is considered an important carbon sink in the Northern Hemisphere [23].

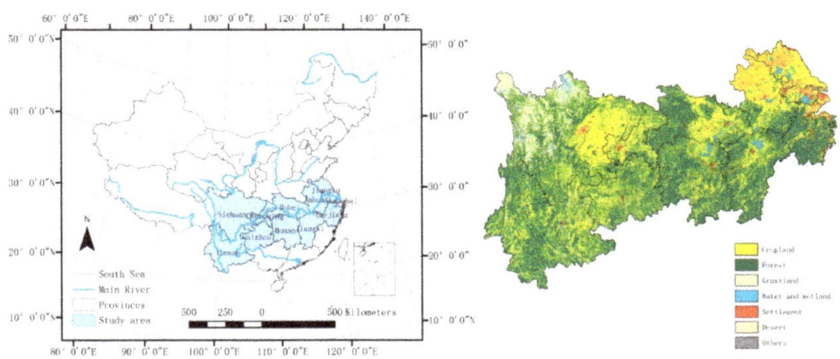

Figure 1. The geographical location and ecological pattern of the YREB of China.

2.2. Model Description and Validation

The Carbon Exchange between Vegetation, Soil, and Atmosphere (CEVSA) model is a process-based terrestrial ecosystem mechanical model that simulates and quantifies energy and the cycles of carbon, nitrogen and water [34]. There are three key modules in the CEVSA model: the biophysical module, which calculates the canopy conductance of plant, soil water supply, and evapotranspiration; the plant growth module, including photosynthesis, autotrophic respiration, carbon allocation in organs, etc., which are key processes of plant growth; and the biochemical module, which simulates how carbon and nitrogen transform and decompose between vegetation and soil. The major parameters and processes of the CEVSA model have been validated and calibrated in different ecosystem types. Gross primary productivity (GPP), NPP, ecosystem heterotrophic respiration (HR), leaf area index (LAI), and net ecosystem exchange outputs by the CEVSA model have exhibited high correlations with the observations by eddy flux tower data and plot-sampling observation datasets of the China Ecosystem Research Network, and the value of remote sensing estimated models [35,36]. The CEVSA model has been widely used to simulate the responses of ecosystems to environmental factors on different scales [18,37]. In this study, we used the latest vision of the CEVSA model—CEVSA2, which incorporates a nitrogen influence module and updated data on nitrogen depositions. The details of the CEVSA2 model are described in Gu et al. (2017) [36]. Comparisons between the observed data of vegetation carbon storage and soil carbon storage showed that the simulation results of the improved CEVSA2 model have stronger agreement with the observed data [37].

2.3. Input Datasets

The gridded datasets included vegetation types, soil parameters, and climate processes, and were inputted to the CEVSA2 model. All input datasets were processed to 10-day averages and interpolated to 0.1 latitude × 0.1 longitude resolutions. Climate data were interpolated with the thin disk spline algorithm in ANUSPLIN software [38]. Vegetation types derived from Global Land Cover (GLC) data were downloaded from the website of the European Commission and resampled to 16 vegetation function types for the CEVSA model. The soil data and parameters were derived from a digitalized soil texture map of China and a soil classifications map, provided by the Institute of Soil Science, Chinese Academy of Science. Climate input data were used for the period of 1961–2099, which constituted observational data for the simulation of ecosystem responses under current climate condition, and climate change scenario data were used to project further ecosystem responses. The observational data were based on daily dataset in situ measurements from about 2400 meteorological stations from 1961 to 2005, which were provided by National Meteorological Information Center. The climate change scenario data were obtained from the BCC_CSM1.1 global climate model, driven by RegCM4.0, with a 25 km × 25 km original spatial resolution, which was released by the National Climate Center of China [21]. The climate projection datasets have been revised on the national scales for China and are widely used to perform impact assessments of climate change across multiple different disciplines [39,40]. In addition, annual mean CO_2 concentration data for historical periods were obtained from the Scripps Institution of Oceanography (SIO). The projections of annual CO_2 concentration varied according to climate scenarios available from the IPCC Data Center. We used two representative concentration pathway (RCP) scenarios: RCP4.5 is a medium greenhouse gas emission scenario, whereas RCP8.5 is a high greenhouse gas emission scenario.

Figure 2 illustrates the recapitulative characteristics of climate change in the YREB. The annual mean air temperature was about 14.8 °C in the period 1986–2005. By the end of the 21st century, the mean temperature is projected to be 15.9 or 17.9 °C under the RCP4.5 or RCP8.5 scenario, respectively. The mean air temperature is projected to increase substantially from 2006 to 2099 (Figure 2a). Precipitation will slightly increase, by 3.4% compared with the present climate, towards the end of the 21st century (2070–2099) under RCP4.5. Similarly, precipitation will increase by 2.3% under the RCP8.5 scenario (Figure 2b). The atmosphere CO_2 concentrations continue to increase, and are projected to continuously increase during the period of 1960 to 2099. The CO_2 concentration will increase strongly under the RCP8.5 scenario; comparatively, it will increase moderately until 2070, and then fluctuate under the RCP4.5 scenario. The magnitudes of CO_2 concentration under RCP4.5 and RCP8.5 are estimated to be slightly over 500 and 900 ppm by the end of the 21st century, respectively (Figure 2c).

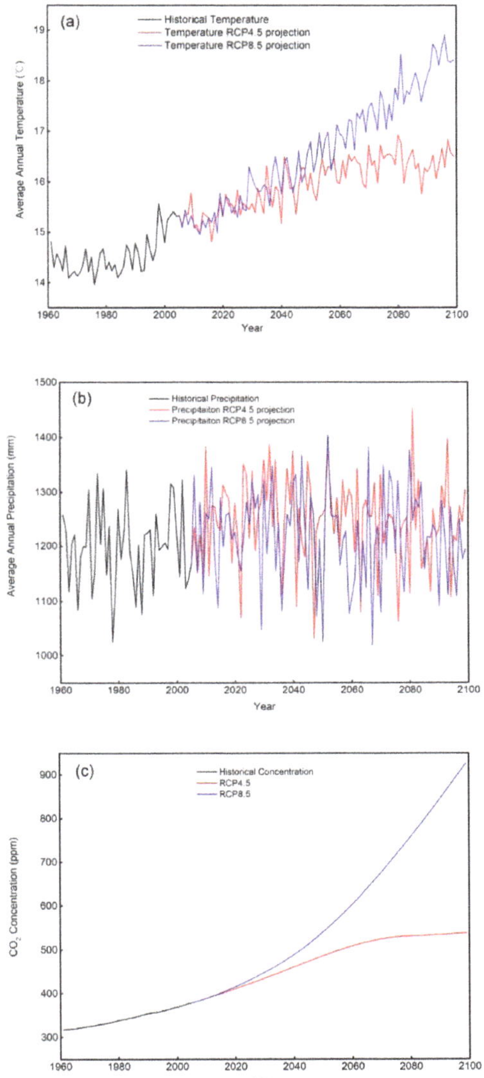

Figure 2. Annual variation in mean air temperature (**a**), precipitation, (**b**) and atmospheric CO_2 concentration (**c**) during the periods 1961–2005/2006–2099 under the RCP4.5 and RCP8.5 scenarios in the YREB.

2.4. Simulations and Analysis

Firstly, the equilibrium CEVSA2 model was performed by average climatic data and average CO_2 concentration in 1971–2000 until equilibrium status of the ecosystem was achieved, which aimed to eliminate the impacts of the initial ecosystem conditions on the simulation. Then, the dynamic model was run using transient climate data and the annual atmospheric CO_2 concentration during the period 1961–2099. The results of the dynamic simulation from 1986 to 2005 were used as a referenced for the current climate, and the outputs of 2020 to 2099 were analyzed to explore the spatiotemporal patterns and changing trends in vegetation NPP over the 21st century.

To show the spatiotemporal characteristics of vegetation NPP to climate change, we defined the period of 2020–2049 as the mid-term 21st century and the period of 2070–2099 as the late-term 21st century. In addition, because of the concerns of the impacts of not achieving global warming targets in key ecological regions, we also analyzed the impacts of 1.5 and 2 °C global warming targets on terrestrial NPP in the YREB. Here, based on the results from Jiang et al. (2016) [41], we extracted the projections of 1.5 and 2 °C warming under RCP4.5 and RCP8.5 scenarios, respectively. All representative periods in this study are shown in Table 1.

Table 1. The representative periods of the 21st century and the periods of 1.5 and 2 °C warming under RCP4.5 and RCP8.5 scenarios.

Scenarios	Mid-Term	Late-Term	1.5 °C Warming Period	2 °C Warming Period
RCP4.5	2020–2049	2070–2099	2020–2039	2040–2059
RCP8.5	2020–2049	2070–2099	2017–2036	2030–2049

To indicate the responses of vegetation NPP to climate change in the YREB, the percentage change of NPP (*R*) was defined as the normal value in different periods against its value in the respective reference period. Here, NPP_{TP} is the mean NPP during the different target periods, and NPP_{RP} is the mean NPP of the reference period. We adopted linear fitting to show the trend in annual NPP spatial and temporal patterns according to climate change under the two RCP scenarios. This was estimated using ordinary least squares based on unitary linear regression. Differences in the NPP slope at different warming targets were used to estimate the distinct responses of a further 0.5 °C increment of warming and the potential risks of plant growth to global warming in the YREB.

3. Results

3.1. Estimated Vegetation NPP in the Current Climate

The spatial distribution of plant NPP varied markedly over the entire YREB region, as shown in Figure 3. The annual NPP was estimated to be about 652.7 ± 177.2 g C m^{-2} $year^{-1}$ during the reference period, which is similar to the results obtained by Miao et al., who used AVIM (674.1 g C m^{-2} $year^{-1}$), and Luo et al., who used statistical data (687.4 g C m^{-2} $year^{-1}$) [42], and the estimate was higher than that in the simulation performed by Zhang et al., who used the LPJ model (530.4 g C m^{-2} $year^{-1}$), and the evaluation carried out by Wu et al., who used remote sensing data (472.6 g C m^{-2} $year^{-1}$). The annual NPP in the YREB is about 1.7 times greater than that of the whole country, the total NPP of the YREB comprises about 33% that of the whole country, although the land area only accounts for about 21% of the total area. Spatially, the higher NPP values, above 750 g C m^{-2} $year^{-1}$ in the YREB, are mainly located in the middle and lower reaches, which account for about 35.5% of the total area of the YREB, covering Jiangxi, Hunan, and Zhejiang provinces, and in some regions of the upper reaches, such as Chongqing municipality, Sichuan Province, and southern Yunnan Province. The lower NPP values, of less than 450 g C m^{-2} $year^{-1}$, are distributed in the northwest of the YREB, which accounts for approximately 13.8% of the total area; these areas are mostly mountainous and plateau areas, including the northwest Sichuan Province and the northern Yunnan Province. Temporally, the tendency of annual mean NPP in the reference period showed significant growth (R^2 = 0.71, $p < 0.01$), which was consistent with the changing tendency of annual NPP on the national scale; however, the growth rate in the YREB was approximately 1.3 times higher than that on the national scale.

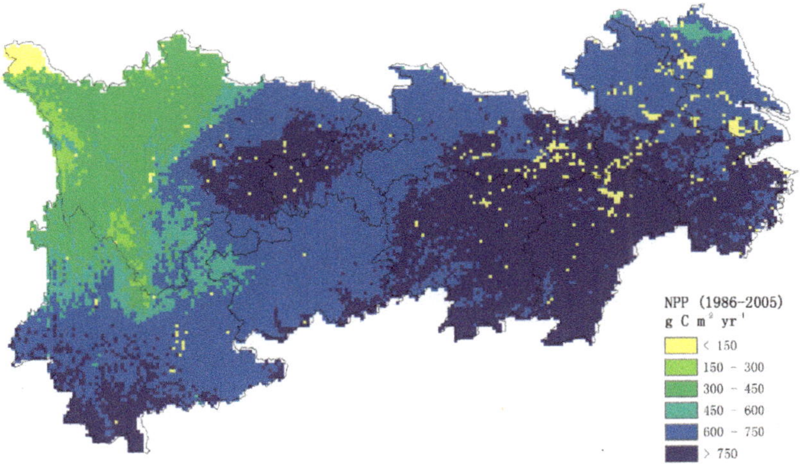

Figure 3. Spatial pattern of NPP in the reference period of 1986–2005 in the YREB.

3.2. Projected Vegetation NPP Due to Future Climate Change

Vegetation NPP is projected to continuously increase in the YREB in the 21st century. The annual mean NPP will increase to 738.7 ± 190.7 g C m^{-2} $year^{-1}$ and 805.6 ± 203.2 g C m^{-2} $year^{-1}$ under the RCP4.5 scenario, and 756.6 ± 193.9 g C m^{-2} $year^{-1}$ and 924.8 ± 241.6 g C m^{-2} $year^{-1}$ under the RCP8.5 scenario in the mid-term and late-term 21st century, respectively. Annual mean NPP was significantly correlated with the mean air temperature (RCP4.5: $R^2 = 0.72$, $p < 0.01$; ECP8.5: $R^2 = 0.93$, $p < 0.01$) for both RCP scenarios (Figure 4). Meanwhile, the NPP growth rate will clearly be higher under the RCP8.5 scenario than that under the RCP4.5 scenario, although both NPP growth rates are higher than that in the reference period.

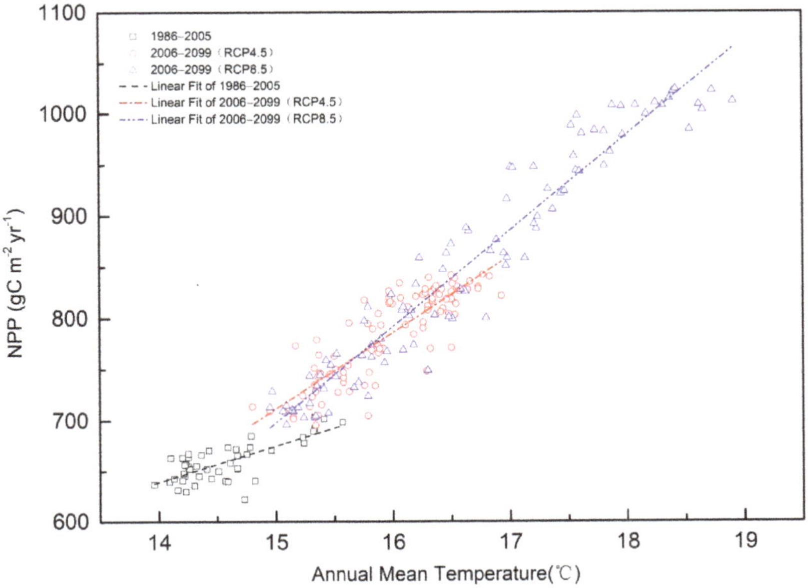

Figure 4. Effects on vegetation NPP and the growth rates of mean air temperature under different scenarios in the YREB.

The trends in annual NPP under the different scenarios indicate that there is a turning point of NPP changing to temperature increase. Before the average temperature rises to 15.7 °C, the rate of NPP change under RCP4.5 is shown to be higher than that under RCP8.5. As the mean air temperature continues to increase, NPP also keeps increasing under the high emission scenario. However, when the mean temperature rises above 17.5 °C under the RCP8.5 scenario, the value of NPP fluctuates at around 1000 g C m^{-2} $year^{-1}$, and the trend in NPP change also shows a remarkable slowing down.

Although global warming prompts vegetation NPP to increase remarkably, our results also reveal that warming does not always have positive effects on the NPP. The interdecadal NPP trend shows that the growth rate of vegetation NPP appears to decrease after 2060 under RCP4.5, and slows down after 2070 under RCP8.5 (Figure 5). In addition, the atmospheric CO_2 concentration shows a positive effect on vegetation NPP based on the interdecadal trends of NPP; the NPP growth rate is consistently greater under higher atmospheric CO_2 concentrations, whether at the same warming level or in the same period. Similarly, the magnitudes and growth rates of NPP under higher CO_2 concentrations are greater than those under low CO_2 concentrations according to the interdecadal tendency of annual NPP.

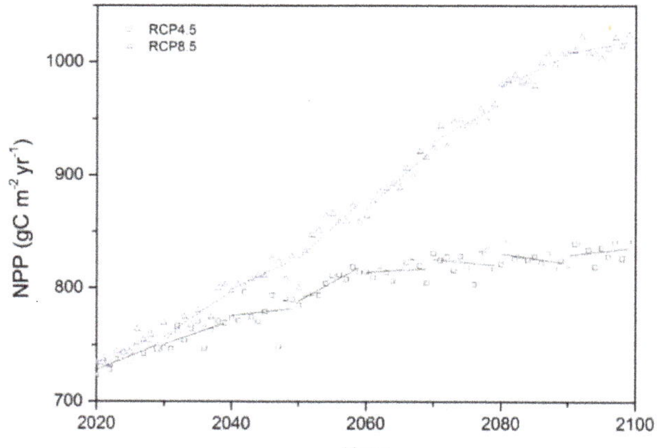

Figure 5. The annual mean NPP under the RCP4.5 and RCP8.5 scenarios during 2020–2099 in the YREB.

3.3. *Spatial Responses of NPP to Future Climate Change*

Figure 5 shows the spatial patterns of NPP variations in the mid-term and late-term 21st century under the two studied RCPs against to the reference period. Spatially, the increments of vegetation NPP in the mid-term period demonstrate a basic pattern of high in the west and low in the east under different RCP scenarios. However, in the late term, the spatial patterns of vegetation NPP show great difference under the RCP8.5 scenario. Generally, the overwhelming majority of areas show an increase in plant NPP in the future; fewer than 1% of the grids show a decrease in NPP, with the exception being that in the late term under the RCP8.5 scenario.

In the mid-term 21st century, vegetation NPP is projected to increase all over the region, the spatial patterns of NPP change will be similar for both RCP scenarios, and the greatest increase in vegetation NPP will be located in the Yunnan and Sichuan provinces, which both have lower NPP values under contemporary climate conditions (Figure 6). The projected vegetation NPP will increases by about 13.5% and 16.4% under the RCP4.5 and RCP8.5 scenarios, respectively. In the late-term 21st century, the magnitude of NPP change in most areas of the YREB will be 20% greater than that in the reference period under the same RCPs. Specifically, NPP will increase by 23% and 35% under the RCP4.5 and RCP8.5

scenario, respectively. Similarly, the magnitude of NPP change is projected to be higher under the RCP8.5 scenario than that under the RCP4.5 scenario. However, vegetation NPP is projected to decline in some areas under the RCP8.5 scenario in the late-term 21st century, which will be concentrated in the central–eastern region of the YREB, including Jiangxi, Hunan, and Hubei, comprising 3.6% of the entire YREB area.

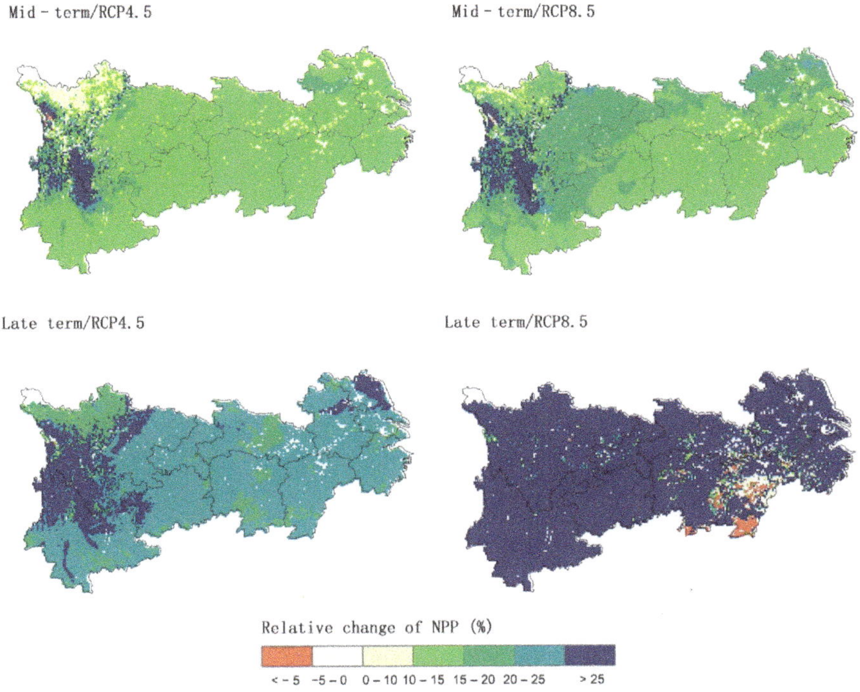

Figure 6. Spatial change patterns of NPP in the mid−term and the late−term 21st century under the RCP4.5 and RCP8.5 scenarios compared with the reference period in the YREB.

3.4. Vegetation NPP Responses to Global Warming

As shown in Figure 7, the spatial change in vegetation NPP presents similar patterns for both 1.5 and 2 °C global warming targets for different RCP scenarios. The magnitude of vegetation NPP increases considerably more in the western YREB than that in the east. Vegetation NPP is projected to increase strongly at both warming targets over the region; fewer than 1% of grids showed it to decrease compared with the reference period. The annual mean NPP is projected to increase to 730 ± 188.1 g C m^{-2} $year^{-1}$ at 1.5 °C warming and to 771 ± 198.3 g C m^{-2} $year^{-1}$ at 2 °C warming. The increase in NPP is approximately 14.7% at 1.5 °C warming and 21% at 2 °C warming compared with the reference period. Additionally, there is no significant difference in NPP variation under different scenarios, although vegetation NPP will increase by approximately 6% more in the 2 °C global warming period than in the 1.5 °C warming period. Generally, warming is projected to promote NPP even at 2 °C warming targets, and the additional 0.5 °C warming shows positive contributions to NPP increasing, especially under the RCP8.5 scenario.

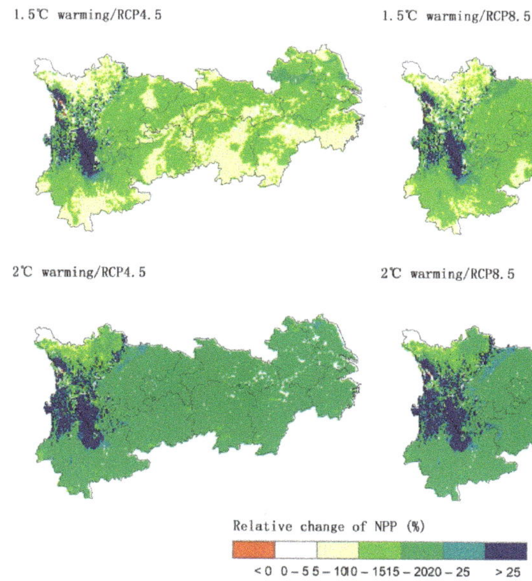

Figure 7. Spatial patterns of vegetation NPP at 1.5 and 2 °C global warming under the RCP4.5 and RCP8.5 scenarios in the YREB.

Although plant NPP is projected to increase in most areas of the YREB with the two global warming targets, the extra 0.5 °C warming will show a substantial impact on the growth rate of vegetation NPP. Figure 8 shows the different NPP tendencies in the 1.5 and 2 °C warming periods under different scenarios. A further 0.5 °C of warming would reduce the growth rate of NPP in about 31% of the regional areas under RCP4.5. Spatially, the growth rate of NPP in about one-third of the YREB, including places such as Sichuan, Yunnan, Guizhou, northern Hubei and Jiangsu, and northwestern Anhui, will be reduced at an enhanced warming target. Conversely, the trend in vegetation NPP is projected to keep growing in over 89% of the region at 2 °C warming under the RCP8.5 scenario. In most places, such as northern Yunnan, northeastern Sichuan, and northern Jiangsu and Anhui, there will be an even further increased growth rate of NPP at 2 °C warming.

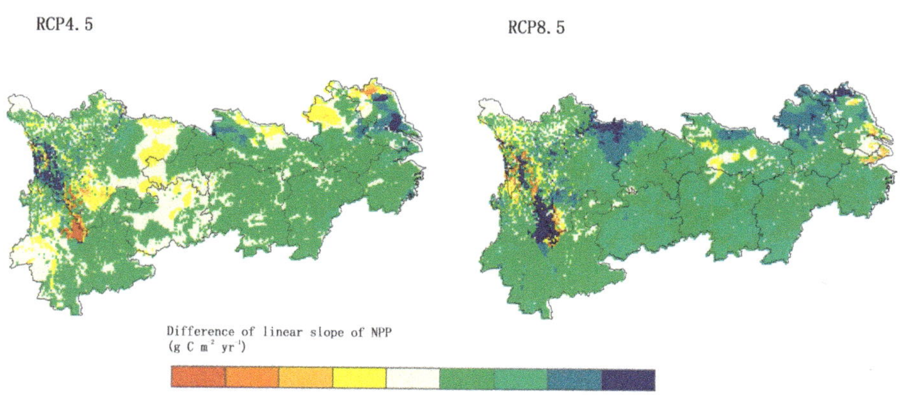

Figure 8. Spatial patterns of NPP trends at 1.5 and 2 °C global warming targets under the RCP4.5 and RCP8.5 scenarios.

4. Discussion

4.1. The Effects of Warming

Global warming is one of the most pervasive characteristics of climate change. Continuous warming has been proven to increase terrestrial vegetation NPP in recent decades; it was projected to increase in the 21st century in mid- and high- latitude regions, as well as alpine regions [1,43]. Previous studies in China also indicated that the rising meant air temperature was the dominant factor accounting for NPP changes on the national scale [14,36]. Based on the projection of high-resolution regional climate models, warming will also be widespread throughout the 21st century in the YREB (Figure 9). Our simulation results show that vegetation NPP is projected to increases in most parts of the YREB. Furthermore, the magnitude of NPP increments will be greater in the west than in the eastern YREB, which is associated with the differences in mean air temperature due to the higher altitude in this region. Our results are similar to those from research on alpine regions in southwestern China, such as the Yunnan–Guizhou Plateau and the Western Sichuan Basin [28]. The effects of different global warming targets on vegetation NPP also show that a rising temperature will promote increased vegetation productivity, which will be beneficial for terrestrial ecosystem carbon storage and achieving carbon neutrality on the regional scale.

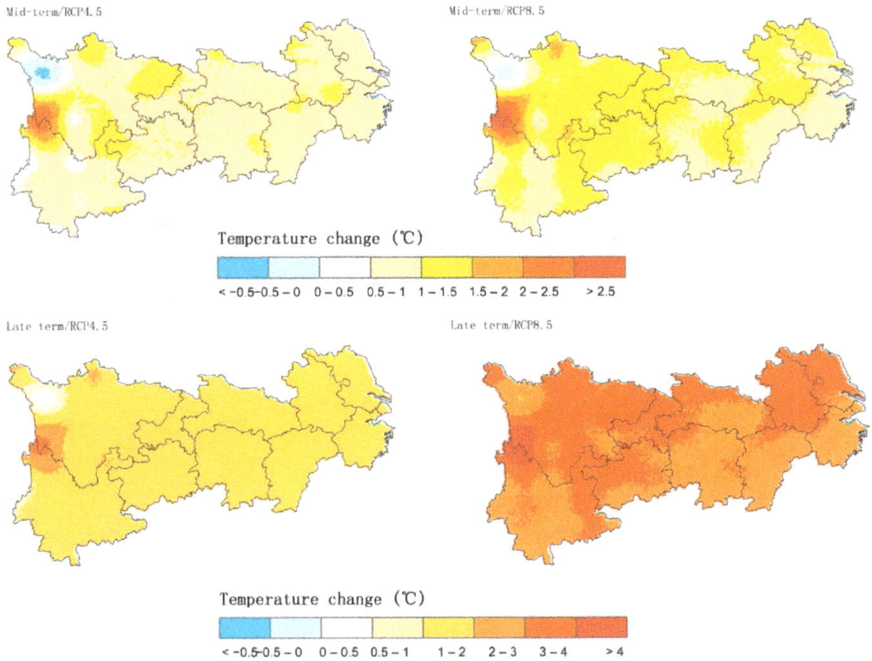

Figure 9. Mean air temperature change patterns in the mid–term and late–term 21st century compared with the reference period under the RCP4.5 and RCP8.5 scenarios.

4.2. The Effects of Precipitation Changes

As one of the most abundant water resource regions in China, the water supply for plant growth in the YREB is usually sufficient, and water resources rarely constrain vegetation productivity [44]. Based on the projections of regional climate models, there are no obvious trends in annual precipitation over the region (Figure 2b), although the variation in precipitation exhibits reasonable spatial heterogeneity, with high uncertainty (Figure 10). Precipitation is projected to increase in the west but generally decrease in the southern and central YREB, and coupled with temperature changes, the spatial hydrothermal patterns will be greatly altered in this region.

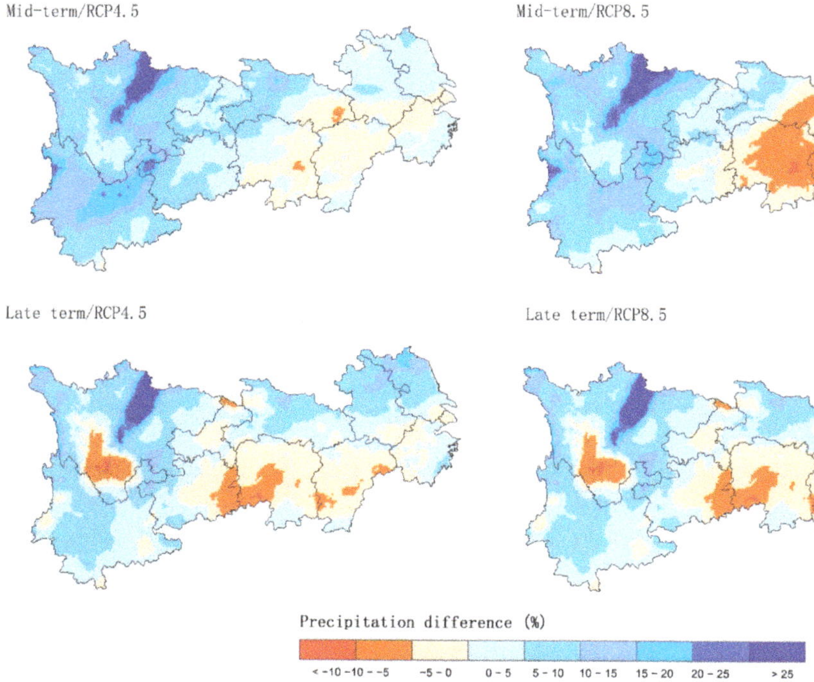

Figure 10. Spatial pattern of annual mean precipitation changes in the mid-term and late-term 21st century relative to the reference period under the RCP4.5 and RCP8.5 scenarios in the YREB.

The detrend fluctuation analysis showed that temperature and precipitation both exhibit positive correlations with the interannual variability of NPP (IAV NPP) during the period of 2021–2099 under RCP4.5, whereas under RCP8.5, only precipitation shows a positive correlation with IVA NPP (Table 2). This suggests that precipitation has a more dominant influence on IVA NPP than temperature, especially under higher emission scenarios, which indicates that as plant demand for water supply increases with rising temperatures, water resources may become a restricted factor for plant growth in this region. Thus, enhanced warming due to anthropogenic CO_2 emission promotes the growth and productivity of plants, but a sufficient water supply is necessary; otherwise, it may lead to stunted vegetation growth and reduced productivity. Vegetation NPP is projected to decline in the central–southern region of the YREB under RCP8.5 in the late-term 21st century maybe for this reason. Studies on other areas of China also found that declines in vegetation NPP will occur in water-deficit areas associated with rising air temperatures [7,45]. Therefore, water availability to terrestrial vegetation is particularly important in the YREB, especially so towards the end of 21st century. However, plant growth and vegetation NPP are always affected by integrated climatic conditions and other environmental factors; thus, the trends and fluctuation in NPP will respond in accordance with certain regional features as well as species diversity and complexity.

Table 2. The coefficients of detrended temperature and precipitation anomalies with IAV NPP under the RCP4.5 and RCP8.5 scenarios in the different periods of 21st century. Statistically significant correlations are marked with ** ($p < 0.01$) and * ($p < 0.05$).

		2021–2049	2070–2099	2021–2099
RCP4.5	T	0.04	0.61 **	0.39 **
	P	0.55 **	0.35	0.38 **
RCP8.5	T	0.27	−0.01	0.15
	P	0.58 **	0.52 *	0.52 *

4.3. Factors Other Than Climate

Except for temperature and precipitation, the elevated atmospheric CO_2 concentrations are a remarkable feature of climate change. The CO_2 concentration will increase even more under the high-emission scenario, and it will be more than twice the current level under RCP8.5 by the end of the 21st century (Figure 2c). Previous studies have confirmed that the fertilization effect of CO_2 contributes to the increase in vegetation NPP, but high CO_2 concentrations will indirectly have a negative impact on biodiversity, which may impact the stability of terrestrial vegetation productivity; thus, the effects of rising atmosphere CO_2 levels are still highly uncertainty [8,46,47]. Our results show that annual NPP and its changing rate under tRCP8.5 are greater than those under RCP4.5, especially at 4 °C warming (Figure 9). Higher CO_2 concentrations and warmer temperature bring further increases in NPP, which means that the elevated CO_2 concentrations will stimulate plant growth as well as temperature increases [48]. Vice versa, the effect of rising air temperature on vegetation would be limited without the cooperative influence of higher CO_2 concentrations. Our results suggest that the increase in CO_2 concentration has a significantly positive effect on NPP. Similar results by Zhu et al. (2018) [43] also indicated that the elevated atmospheric CO_2 concentration is the dominant factor on increasing NPP, particularly in mid- and low-latitude regions.

Land use and land cover change (LUCC) present realistic pressures on terrestrial ecosystems in addition to the impacts of climate change, and also have notable impacts on species and productivity [49,50]. Anthropogenic LUCC is a substantial factor that impacts terrestrial ecosystem functions and services associated with biodiversity and productivity, especially in areas with extensive human activity [17]. In this study, the CEVSA2 model did not consider the effects of LUCC, which would bring uncertainty in the projection. In the YREB, urban and construction areas have continuously increased in recent decades [51,52]. Additionally, wetlands have been shrinking since the 1970s [53]. However, forestland and grassland have shown an increasing tendency since the 1990s, due to a series of major ecological conservation and restoration projects, such as "the Grain for Green Program" and "Natural Forest Protection Projects". Now, there are high forest coverage rates in the YREB, significantly above the national average [25,51,54]. Some studies have suggested that the increases in NPP due to climate change are not enough to offset the decreases in plant NPP due to anthropogenic LUCC [55], although other studies have suggested that LUCC has already brought benefits for vegetation and terrestrial ecosystems in the YREB, for ecological restoration and protection projects in this region [26,56]. Furthermore, these ecological projects have been implemented and the effects have continually been improved, with the target of carbon neutrality. Therefore, ecosystem protection combined with climate change would be more conducive to the promotion of vegetation NPP. Thus, it would be expected that the terrestrial carbon storage in this region has great potential to improve, especially for the forest ecosystem [57]. Another study found that the carbon sequestration by terrestrial ecosystems in the YREB is projected to comprise about 42% of China's total capacity by the end of the 21st century; the magnitude will be higher because of ecological policies and projects in this region [37]. However, LUCC intensively affects ecosystems, including species, vegetation types, biodiversity, and productivity, and it is a great challenge to simulate the effects of both anthropogenic LUCC and climate change in ecosystem models [1]. Model development is ongoing and could consider LUCC effects by coupling a module of climate change with land use, and it could consider the impacts of human management measures on terrestrial ecosystems, such as natural forest protection and the management of nature reserves, in order to reduce uncertainties in modeling vegetation growth on the regional scale.

5. Conclusions

In this study, we used a process-based ecosystem model, CEVSA2, driven by high-resolution climate data, to project future vegetation NPP, analyze the spatial and temporal changing patterns of NPP, and explore the responses of NPP to key climate factors in the YREB. Our study indicates that global warming will promote vegetation NPP, and vegetation NPP will increase strongly in the 21st century in the YREB; it is estimated to increase by about 14% under RCP4.5 and 16% under RCP8.5 in the mid-term, and 23% under RCP4.5 and 35% under RCP8.5 in the late-term. Generally, NPP is projected to be higher under the RCP8.5 scenario than that under the RCP4.5 scenario, associated with the positive effects of high atmospheric CO_2 concentrations. NPP is projected to increase at both global warming targets as well. The magnitude of the NPP increase will be approximately 15% at 1.5 °C and 21% at 2 °C global warming compared with the reference period, which are higher values than those expected on the national scale. However, continuous warming does not always result in positive contributions to vegetation NPP; the trends and variations in NPP change show considerable differences for different warming levels and climate change scenarios. In addition, precipitation exerts a more dominant role in NPP changes than air temperature under the higher emission scenario, especially towards the end of the 21st century. Our findings on vegetation NPP change and its responses to climate change suggest that climate change may be beneficial for vegetation growth and the carbon storage of terrestrial ecosystems in the YREB, but the projections of and responses to NPP exhibit high uncertainty and complexity. Our research contributes a detailed modeling study on a significant river-basin scale, which covers diverse terrestrial ecosystems and rich biodiversity. Our conclusions contribute to the knowledge of sustainable forest management and terrestrial ecosystem adaptations to climate change on river-basin scales and in subtropical monsoon zones.

Author Contributions: Conceptualization, L.Y. and F.G.; methodology, L.Y. and B.Z.; software, F.G. and L.Y.; validation, Y.X. and Y.Z.; formal analysis, L.Y.; writing, L.Y. and Y.X.; review and editing, Y.Z. and Y.X.; visualization, L.Y.; supervision, L.Y. and B.Z. All authors have read and agreed to the published version of the manuscript.

Funding: This research was funded by National Natural Science Foundation of China grant number [41991285].

Institutional Review Board Statement: Not applicable.

Data Availability Statement: Data are available from the authors upon reasonable request as the data need further use.

Acknowledgments: We acknowledge Shi Ying provided the projection climate data of RCM in China.

Conflicts of Interest: No conflict of interest exists in the submission of this manuscript, and the manuscript is approved by all authors for publication.

References

1. Settele, J.; Scholes, R.; Betts, R.; Bunn, S.; Leadley, P.; Nepstad, D.; Overpeck, J.T.; Taboada, M.A. Terrestrial and inland water systems. In *Climate Change 2014: Impacts, Adaptation, and Vulnerability. Part A: Global and Sectorial Aspects. Contribution of Working Group II to the Fifth Assessment Report of the Intergovernmental Panel on Climate Change*; Cambridge University Press: Cambridge, UK; New York, NY, USA, 2014; pp. 271–359.
2. Vitousek, M.P.; Aber, D.J.; Howarth, W.R.; Likens, E.G.; Matson, A.P.; Schindler, W.D.; Schlesinger, H.W.; Tilman, G.D. Human aletation of the global nitrogen cycle: Sources and consequences. *Ecol. Appl.* **1997**, *7*, 737–750.
3. Ito, A. A historical meta-analysis of global terrestrial net primary productivity: Are estimates converging? *Glob. Chang. Biol.* **2011**, *17*, 3161–3175. [CrossRef]
4. Bonan, B.G.; Doney, C.S. Climate, ecosystems, and planetary futures: The challenge to predict life in Earth system models. *Science* **2018**, *359*, 6375. [CrossRef] [PubMed]

5. Cramer, W.; Bondeau, A.; Woodward, I.F.; Prentice, I.C.; Betts, R.A.; Brovkin, V.; Cox, P.M.; Fisher, V.; Foley, J.A.; Friend, A.D. Global response of terrestrial ecosystem structure and function to CO_2 and climate change: Results from six dynamic global vegetation models. *Glob. Chang. Biol.* **2001**, *7*, 357–373. [CrossRef]
6. Bachelet, D.; Neilson, R.P.; Hickler, T.; Drapek, R.J.; Lenihan, J.M.; Sykes, M.T.; Smith, B.; Sitch, S.; Thonicke, K. Simulating past and future dynamics of natural ecosystems in the United States. *Glob. Biogeochem. Cycles* **2003**, *17*, 1045. [CrossRef]
7. Wang, Z.; Chang, J.; Peng, S.; Piao, S.; Ciais, P.; Betts, R. Changes in productivity and carbon storage of grasslands in China under future global warming scenarios of 1.5 °C and 2 °C. *J. Plant. Ecol.* **2019**, *12*, 804–814. [CrossRef]
8. Gerber, S.; Joos, F.; Prentice, I.C. Sensitivity of a dynamic global vegetation model to climate and atmospheric CO_2. *Glob. Chang. Biol.* **2004**, *10*, 1223–1239. [CrossRef]
9. Gang, C.C.; Zhang, Y.Z.; Wang, Z.Q.; Chen, Y.Z.; Yang, Y.; Li, J.L.; Cheng, J.M.; Qi, J.G.; Odeh, I. Modeling the dynamics of distribution, extent, and NPP of global terrestrial ecosystems in response to future climate change. *Global. Planet. Chang.* **2017**, *148*, 153–165. [CrossRef]
10. Alo, C.A.; Wang, G. Potential future changes of the terrestrial ecosystem based on climate projections by eight general circulation models. *J. Geophys. Res.* **2008**, *113*, G01004. [CrossRef]
11. Sun, G.D.; Mu, M. Assessing the characteristic of net primary production due to future climate change and CO_2 under RCP4.5 in China. *Ecol. Complex.* **2018**, *34*, 58–68. [CrossRef]
12. Piao, S.L.; Fang, J.Y.; Zhou, L.M.; Zhu, B.; Tan, K.; Tao, S. Changes in vegetation net primary productivity from 1982 to 1999 in China. *Global. Biogeochem. Cycles* **2005**, *19*, 1–16. [CrossRef]
13. Wang, J.Y.; Delang, O.C.; Hou, G.L.; Gao, L.; Lu, X.X. Net Primary production increase in the Yangtze River Basin within the latest two decades. *Glob. Ecol. Coserv.* **2021**, *26*, e01497. [CrossRef]
14. Zhao, D.; Wu, S.H.; Yin, Y.H. Responses of terrestrial ecosystems' net primary productivity to future regional climate change in China. *PLoS ONE* **2013**, *8*, e60849. [CrossRef] [PubMed]
15. Ren, W.; Tian, H.Q.; Tao, B.; Chappelka, A.; Sun, G.; Lu, C.Q.; Liu, M.; Chen, G.; Xu, X.F. Impacts of tropospheric ozone and climate change on net primary productivity and net carbon exchange of China's forest ecosystems. *Global Ecol. Biogeogr.* **2011**, *20*, 391–406. [CrossRef]
16. Liang, W.; Yang, Y.; Fan, D.; Guan, H.; Zhang, T.; Long, D.; Zhou, Y.; Bai, D. Analysis of spatial and temporal patterns of net primary production and their climate controls in China from 1982 to 2010. *Agric. Forest Meteorol.* **2015**, *204*, 22–36. [CrossRef]
17. Li, J.; Wang, Z.; Lai, C.; Wu, X.; Zeng, Z.; Chen, X.; Liao, Y. Response of net primary production to land use and land cover change in mainland China since the late 1980s. *Sci. Total. Environ.* **2018**, *639*, 237–247. [CrossRef] [PubMed]
18. Yu, L.; Gu, F.X.; Huang, M.; Tao, B.; Hao, M.; Wang, Z.S. Impacts of 1.5 °C and 2 °C Global Warming on Net Primary Productivity and Carbon Balance in China's Terrestrial Ecosystems. *Sustainability* **2020**, *12*, 2849. [CrossRef]
19. Yan, Z.W.; Ding, Y.H.; Zhai, P.M.; Song, L.C.; Cao, L.J.; Li, Z. Re-assessing climatic warming in China since the last century. *Acta Meteorol. Sin.* **2020**, *78*, 370–378.
20. Dai, A. Increasing drought under global warming in observations and models. *Nat. Clim. Chang.* **2012**, *3*, 52–58. [CrossRef]
21. Chen, N.; Gao, X. Climate change in the twenty-first century over China: Projections by an RCM and the driving GCM. *Atmos Ocean. Sci. Lett.* **2019**, *4*, 270–277. [CrossRef]
22. He, H.L.; Wang, S.Q.; Zhang, L.; Wang, J.B.; Ren, X.L.; Zhou, L.; Piao, S.L.; Yan, H.; Ju, W.M.; Gu, F.X.; et al. Altered trends in carbon uptake in China's terrestrial ecosystems under the enhanced summer monsoon and warming hiatus. *Natl. Sci. Rev.* **2019**, *6*, 505–514. [CrossRef]
23. Yu, G.R.; Chen, Z.; Piao, S.L.; Peng, C.H.; Ciais, P.; Wang, Q.F.; Li, X.R.; Zhu, X.J. High carbon dioxide uptake by subtropical forest ecosystems in the East Asian monsoon region. *Proc. Natl. Acad. Sci. USA* **2014**, *111*, 4910–4915. [CrossRef]
24. Li, G.J. Adhere to the ecological priority, green development and advance the Yangtze River Economic Belt of the ecological environment protection work. *Environ. Prot.* **2016**, *44*, 7–13. (In Chinese)
25. Luo, Q.L.; Luo, Y.L.; Zhou, Q.F.; Song, Y. Does China's Yangtze River Economic Belt policy impact on local ecosystem services? *Sci. Total. Environ.* **2019**, *676*, 231–241. [CrossRef] [PubMed]
26. Xu, X.B.; Yang, G.S.; Tan, Y.; Liu, J.P.; Hu, H.Z. Ecosystem services trade-offs and determinants in China's Yangtze River Economic Belt from 2000 to 2015. *Sci. Total. Environ.* **2018**, *634*, 1601–1614. [CrossRef] [PubMed]
27. Ke, J.H.; Piao, S.L.; Fang, J.Y. NPP and its spatiotemporal patterns in the Yangtze River watershed. *Acta Phytoecol. Sin.* **2003**, *27*, 764–770.
28. Zhang, F.Y.; Zhang, Z.X.; Tian, J.X. Forest NPP simulation in the Yangtze River Basin and its response to climate change. *J. Nanjing For. Univ. (Nat. Sci. Ed.)* **2021**, *45*, 175–181. (In Chinese)
29. Wu, N.; Gao, J.X.; Sudebilige, L.U.O.; Li, D.Q. Inter-annual variation trends of vegetation net primary productivity and its responses to climate factors in the upper Yangtze River basin. *Resour. Environ. Yangtze Basin* **2010**, *19*, 296–389. (In Chinese)
30. Miao, Q.; Huang, M.; Li, R.Q. Impacts of climate change on vegetation net primary productivity of the Yangtze River Basin. *J. Nat. Resour.* **2010**, *25*, 1296–1305. (In Chinese)
31. Wu, S.; Liu, L.; Gao, J.; Wang, W. Integrate Risk From Climate Change in China Under Global Warming of 1.5 °C and 2.0 °C. *Earths Future* **2019**, *7*, 1301–1322. [CrossRef]

32. Ministry of Ecology and Environment of the People's Republic of China; National Development and Reform Commission; Ministry of Water Resources. *Environmental Protection Plan for the Yangtze River Economic Belt*; Beijing; 2017. Available online: https://www.mee.gov.cn/gkm/hbb/bwj/201707/t20170118_418053.htm (accessed on 19 January 2017).
33. Huang, M.T.; Piao, S.L.; Sun, Y.; Ciais, P.; Cheng, L.; Mao, J.F.; Poulter, B. Change in terrestrial ecosystem water-use efficiency over the last three decades. *Glob. Chang. Biol.* **2015**, *21*, 2366–2378. [CrossRef] [PubMed]
34. Cao, M.K.; Woodward, I.F. Dynamic responses of terrestrial ecosystem carbon cycling to global climate change. *Nature* **1998**, *393*, 249–252. [CrossRef]
35. Tao, B.; Cao, M.K.; Li, K.R.; Gu, F.F.; Ji, J.J.; Huang, M.; Zhang, L.M. Spatial patterns of terrestrial net ecosystem producticity in China during 1981–2000. *Sci. China Earth Sci.* **2007**, *50*, 745–753. [CrossRef]
36. Gu, F.X.; Zhang, Y.D.; Huang, M.; Tao, B.; Liu, Z.J.; Hao, M.; Guo, R. Climate-driven uncertainties in modeling terrestrial ecosystem net primary productivity in China. *Agric. Forest Meteorol.* **2017**, *246*, 123–132. [CrossRef]
37. Gu, F.; Zhang, Y.; Huang, M.; Yu, L.; Yan, H.M.; Guo, R.; Zhang, L.; Zhong, X.L.; Yan, C.R. Climate-induced increase in terrestrial carbon storage of Yangtze River Economic Belt. *Ecol. Evol.* **2021**, *11*, 7211–7225. [CrossRef] [PubMed]
38. Hutchinson, M.F. *Anusplin Version 4.36 User Guide*; The Austrian National University: Canberra, Australia, 2006; pp. 14–18.
39. Han, Z.; Zhou, B.; Xu, Y.; Wu, J.; Shi, Y. Projected changes in haze pollution potential in China: An ensemble of regional climate model simulations. *Atmos. Chem. Phys.* **2017**, *17*, 10109–10123. [CrossRef]
40. Gao, X.J.; Wu, J.; Shi, Y.; Wu, J.; Han, Z.Y.; Zhang, D.F. Future changes in thermal comfort conditions over China based on multi-RegCM4 simulations. *Atmos. Ocean. Sci. Lett.* **2018**, *11*, 291–299. [CrossRef]
41. Jiang, D.B.; Sui, Y.; Lang, X.M. Timing and associated climate change of a 2 °C global warming. *Int. J. Climatol.* **2016**, *36*, 4512–4522. [CrossRef]
42. Luo, T.X. Observation and Modeling of the Distribution of Biological Productivities of China's Forest. Ph.D. Thesis, Commission for Integrated Survey of National Resources, Chinese Academy of Science, Beijing, China, 1997.
43. Zhu, Z.; Liu, Y.; Liu, Z.; Piao, S. Projection of changes in terrestrial ecosystem net primary productivity under future global warming scenarios based on CMIP5 models. *Clim. Chang. Res.* **2018**, *14*, 31–39. (In Chinese)
44. Zhang, Y.X.; Tao, B.; Yu, L. Assessment on the vulnerability of different ecosystems to extreme rainfalls in the middle and lower reaches of Yangtze River. *Theor. Appl. Climatol.* **2014**, *121*, 157–166.
45. Li, H.W.; Wu, Y.P.; Liu, S.G.; Xiao, J.F. Regional contributions to interannual variability of net primary production and climatic attributions. *Agric. Forest Meteorol.* **2021**, *303*, 1080384. [CrossRef]
46. Piao, S.; Ciais, P.; Lomas, M.; Beer, C.; Liu, H.; Fang, J.; Friedlingstein, P.; Huang, Y.; Muraoka, H.; Son, Y.; et al. Contribution of climate change and rising CO_2 to terrestrial carbon balance in East Asia: A multi-model analysis. *Glob. Planet. Chang.* **2011**, *75*, 133–142. [CrossRef]
47. Rinawati, F.; Stein, K.; Lindner, A. Climate change impacts on biodiversity—The setting of a lingering global crisis. *Diversity* **2013**, *5*, 114–123. [CrossRef]
48. Schimel, D.; Stephens, B.B.; Fisher, J.B. Effect of increasing CO_2 on the terrestrial carbon cycle. *Proc. Natl. Acad. Sci. USA* **2015**, *112*, 436–441. [CrossRef]
49. Lucht, W.G.; Schaphoff, S.; Erbrecht, T.; Heyder, U.; Cramer, W. Terrestrial vegetation redistribution and carbon balance under climate change. *Carbon Balance Manag.* **2006**, *1*, 6. [CrossRef] [PubMed]
50. Habibullah, S.M.; Din, H.B.; Tan, S.H.; Zahid, H. Impact of climate change on biodiversity loss: Global evidence. *Environ. Sci. Pollut. R* **2022**, *29*, 1073–1086. [CrossRef] [PubMed]
51. Liu, J.; Kuang, W.; Zhang, Z.; Xu, X. Spatiotemporal characteristics, patterns, and causes of land-use changes in China since the late 1980s. *J. Geogr. Sci.* **2014**, *24*, 195–210. [CrossRef]
52. Jin, G.; Deng, X.Z.; Zhao, X.D.; Guo, B.S.; Yang, J. Spatiotemporal patterns in urbanization efficiency within the Yangtze River Economic Belt between 2005 and 2014. *J. Geogr. Sci.* **2018**, *28*, 1113–1126. [CrossRef]
53. Yu, Z.L.; Qin, T.L.; Yan, D.M.; Yang, M.J.; Yu, H.X.; Shi, W.L. The impact on the ecosystem services value of the ecological shelter zone reconstruction in the upper reaches basin of the Yangtze River in China. *Int. J. Environ. Res. Public Health* **2018**, *15*, 2273. [CrossRef] [PubMed]
54. Wu, X.; Shen, Z.Y.; Liu, R.M.; Ding, X.W. Land Use/Cover Dynamics in Response to Changes in Environmental and Socio-Political Forces in the Upper Reaches of the Yangtze River, China. *Sensors* **2008**, *8*, 8104–8122. [CrossRef] [PubMed]
55. Jiang, C.; Wu, Z.F.; Cheng, J.; Yu, Q.; Rao, X.Q. Impacts of urbanization on net primary productivity in the Pearl River Delta, China. *Int. J. Plant Prod.* **2015**, *9*, 581–598.
56. Yang, H.F.; Zhong, X.N.; Deng, S.Q.; Xu, H. Assessment of the impact of LUCC on NPP and its influencing factors in the Yangtze River basin, China. *Catena* **2021**, *206*, 105542. [CrossRef]
57. Zhang, C.; Park, T.; Wang, X.; Piao, S.; Xu, B.; Chaturvedi, R.K.; Fuchs, R.; Brovkin, V.; Ciais, P.; Fensholt, R. China and India lead in greening of the world through land-use management. *Nat. Sustain.* **2019**, *2*, 122–129.

diversity

MDPI

Article

Habitat Suitability Assessment of Black-Necked Crane (*Grus nigricollis*) in the Zoige Grassland Wetland Ecological Function Zone on the Eastern Tibetan Plateau

Junjun Bai [1,2], Peng Hou [1,2,*], Diandian Jin [2], Jun Zhai [2], Yutiao Ma [2,3] and Jiajun Zhao [1,2]

1 Chinese Research Academy of Environmental Sciences, Beijing 100012, China; junj_bai@163.com (J.B.); zhaojiajun2345@163.com (J.Z.)
2 Satellite Environment Application Center, Ministry of Ecology and Environment, Beijing 100094, China; jin_diandian@163.com (D.J.); zhaij@lreis.ac.cn (J.Z.); qwertbng@126.com (Y.M.)
3 School of Geomatics and Spatial Information, Shandong University of Science and Technology, Qingdao 266000, China
* Correspondence: houpcy@163.com

Abstract: Habitat suitability assessment is critical for wildlife population conservation and management planning. The MaxEnt model is widely used in species habitat suitability modeling. In order to investigate the habitat status of the black-necked crane in the Zoige grassland wetland ecological function area, this study evaluated the habitat suitability of the black-necked crane using the MaxEnt model with 152 occurrence records and 13 environmental variables. Based on the Akaike information criterion corrected for the small sample size, the best optimal parameter combination (feature class LQPHT, regularization multiplier 3.0) was selected. The results show that the Maxent model had good accuracy with an area under the curve (AUC) value of 0.895. Distance to roads, average summer precipitation, distance to lakes, and altitude are the dominant environmental variables affecting the potential distribution of black-necked cranes, with the contribution rates of 31.6%, 29.8%, 14.3%, and 8.2%, respectively. The suitable habitat area of black-necked cranes is mainly distributed in and around the Zoige Wetland National Nature Reserve, accounting for 46.49% of the Zoige Grassland Wetland National Key Ecological Function Zone. The potential distribution area has a tendency to spread to Hongyuan County in the south of the functional zone, and the unsuitable habitat is mainly distributed in the high-altitude area in the southwest of the functional zone. This study recommends focusing on the distribution area of black-necked cranes around Zoige Wetland National Nature Reserve and Hongyuan County to improve conservation strategies and strengthen protection efforts.

Keywords: Zoige grassland wetland; black-necked crane; species distribution model; MaxEnt; habitat suitability assessment; model optimization

Citation: Bai, J.; Hou, P.; Jin, D.; Zhai, J.; Ma, Y.; Zhao, J. Habitat Suitability Assessment of Black-Necked Crane (*Grus nigricollis*) in the Zoige Grassland Wetland Ecological Function Zone on the Eastern Tibetan Plateau. *Diversity* **2022**, *14*, 579. https://doi.org/10.3390/d14070579

Academic Editor: Corrado Battisti

Received: 23 May 2022
Accepted: 19 July 2022
Published: 20 July 2022

Publisher's Note: MDPI stays neutral with regard to jurisdictional claims in published maps and institutional affiliations.

1. Introduction

Biodiversity refers to the diversity of all living things on earth and is the basis for human survival and development. Furthermore, biodiversity conservation can enhance human well-being. Habitat loss and fragmentation are the major cause of biodiversity loss [1]. Habitats, as places where species can live and reproduce, can provide for individuals, populations, and communities to complete their cycle. Habitat suitability assessment is a critical component of species conservation research and an important indication of habitat quality [2,3]. The assessment of the habitat suitability of wildlife has become a hot issue in the study of the changes in the global species spatial pattern [4–7]. Therefore, understanding the habitat conditions of the studied species, that is, exploring and analyzing the environmental variables affecting the distribution of species and identifying potential geographic distribution areas, is required in order to provide scientific theories for effective species conservation and protected area management planning [8].

Species distribution models are important tools for studying the habitat suitability of species, identifying potential species distribution areas, revealing possible influencing factors, and providing an important scientific basis for biodiversity conservation [9,10]. Currently, there is extensive scientific literature on species distribution models [11]. Based on whether species occurrence records are necessary when the model is created, habitat suitability models are classified into three categories: mechanistic models, statistical models, and niche models [7,10,12–14]. The commonly used model are ecological niche factor analysis (ENFA) [15], random forest (RF) [16], Maximum Entropy Models (MaxEnt) [9,17], generalized linear model (GLM) [18], generalized additive model (GAM) [18], and artificial neural networks (ANN) [19]. Among them, the MaxEnt model relies on species occurrence records and environmental variables and has broad applicability, allowing for less bias and more accurate results [20–22]. At the same time, it can still obtain better results when compared with other species distribution models and is widely used to assess the distribution of wildlife habitats [17,23].

The black-necked crane (*Grus nigricollis*) is listed as a national I-class protected animal by China and a near-threatened species (NT) by the International Union for Conservation of Nature [24,25]. Only black-necked cranes inhabit the plateau, which is mainly found on the Qinghai–Tibet Plateau and the Yunnan–Guizhou Plateau. So far, the total number of black-necked cranes around the globe has reached 10,000–10,200 [26]. The majority of the black-necked cranes overwinter in the low-altitude areas of the Qinghai–Tibet Plateau, the Yunnan–Guizhou Plateau, Bhutan, and southern Tibet, and breed in the Zoige wetland at the northeastern end of the Qinghai–Tibet Plateau [26]. With around 2600 black-necked cranes [27], the Zoige Wetland is the largest swamp wetland on the Eastern Tibetan Plateau and one of the most important nesting places for them [28]. The Zoige Wetland National Nature Reserve was established to protect local biodiversity, while the Zoige Grassland Wetland Ecological Function Zone was established to provide regional ecological security. Although the black-necked crane population is increasing, it is also threatened by the reduction in wetland area during the 21st century, as melting glaciers and permafrost degradation caused by local economic development, as well as future global climate change, may negatively affect shallow wetlands [29,30].

Researchers conducted surveys on the population abundance, distribution [31–33], and migration patterns of black-necked cranes [34,35]. In terms of the behavioral ecology of black-necked cranes, the feeding time during the overwintering period is mainly regulated by humidity indirectly [36,37], and the breeding season is mainly distributed in meadows and marsh meadows [38]. The study of Kong et al. suggested that the impact of predator threat and human disturbance on black-necked cranes should be considered in future tourism management, and a safe distance should be planned reasonably [39]. Human disturbance, food, and water conditions are the key environmental variables impacting the habitat quality of black-necked cranes in the Napa Sea wetland [40], according to studies on their habitat choices. Furthermore, land-use change influences the feeding and nocturnal habitat selection of black-necked cranes [41], and precipitation is another key factor impacting their habitat [42]. The black-necked cranes distributed in the Zoige Wetland are mainly influenced by altitude and autumn climate [43]. The distance from the cultivated land, the distance from the water, and the dominant vegetation are the main environmental factors affecting the distribution of black-necked cranes in the Caohai National Nature Reserve, Guizhou, China [44].

Previous studies have mostly focused on the migration routes, population changes, behavioral ecology, and habitat quality of black-necked cranes. At the spatial scale, the focus has been on the global distribution of black-necked cranes, especially in China, but most studies at the regional level have focused on the distribution, influencing factors, and the conservation status of the wintering areas in Yunnan and Guizhou, with insufficient attention to the breeding sites in Zoige, Sichuan [31–44]. The Zoige Wetland is the largest breeding place for black-necked cranes, and it is a key element of their life cycle. To conserve species and ecosystems, nature reserves are defined and zoned. Therefore, further research on the distribution and habitat of black-necked cranes in breeding grounds is needed to narrow the gap with actual local conservation actions. Thus, in this study, we selected the Zoige Grassland Wetland Ecological Function Area, which is located in the core area of the Zoige wetland. The optimized maximum entropy model was used to predict the distribution of the black-necked cranes, analyze the main environmental factors affecting the distribution of black-necked cranes and their habitat distribution characteristics, and provide a scientific basis for the formulation of future measures for the efficient conservation and management of black-necked cranes.

2. Materials and Methods

2.1. Study Area

Zoige Grassland Wetland National Key Ecological Function Areas (Figure 1) (31°51′—34°18′ N, 101°6′—103°38′ N), located in the center of the Zoige Wetland, the largest marsh wetland on the Qinghai–Tibet Plateau, is an important part of the conservation land system of the Qinghai–Tibet Plateau. Meanwhile, the area is one of the world's most important alpine marsh wetlands, with a unique role in global climate change and regional ecological security [28,45]. There are about 2600 black-necked cranes in the entire Zoige wetland, which has the largest breeding population of black-necked cranes in the world. The region's unique geological, climatic, and hydrological natural conditions provide a favorable environment for black-necked crane survival and reproduction. The functional zone is at an altitude of 2442–4921 m, including Aba County, Zoige County, and Hongyuan County, with a total area of about 28,500 km^2. It is located in the watershed of the Yellow River and Yangtze River system, with abundant wetland peat resources, which play an important role in water conservation, hydrological regulation, and biodiversity maintenance of the Yellow River basin [27]. The National Main Functional Zone Plan, which defined 25 national key ecological functional zones, was promulgated and implemented in China in 2010 [46]. Among them, the Zoige Grassland Wetland National Key Ecological Function Zone is an important water conservation type zone in China, serving as a demonstration environment for people living in harmony with nature. The Zoige Wetland has an annual average temperature of 0.7–1.1 °C, with January temperatures of −10.5–7.9 °C, July temperatures of 10.9–11.4 °C, and annual average precipitation of 650–750 mm [47]. The Zoige wetland primarily protects rare wild species such as black-necked cranes, white storks (*Ciconia ciconia*), and the plateau swamp wetland habitat [48].

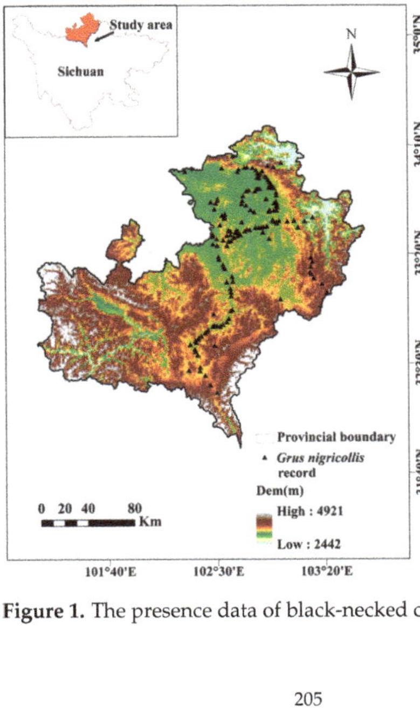

Figure 1. The presence data of black-necked crane.

2.2. Collection and Processing of Sample Data

Species distribution data were obtained from the published literature [49], field survey sites provided by Li et al. [50], and downloaded from the Global Species Diversity Information Database (http://www.gbif.org, accessed on 1 April 2022) and China Citizen Science Platform (http://www.birdreport.cn, accessed on 5 March 2022) [51,52], with record sites prior to 2010 removed. To eliminate duplicate, geographically and temporally unclear black-necked crane occurrence records and prevent covariance in environmental variables from affecting the model's accuracy, the SDM Toolbox was used, the GIS toolkit operating based on Python, version 2.5, developed by the Brown Lab et al., in Durham, America. Spatial filtering using the tool to ensure that only one point per grid (1 km $\times$ 1 km) was retained, resulting in 152 occurrence records (Figure 1, Table S1).

2.3. Screening and Determination of Environment Variables

Wildlife habitats must have the ability to provide their habitat, water, and food, so the distribution of wildlife is subject to a combination of climatic factors and human activities. Based on the principles of habitat suitability evaluation, combined with the previous studies analyzed above, the characteristics of the study area and the behavioral and ecological characteristics of black-necked cranes [36–38,40–44], the paper identified four major categories of factors related to the distribution of suitable habitats for black-necked cranes: climatic conditions, geomorphic types, foraging conditions, and human activities. First of all, in terms of climatic conditions, black-necked cranes return from migration in March and prefer to nest and breed near wetland marshes or in the center of shallow lakes, followed by a vital period of growth and development from June to August. The spring and summer precipitation resources are sufficient for vegetation growth as well as fish and shrimp spawning, which provides abundant food for the black-necked cranes. Considering the changes in behavioral habits of black-necked cranes in different seasons, factors such as average temperature and average precipitation from 2015 to 2020 were selected (bio1–bio10). The data were obtained from the Chinese 1 km resolution monthly precipitation dataset (1901–2020) of the National Tibetan Plateau Science Data Center [53]. Secondly, the geomorphic types (bio11–bio13) include elevation, slope, and aspect, and the data were downloaded from the Resource and Environment Science and Data Center (https://www.resdc.cn/, accessed on 4 March 2022). Thirdly, the foraging conditions (bio14–bio15, bio19) included the normalized difference vegetation index (NDVI), distance to rivers, and distance to lakes. The former data were obtained from the 2015–2020 NASA MODIS product data MODIS09A1 (http://ladsweb.nascom.nasa.gov/, accessed on 3 June 2021), and the latter two data were obtained from the China National Catalogue Service For Geographic Information (http://www.webmap.cn/, accessed on 22 January 2022). Finally, human activities (bio16–bio18) include the distance from roads, the distance from settlements, and land-use types. The first two data come from the National Geographic Information Resource Catalog Service System (http://www.webmap.cn/, accessed on 22 January 2022), and the latter data come from the Satellite Environment Application Center of the Ministry of Ecology and Environment of China. The environment variable details and sources are in Table 1.

Table 1. Variables used for modeling.

Code	Environmental Variable	Source
Bio1	Average spring precipitation	
Bio2	Average summer precipitation	
Bio3	Average autumn precipitation	
Bio4	Average winter precipitation	
Bio5	Average precipitation	
Bio6	Average spring temperature	
Bio7	Average summer temperature	
Bio8	Average autumn temperature	

Table 1. *Cont.*

Code	Environmental Variable	Source
Bio9	Average winter temperature	
Bio10	Average temperature	http://data.tpdc.ac.cn/, accessed on 2 March 2022 (Bio1-Bio10)
Bio11	Aspect (°)	
Bio12	Altitude (m)	
Bio13	Slope (°)	https://www.resdc.cn/, accessed on 4 March 2022 (Bio11-Bio13)
Bio14	Distance to lakes (m)	
Bio15	Distance to rivers (m)	
Bio16	Distance to roads (m)	
Bio17	Distance to settlements (m)	http://www.webmap.cn/, accessed on 22 January 2022 (Bio14-Bio17)
Bio18	Land use	http://www.secmep.cn/, accessed on 11 August 2021
Bio19	Normalized difference vegetation index	http://ladsweb.nascom.nasa.gov/, accessed on 3 June 2021

Applying all of the environmental factors to the model modeling would result in overfitting due to the possible correlation between them. Therefore, the study used the ENMTools, version 1.0.6, developed by Warren et al., an R package for correlation analysis of each environmental factor, which does not depend on the distribution data and is able to obtain reliable results [54]. Figure 2 shows the environmental factor correlation heat map after processing the correlation plot with the corrplot, an R package, version 0.92, developed by Wei et al.; the minor environmental variables with $|R| \geq 0.9$ between the two environmental factors were excluded. Finally, the Jackknife method was used to screen again to remove the environmental factors with zero contribution rate, and only 13 environmental factors (bio1–bio4, bio10–bio16, bio18–bio19) were retained.

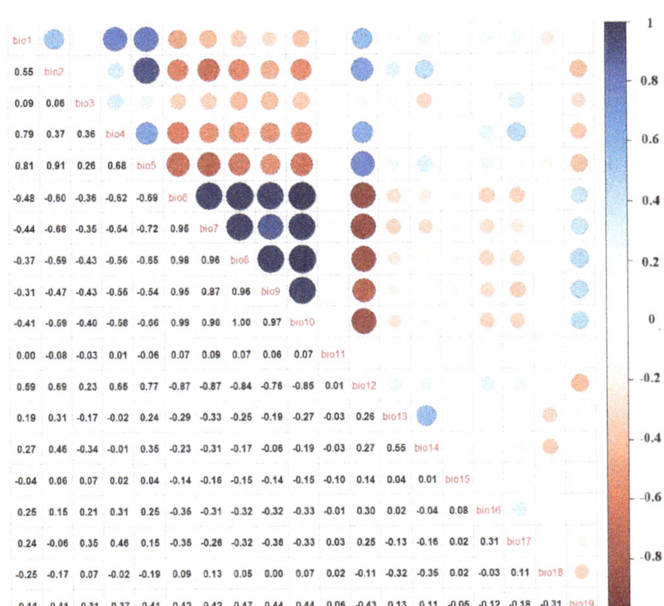

Figure 2. Heat map for correlation analysis of environmental factors. The darker the blue and red circles, the greater the correlation between the two environmental factors.

2.4. MaxEnt Model Optimization and Selection

The MaxEnt model has strong applicability and can use the area enclosed by the receiver operating characteristic curve (ROC) and the abscissa, that is, the AUC value, to evaluate the model prediction accuracy [9,20–22]. Moreover, the MaxEnt model's predictions based on default parameters are conservative, and the complexity of MaxEnt can be changed by altering the model parameter settings to predict the potential distribution of species more reasonably [7,55]. The study utilized the Enmeval, version 2.0.3, developed by Muscarella et al., an R package in R (v4.0.5) to optimize the MaxEnt model [56], which contains feature combination (FC) and regularization multiplier (RM). The feature class included Linear features (L), Quadratic features (Q), Product features (P), Hinge features (H), and Threshold features (T). The regularization multipliers were set to 1–4 with a 1 interval each, and the six feature combinations offered by the MaxEnt model (L, H, LQ, LQH, LQHP, and LQHPT) were merged to generate 24 combinations. The Akaike information criterion corrected for small sample size (AICc) was used as an indicator to determine the RM and FC of the model [57].

The black-necked crane occurrence records and the above 13 environmental factors were imported into the MaxEnt model, and the other settings were as follows: RM and FC values under the optimal parameters were input, 10-fold cross-validation was selected, the number of repetitions was 10, and the Jackknife method was chosen to test the importance of each environmental factor, and the output results were Logistic format. The model prediction results were examined using the area AUC under the ROC curve, and the value of AUC was taken in the range of 0–1, and the closer the value was to 1, the higher the model prediction accuracy. The AUC values are 0.5–0.6, unqualified; 0.6–0.7, poor; 0.7–0.8, fair; 0.8–0.9, good; and 0.9–1.0, excellent.

2.5. Habitat Suitability Classification of Black-Necked Cranes

Species predictive distribution maps show species preferences for habitat as probabilities (0–1), with the closer the value to 1, the more suitable the species distribution. The selection of thresholds generally follows three principles: objectivity, equivalence, and discriminative power [58]. The threshold is generally determined based on the omission error or based on the sensitivity and specificity of the prediction results. The former does not consider commission error, while the latter comprehensively considers omission error and commission error. The model maximum training sensitivity plus specificity (MTSS) belongs to the latter and satisfies the three principles of threshold selection [28]. MTSS and balance training omission and predicted area and threshold value (TPT) were selected as classification thresholds for suitable and low suitable habitats, respectively, to reclassify the MaxEnt model outputs into unsuitable, low suitable, moderately suitable, and highly suitable habitats [58–60]. Finally, the Reclassify tool of ArcGIS software, version 10.8, developed by Environmental Systems Research Institute, in RedLands, America, was used to count and calculate the area of the corresponding distribution area for each class.

3. Results and Analysis

3.1. MaxEnt Optimal Model and Accuracy Evaluation

Based on 152 occurrence records and 13 environmental factors, this study used the Enmeval package to invoke MaxEnt to predict the potential distribution area of black-necked cranes. The model with the lowest AICc value (i.e., ΔAICc = 0) is considered the best model out of the current suite of models [55,56]. When the model was the default parameter, FC = LQHPT, RM = 1, ΔAICc = 296.00, and when the model parameter was set to FC = LQHPT, RM = 4, ΔAICc = 0 (Table 2), the AICc value was the smallest and the model with this parameter was the optimal model. Figure 3 shows the results of comparing different parameters in the model. When the model parameters are set to FC = LQHPT, RM = 4, compared with the default parameters, Mean.AUC is nearly the same (decreased by 4.00%), but the difference between the AUC Values (Auc.diff.avg) decreased by 45.56% and 10%, and the training omission rate (OR10) decreased by 60.07%, the latter two are lower than the default values, indicating that the optimized model reduced overfitting, so FC = LQHPT, RM = 4 was set as the modeling parameter. The model was reconstructed using the optimized parameters to simulate the suitable area for black-necked cranes, and the model was repeated 10 times, obtaining a mean value of 0.895 for the test AUC (Figure 4), indicating that the prediction accuracy of the MaxEnt model reached a good level.

Table 2. Evaluation metrics of MaxEnt model generated by Enmeval.

Type	Feature Combination	Regularization Multiplier	ΔAICc	Avg.diff.avg
Default	LQPHT	1	296.00	0.0413
Optimized	LQPHT	4	0	0.0225

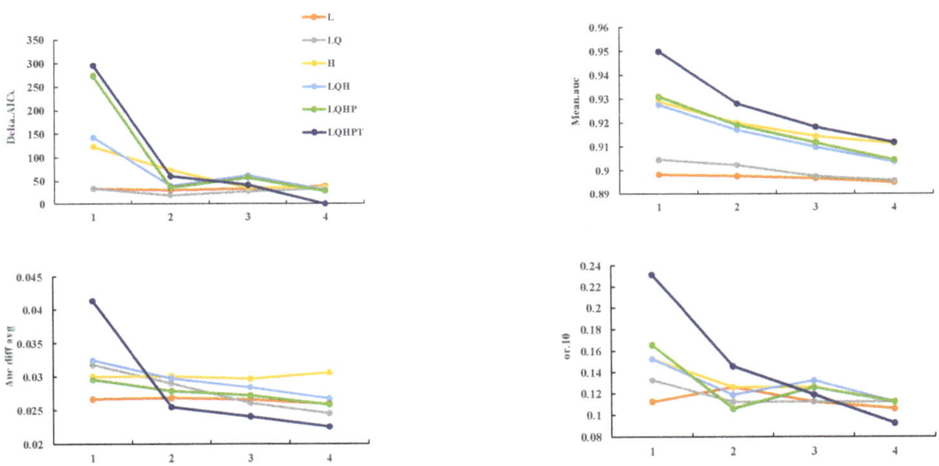

Figure 3. Performances of the maximum entropy model under different settings.

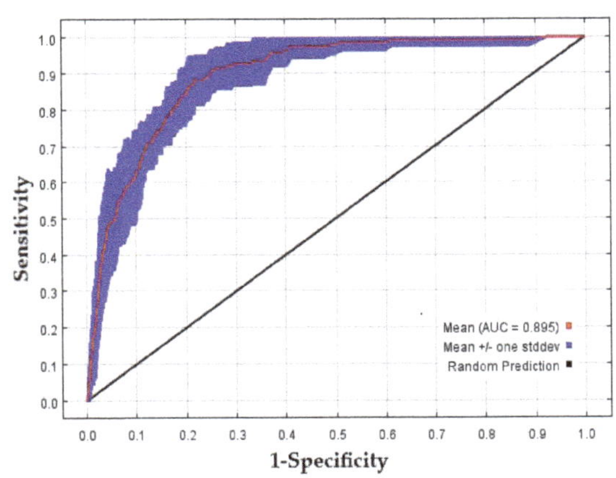

Figure 4. Performances of MaxEnt model under different settings.

3.2. The Influence of Environmental Factors on the Distribution of Black-Necked Cranes

The research used the Jackknife method to analyze the importance of 13 environmental factors affecting the habitat selection of black-necked cranes (Figure 5). The results of the contribution of environmental variables showed (Table 3) that distance to roads, average summer precipitation, distance to lakes, and elevation may be the major environmental factors affecting black-necked cranes, where the contribution rates were 31.6%, 29.8%, 14.3%, and 8.2%, respectively, with a cumulative contribution rate of 83.9%; the secondary variables affecting the distribution of black-necked cranes were average summer precipitation, slope direction, NDVI, and the contribution rates were 5.6%, 2.5%, 2.3%, and 1.9%, respectively; and the percentages of other environmental factors were around 1%, indicating that the influence on the habitat suitability of black-necked cranes was small.

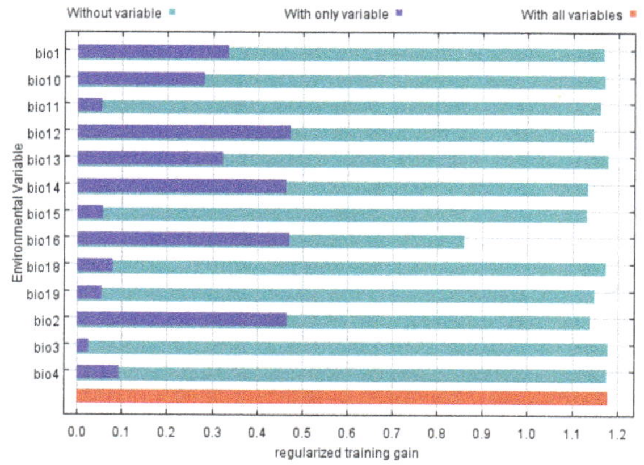

Figure 5. The importance of environmental variables evaluated by Jackknife testing.

Table 3. Contributions of the environmental variables to the MaxEnt model.

Variable	Contribution/%	Variable	Contribution/%
Distance to roads (m)	31.6	Distance to rivers (m)	1.9
Average summer precipitation	29.8	Average winter precipitation	1.1
Distance to lakes (m)	14.3	Slope (°)	1.1
Altitude (m)	8.2	Average autumn precipitation	0.8
Average spring precipitation	5.6	Land use	0.4
Aspect (°)	2.5	Average temperature	0.3
Normalized difference vegetation index	2.3		

3.3. *Habitat Suitability Distribution of Black-Necked Cranes in Zoige Grassland Wetland Ecological Function Zone*

In the MaxEnt model results, MTSS = 0.304 and TPT = 0.0623, so the thresholds for classifying more suitable and less suitable habitats for black-necked cranes are 0.304 and 0.0623, that is. 1–0.5 is the highly suitable habitat, 0.5–0.304 is the moderately suitable habitat, 0.304–0.0623 is the low suitable habitat, and 0.0623–0 is an unsuitable habitat. To obtain the suitable habitat distribution map of black-necked cranes in the functional zone (Figure 6), the above thresholds were applied to reclassify the model outputs into different habitat classes, and the area of each suitable distribution area was calculated separately. The statistics reveal that the highly suitable habitat for black-necked cranes in the Zoige grassland wetland ecological function zone is about 2356.17 km^2, accounting for 8.27% of the total function area, mainly in the Zoige National Nature Reserve in the northern part of the function zone. Low suitable habitat covers approximately 7899.43 km^2, accounting for 27.72% of the total functional zone, mainly in Zoige country and Hongyuan country; unsuitable habitat covers about 0.98 km^2, accounting for 53.51% of the total functional zone, mainly in Aba County, a high-altitude area in the southwest. Moreover, the study indicated that the highly suitable habitat for black-necked cranes in the Zoige National Nature Reserve was nearly 668.70 km^2, accounting for 39.22% of the whole nature reserve area and 28.38% of the highly suitable habitat area in the total functional zone. The highly suitable habitat is mainly in the Zoige National Nature Reserve, which is primarily composed of marshes, wetlands, grasslands, lakes, and rivers. The reserve has abundant summer rainfall and is covered with small lakes, creating good foraging conditions for black-necked cranes. Black-necked cranes prefer to be distributed close to water sources, which can provide good nesting conditions. The moderately suitable habitat was approximately 475.04 km^2, accounting for 27.86% of the whole nature reserve area and 15.87% of the moderately suitable habitat area in the total functional zone. The moderately suitable habitat was distributed in and around the nature reserve area, where there are undulating hills that cause differences in precipitation and temperature. All suitable habitats for black-necked cranes took up 97.48% of the whole nature reserve area and 46.49% of the total functional zone. In contrast to the nature reserve, the land use types of Aba County are mostly woodland and grassland in the southwest of the functional zone, which does not meet the demands of black-necked cranes for foraging and breeding. The spatial distribution of environmental variables affecting the distribution of black-necked cranes is more concentrated in the nature reserve but is not consistent in the entire functional zone. Therefore, the suitable distribution area of black-necked cranes is largely distributed in the nature reserve area.

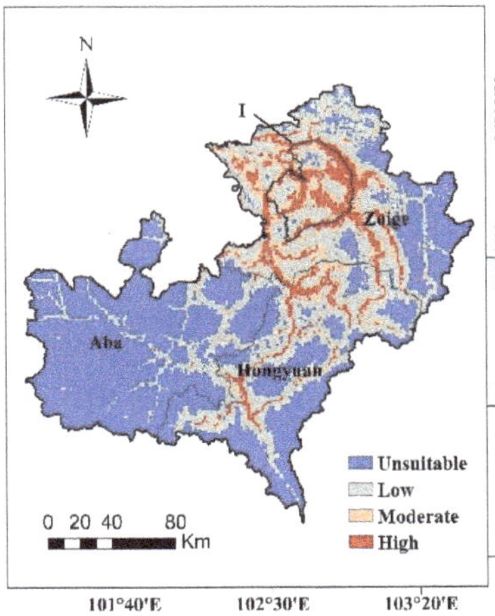

Figure 6. Map of the habitat suitability of black-necked crane in the Zoige Grassland Wetland National Key Ecological Function Zone (I. Zoige Wetland National Nature Reserve).

4. Discussion

MaxEnt model optimization generally includes correction of sampling bias, selection of environmental factors closely related to species distribution, and the optimization of model parameters [55,59,61–63]. Firstly, for the correction of sampling bias, the spatial filtering method used in this research ensures only one occurrence record in every 1 km × 1 km grid to reduce the overfitting phenomenon existing in model modeling [22], but this method may also overlook the ecological value of high-density areas of species distribution.

Second, habitat selection determinants for black-necked cranes may differ by region; thus, we recommend choosing environmental factors that are closely related to the species' distribution. Previous studies on the habitat selection of black-necked cranes show that human disturbance, food, and water conditions are the significant environmental variables determining the habitat quality of black-necked cranes [40]. Black-necked cranes distributed in the Zoige wetlands are mainly affected by altitude and autumn climate [43]. Distance to cultivated land, distance to water, and dominant vegetation are the impacting environmental factors affecting the distribution of black-necked cranes in the Caohai National Nature Reserve [44]. As a result, the research on black-necked crane habitat selection should be adapted to the study area's conservation situation and needs. After correlation and Jackknife analysis, only 13 environmental components were maintained in this study to reduce model overfitting. The results of the MaxEnt model prediction showed that the habitat distribution of black-necked cranes was mainly influenced by distance to roads, summer precipitation, distance to lakes, and elevation. The distance to roads is closely related to the distribution of black-necked cranes, and by observing the occurrence records of black-necked cranes, their collection routes may be mostly distributed along the road, resulting in a strong contribution of the distance to roads factor, so the prediction results may be biased. The distribution of black-necked cranes is also mainly affected by spring precipitation, summer precipitation, and the distance to lakes, indicating that water and food are considered to be the main environmental factors, which is consistent with the results of previous investigations [40,42]. Black-necked cranes return from migration in March, preferring to nest and breed near wetlands and swamps or in the center of shallow lakes. After that, June-August is a critical period for the growth and development of black-necked cranes. The precipitation resources in spring and summer just provide sufficient conditions for vegetation growth and fish and shrimp reproduction and provide a rich food source for black-necked cranes [64]. However, the distance to cultivated land was discovered to be the primary determinant for the distribution of black-necked cranes in Dashanbao, Yunnan, and Caohai, Guizhou, China, because it makes up a significant portion of these research areas and offers more food than other environmental factors [41,44]. The majority of the swamp and lakes in our study region can offer the optimal environment for black-necked cranes to breed. From previous studies on the habitat of black-necked cranes, it can also be found that for other similar waterfowl, such as red-crowned cranes, Cao et al. used the MaxEnt model to reveal that the distance to roads, fishponds, and smooth cordgrass (*Spartina alterniflora*), and the distance to reed shoals and seepweed shoals, as well, were the main factors that influenced the selection of a wintering habitat by red-crowned cranes in the Yancheng Nature Reserve [65]. Na et al. found that habitat composition, water depth, and distance to roads and ditches were the most important habitat features for Red-crowned cranes in the Zhalong National Nature Reserve during the breeding season [66]. This also demonstrates that water and food are considered to be the main environmental factors in the distribution of waterfowl species. Black-necked cranes are the only cranes that live on plateaus among the 15 species of cranes in the world, mainly distributed in the Qinghai–Tibet Plateau and the Yunnan–Guizhou Plateau [26]. Therefore, compared with other cranes, Black-necked cranes have evolved good physiological adaptation characteristics and perfectly adapted to the living environment of high altitude and low temperature, which can explain that altitude is the main environmental factor for black-necked cranes distribution. In reality, the Zoige National Nature Reserve, the core area of the high suitable distribution of black-necked cranes, is strictly controlled, while the functional zone outside the nature reserve is a bit weaker and more vulnerable to human activities. For a more targeted analysis of the suitable distribution of black-necked cranes, it might be useful to further distinguish in detail the environmental variables affecting the distribution of black-necked cranes in the nature reserve and in the functional zone outside of the nature reserve, i.e., to select environmental factors separately for the nature reserve and functional zone, and then to superimpose these two results. In the future, we also need to supplement black-necked crane occurrence records to reduce data sampling bias and simulate the real black-necked crane distribution to provide a more scientific and theoretical basis for the nature reserve and functional zone planning and management.

Finally, in order to make the model findings more ecologically interpretable, a balance of model complexity is required for optimizing the model parameters [22]. Adjusting model complexity through FC and RM settings is a hot area in MaxEnt model research. With different sample sizes and feature combinations, the results of FC selection will be reflected in response curve plotting, making simple linear correlations or complex nonlinear correlations between environmental variables and distribution, which can have an impact on model fitting and prediction. Generally, simple models are simpler to understand ecologically, but if too few features are selected, such as selecting only the L function, as the sample size increases, the sampling bias will increase, resulting in a lower AUC value [67]. It is also argued that FC has little effect on the predictive ability of the model and that a complex model will only slightly increase the AUC value [63]. The setting of RM is in order to balance the model fitting degree and extrapolation ability. When the RM value is set too low, the model is more likely to overfit and raise the omission error; when the RM value is set too high, the model becomes smooth, increases the misjudgment error, and loses its ability to discriminate in unsuitable areas. In this paper, the AUC values after model optimization are approximate to the default case, but the response curves obtained for some environmental factors are not particularly flat. Moreover, the increase in RM expands the error boundary range compared to the default setting. In practice, the selection of both FC and RM needs further judgment, and more studies are needed to show the relationship between the model parameter optimization results, sample size, and study subjects.

5. Conclusions

This study evaluated the habitat suitability of black-necked cranes in the Zoige grassland wetland ecological function area based on the optimized MaxEnt model and investigated the key environmental factors and suitable ranges affecting their distribution. The complexity and overfitting of the optimized model were minimized compared to the default parameters by optimizing two MaxEnt model parameters: FC and RM, establishing 24 combinations, and utilizing Akaike's small sample corrected information criterion (AICc) as an indication. The AUC value was 0.895, and the prediction results reached a good level. MaxEnt model prediction results show that the habitat distribution of black-necked cranes may be mainly influenced by distance to roads, summer precipitation, distance to lakes, and elevation. The suitable habitat area for black-necked cranes accounts for 46.49% of the total functional zone; with the suitable distribution area for black-necked cranes mainly located in the Zoige National Nature Reserve in the functional zone's north, the potential distribution area tends to spread to Hongyuan County in the functional zone's south, and the unsuitable habitat is mainly distributed in the high-altitude area in the southwest of the functional zone. The core distribution area is located in Zoige National Nature Reserve, so the delineation of the nature reserve can better protect black-necked cranes to some extent. The results of this study indicate that the summer precipitation and the spatial distribution of water resources have a major impact on black-necked crane distribution. It is recommended to strengthen the management of river and lake shorelines, protect the water ecological environment, actively communicate with local residents to raise their awareness of protection, and reasonably formulate the summer grazing intensity standards for grasslands to ensure the energy required by the black-necked cranes during the breeding and growth periods. In view of the impact of future climate change on wetland ecosystems and biodiversity, it is also necessary to strengthen the assessment and construction of protected area management capacity and to design evaluation indicators and management effectiveness assessment tools [68]. It is suggested to regularly monitor the population size of black-necked cranes and other wildlife, changes in the wetland area and human disturbance, assess changes in habitat quality of black-necked cranes, and adjust management activities in time to achieve conservation goals. At the same time, the contradiction between protection and development is still prominent, so the study recommends that attention be paid to the suitable distribution range of black-necked cranes on the vulnerable edge of the Zoige National Nature Reserve and Hongyuan County, as well as the impact of human activities on the distribution of black-necked cranes in these areas.

Supplementary Materials: The following supporting information can be downloaded at: https://www.mdpi.com/article/10.3390/d14070579/s1, Table S1: Occurrence records used for MaxEnt Model of black-necked crane in Zoige Grassland Wetland Ecological Function Zone.

Author Contributions: P.H., D.J., J.Z. (Jun Zhai), Y.M., J.Z. (Jiajun Zhao) and J.B. all contributed to the data analyses; J.B. performed analyses and led the writing; P.H. assisted with collect and check the data, and provided valuable comments in the paper writing. All authors have read and agreed to the published version of the manuscript.

Funding: This research was funded by National Key R&D Program of China (grant number 2021YFF0703903) and Major Projects of High-Resolution Earth Observation Systems of National Science and Technology (grant number 05-Y30B01-9001-19/20-4).

Institutional Review Board Statement: Not applicable.

Data Availability Statement: All occurrence records and environmental variables used in the manuscript are already publicly accessible, and we provided the download address in the manuscript.

Acknowledgments: Thanks to the Zoige National Nature Reserve Administration of Sichuan Province for providing data support.

Conflicts of Interest: The authors declare no conflict of interest.

References

1. Pereira, H.M.; Navarro, L.M.; Martins, I.S. Global Biodiversity Change: The Bad, the Good, and the Unknown. *Annu. Rev. Environ. Resour.* **2012**, *37*, 25–50. [CrossRef]
2. Grinnell, J. The niche-relationships of the California Thrasher. *Auk* **1917**, *34*, 427–433. [CrossRef]
3. OuYang, Z.Y.; Liu, J.G.; Xiao, H.; Tan, Y.C.; Zhang, H.M. An assessment of giant panda habitat in Wolong Nature Reserve. *Acta Ecol. Sin.* **2001**, *21*, 1869–1874.
4. Gurnell, J.; Clark, M.J.; Lurz, P.W.; Shirley, M.D.; Rushton, S.P. Conserving red squirrels (*Sciurus vulgaris*): Mapping and forecasting habitat suitability using a Geographic Information Systems Approach. *Biol. Conserv.* **2002**, *105*, 53–64. [CrossRef]
5. Peterson, A.T. Uses and requirements of ecological niche models and related distributional models. *Biodivers. Inform.* **2006**, *3*, 59–72. [CrossRef]
6. Hooper, D.U.; Adair, E.C.; Cardinale, B.J.; Byrnes, J.E.K.; Hungate, B.A.; Matulich, K.L.; Gonzalez, A.; Duffy, J.E.; Gamfeldt, L.; O'Connor, M.I. A global synthesis reveals biodiversity loss as a major driver of ecosystem change. *Nature* **2012**, *486*, 105–129. [CrossRef] [PubMed]
7. Qiao, H.J.; Hu, H.J.; Huang, J.H. Theoretical Basis, Future Directions, and Challenges for Ecological Niche Models. *Sci. Sin. Vitae* **2013**, *43*, 915–927. [CrossRef]
8. Hou, P.; Gao, J.X.; Wan, H.W.; Shi, P.R.; Wang, Y.C.; Sun, C.X. Progress and some scientific issues on effectiveness assessment of terrestrial ecosystem conservation and restoration. *Environ. Ecol.* **2021**, *3*, 1–7.
9. Phillips, S.J.; Anderson, R.P.; Schapire, R.E. Maximum entropy modeling of species geographic distributions. *Ecol. Model.* **2006**, *190*, 231–259. [CrossRef]
10. Elith, J.; Leathwick, J.R. Species Distribution Models: Ecological Explanation and Prediction Across Space and Time. *Annu. Rev. Ecol. Evol. Syst.* **2009**, *40*, 677–697. [CrossRef]
11. Norberg, A.; Abrego, N.; Blanchet, F.G.; Adler, F.R.; Anderson, B.J.; Anttila, J.; Araujo, M.B.; Dallas, T.; Dunson, D.; Elith, J.; et al. A comprehensive evaluation of predictive performance of 33 species distribution models at species and community levels. *Ecol. Monogr.* **2019**, *89*, e01370. [CrossRef]
12. Liu, C.; White, M.; Newell, G. Measuring and comparing the accuracy of species distribution models with presence-absence data. *Ecography* **2010**, *34*, 232–243. [CrossRef]
13. Liu, F.; Li, S.; Li, D.Q. The Review of Methods for Mapping Species Spatial Distribution Using Presence/Absence Data. *Acta Ecol. Sin.* **2013**, *33*, 7047–7057. [CrossRef]
14. Bai, J.J.; Hou, P.; Zhao, Y.H.; Xu, H.T.; Zhang, B. The main research progress of species habitat suitability models and verification. *Chin. J. Ecol.* **2022**, *41*, 1423–1432. [CrossRef]
15. Hirzel, A.H.; Hausser, J.; Chessel, D.; Perrin, N. Ecological-niche factor analysis: How to compute habitat-suitability maps without absence data? *Ecology* **2002**, *83*, 2027–2036. [CrossRef]
16. Breiman, L. Random forests. *Mach. Learn.* **2001**, *45*, 5–32. [CrossRef]
17. Elith, J.; Graham, C.H.; Anderson, R.P.; Dudík, M.; Ferrier, S.; Guisan, A.; Hijmans, R.J.; Huettmann, F.; Leathwick, J.R.; Lehmann, A. Novel methods improve prediction of species' distributions from occurrence data. *Ecography* **2006**, *29*, 129–151. [CrossRef]
18. Guisan, A.; Edwards, T.C.; Hastie, T. Generalized linear and generalized additive models in studies of species distributions: Setting the scene. *Ecol. Model.* **2002**, *157*, 89–100. [CrossRef]
19. Olden, J.D.; Jackson, D.A. Illuminating the "black box": A randomization approach for understanding variable contributions in artificial neural networks. *Ecol. Model.* **2002**, *154*, 135–150. [CrossRef]
20. Phillips, S.J.; Dudík, M. Modeling of species distributions with Maxent: New extensions and a comprehensive evaluation. *Ecography* **2008**, *31*, 161–175. [CrossRef]
21. Merow, C.; Smith, M.J.; Silander, J.A., Jr. A practical guide to MaxEnt for modeling species' distributions: What it does, and why inputs and settings matter. *Ecography* **2013**, *36*, 1058–1069. [CrossRef]
22. Radosavljevic, A.; Anderson, R.P. Making better MAXENT models of species distributions: Complexity, overfitting and evaluation. *J. Biogeogr.* **2014**, *41*, 629–643. [CrossRef]
23. Hernandez, P.A.; Graham, C.H.; Master, L.L.; Albert, D.L. The effect of sample size and species characteristics on performance of different species distribution modeling methods. *Ecography* **2006**, *29*, 773–785. [CrossRef]

24. Wang, S.; Zheng, G.M.; Wang, Q.S. *China Red Data Book of Endangered Animals*; Science Press: Beijing, China, 1998.
25. BirdLife International. *Grus nigricollis*. The IUCN Red List of Threatened Species 2020: e.T22692162A180030167. 2020. Available online: https://dx.doi.org/10.2305/IUCN.UK.2020-3.RLTS.T22692162A180030167.en (accessed on 1 April 2022).
26. Li, F.S. IUCN black-necked crane (*Grus nigricollis*) conservation plan. *Zool. Res.* **2014**, *35*, 7. [CrossRef]
27. Liu, Q.; Wang, K.; Li, F.S.; Kong, D.J.; Yang, X.J. Numbers and distribution of Black-necked Cranes (*Grus nigricollis*) at Ruoergai Wetland on the Eastern Qinghai-Tibet Plateau. *Zool. Res.* **2014**, *35*, 134. [CrossRef]
28. Chai, Y. *The Swamps of the Zoige Plateau*; Science Press: Beijing, China, 1965.
29. Qiu, J. Thawing permafrost reduces river runoff. *Nature* **2012**, 6. [CrossRef]
30. Li, F. Species review: Black-necked Crane. In *Crane Conservation Strategy*; Mirande, C.M., Harris, J.T., Eds.; International Crane Foundation: Baraboo, WI, USA, 2002.
31. Meine, C.; Archibald, G. *The Cranes: Status Survey and Conservation Action Plan*; World Conservation Union: Gland, Switzerland; Cambridge, UK,, 1996.
32. Dou, L.; Li, H.; Li, F.S.; Zhang, M.; Zheng, Z.R.; Ran, H.J. Survey on the black-necked cranes during the breeding jperiod at Sichuan Ruoergai Wetland National Nature Reserve. *Sichuan J. Zool.* **2013**, *32*, 770–773.
33. Mi, C.R.; Guo, M.Y.; Falk, H.; Han, X.S. Species distribution model sampling contributes to the identification of target species: Take black-necked crane and hooded crane as two cases the modelbased sampling approach could help to reduce areas to be investigated and it can find target species more effectively re. cost and effort. *Acta Ecol. Sin.* **2017**, *037*, 4476–4482. [CrossRef]
34. Wu, Z.K.; Li, Z.M.; Wang, Y.H.; Jiang, Y.M.; Li, R.X.; Li, D.H.; Zhou, Z.J.; Li, L.X. Migration of black-necked crane in China. *Curr. Zool.* **1993**, *39*, 105–106.
35. Jiang, Z.Q.; Li, F.S.; Ran, J.H.; Tang, T.; Zhang, B.; Wang, B.; Li, H. Survey on the population and breeding conditions of black-necked crane (*Grus nigricollis*) in Ruoergai international important wetland. In Proceedings of the Abstract Collection of Papers of the 3rd Western China Zoology Symposium, Xining, China, 3 July 2014.
36. Kong, D.J.; Yang, X.Y.; Zhong, X.Y.; Dao, M.B.; Zhu, Y. Diurnal time budget and behavior rhythm of wintering black-necked crane (*Grus nigricollis*) at Dashanbao in Yunnan. *Zool. Res.* **2008**, *29*, 195–202. [CrossRef]
37. Zhang, Y.N. *How Wintering Habitat Structure Affects Number and Distribution of Black-Necked Cranes (Grus nigricollis): A Case in Xundian Mountains*; Yunnan University: Kunming, China, 2019.
38. Liu, W.; Jin, Y.; Wu, Y.; Zhao, C.; He, X.; Wang, B.; Ran, J. Home range and habitat use of breeding black-necked cranes. *Animals* **2020**, *10*, 1975. [CrossRef]
39. Kong, D.; Møller, A.P.; Zhang, Y. Disturbance and predation risk influence vigilance synchrony of black-necked cranes (*Grus nigricollis*), but not as strongly as expected. *Ecol. Evol.* **2021**, *11*, 2289–2298. [CrossRef] [PubMed]
40. Wang, L. *Landscape Dynamics and It Is Impacts on the Habitat of Black-Necked Crane in Napa Wetland in the Last two Decades*; Yunnan University: Kunming, China, 2015.
41. Yang, Y.X.; Bai, H.T.; Wei, F.; Peng, M.C.; Zhao, Z.J.; Wu, T.P.; Ran, Q.Y.; Wu, Z.L. Bird diversity and its correlation with the habitat structure in Dashanbao Black-necked Cranes Nature Reserve. *Chin. J. Ecol.* **2018**, *37*, 147–156. [CrossRef]
42. Yan, R.Q.; Shen, W.S.; Zhao, W.; Ou Yang, Y. Dynamics of suitability and conservation of the black-necked crane habitats in the midstream of the Yarlung Zangbo River. *J. Ecol. Rual Environ.* **2018**, *34*, 146–152. [CrossRef]
43. Fang, Y.; Zhang, X.X.; Guo, C.Q. Habitat suitability of black-necked cranes (*Grus nigricollis*) in Zoige wetland. *Chin. J. Ecol.* **2020**, *39*, 203–213. [CrossRef]
44. Ruan, O.; Liu, S.H.; Chen, F.; Luo, J.; Hu, H.T. Habitat suitability evaluation of black-necked based on multi-source remote in Caohai National Nature Reserve, Guizhou. *Acta Ecol. Sin.* **2022**, *42*, 1947–1957. [CrossRef]
45. Zhao, Z.L.; Zhang, Y.L.; Liu, L.S.; Liu, F.G.; Zhang, H.F. Recent changes in wetlands on the Tibetan Plateau: A review. *J. Geogr. Sci.* **2015**, *25*, 879–896. [CrossRef]
46. Circular of the State Council on Printing and Distributing the National Main Function Zone Planning. Available online: http://www.gov.cn/zhengce/content/2011-06/08/content_1441.htm (accessed on 1 December 2021).
47. Bai, J.H.; OuYang, H.; Cui, B.S.; Wang, Q.G.; Chen, H. Changes in landscape pattern of alpine wetlands on the Zoihe Plateau in th past four decades. *Acta Ecol. Sin.* **2008**, *28*, 2245–2252.
48. Chengdu Institute of Biology, Chinese Academy of Sciences. *A Survey Report on Animal Diversity in Sichuan Ruoergai Wetland National Nature Reserve*; Chengdu Institute of Biology: Chengdu, China, 2016.
49. Mi, C.; Huettmann, F.; Guo, Y.; Han, X.; Wen, L. Why choose Random Forest to predict rare species distribution with few samples in large undersampled areas? Three Asian crane species models provide supporting evidence. *PeerJ* **2017**, *5*, e2849. [CrossRef]
50. Li, F.S.; Suo Lang, D.E.J.; Brian, B.; Dorn, M.; Ran, H.J.; Jiang, Z.Q.; Phoebe, K. *Monitoring of Blacke-Necked Canes of Ruoergai in 2016–2020*, 2020.
51. The Global Biodiversity Information Facility. Occurrence Download. Available online: https://www.gbif.org/occurrence/download/0201219-210914110416597 (accessed on 1 April 2022).
52. China Citizen Science. Available online: http://www.birdreport.cn/home/search/page.html (accessed on 5 March 2022).
53. Peng, S.Z. 1-km Monthly Precipitation Dataset for China (1901–2020). Available online: https://doi.org/10.5281/zenodo.3185722 (accessed on 1 January 2021). [CrossRef]
54. Warren, D.L.; Matzke, N.J.; Cardillo, M.; Baumgartner, J.B.; Beaumont, L.J.; Turelli, M.; Glor, R.E.; Huron, N.A.; Simões, M.; Iglesias, T.L. ENMTools 1.0: An R package for comparative ecological biogeography. *Ecography* **2021**, *44*, 504–511. [CrossRef]

55. Warren, D.L.; Seifert, S.N. Ecological niche modeling in Maxent: The importance of model complexity and the performance of model selection criteria. *Ecol. Appl.* **2011**, *21*, 335–342. [CrossRef] [PubMed]
56. Muscarella, R.; Galante, P.J.; Soley-Guardia, M.; Boria, R.A.; Kass, J.M.; Uriarte, M.; Anderson, R.P. ENM eval: An R package for conducting spatially independent evaluations and estimating optimal model complexity for Maxent ecological niche models. *Methods. Ecol. Evol.* **2014**, *5*, 1198–1205. [CrossRef]
57. Phillips, S.J.; Anderson, R.P.; Dudík, M.; Schapire, R.E.; Blair, M.E. Opening the black box: An open-source release of Maxent. *Ecography* **2017**, *40*, 887–893. [CrossRef]
58. Liu, C.; White, M.; Newell, G. Selecting thresholds for the prediction of species occurrence with presence-only data. *J. Biogeogr.* **2013**, *40*, 778–789. [CrossRef]
59. Kong, W.Y.; Li, X.H.; Zhou, H.F. Optimizing MaxEnt model in the prediction of species distribution. *J. Appl. Ecol.* **2019**, *30*, 2116–2128.
60. Liu, H.C.; Zhao, N.X.; Zhuang, Y.Q.; Yang, M.L.; Zhao, H.R.; Ye, X.P. Assessment of habitat suitability for *Naemorhedusn grieus* in the Qinling Mountains with MaxEnt model. *Acta Ecol. Sin.* **2022**, *42*, 1–8. [CrossRef]
61. Zadrozny, B. Learning and evaluating classifiers under sample selection bias. In Proceedings of the Twenty-First International Conference on Machine Learning, Banff, AB, Canada, 4 July 2004. [CrossRef]
62. Phillips, S.J.; Dudik, M.; Elith, J.; Graham, C.H.; Lehmann, A.; Leathwick, J.; Ferrier, S. Sample selection bias and presence-only distribution models: Implications for background and pseudo-absence data. *Ecol. Appl.* **2009**, *19*, 181–197. [CrossRef]
63. Syfert, M.M.; Smith, M.J.; Coomes, D.A. The Effects of Sampling Bias and Model Complexity on the Predictive Performance of MaxEnt Species Distribution Models. *PLoS ONE* **2013**, *8*, e55158. [CrossRef]
64. Li, Z.M.; Li, F.S. *Black-Necked Crane Research*; Shanghai Scientific & Technological Education Publishing House: Shanghai, China, 2005.
65. Cao, M.C.; Sun, X.P.; Le, Z.F.; Cui, P.; Gong, X.; Xu, H.G. Analysis of Changes in Wintering Habitat of Red-Crowned Cranes Based on MaxEnt Model: A Case Study ofYancheng Nature Reserve. *J. Ecol. Rural Environ.* **2016**, *32*, 964–970. [CrossRef]
66. Na, X.D.; Zhou, H.T.; Zang, S.Y.; Wu, C.S.; Li, W.L.; Li, M. Maximum Entropy modeling for habitat suitability assessment of Red-crowned crane. *Ecol. Indic.* **2018**, *91*, 439–446. [CrossRef]
67. Anderson, R.P.; Gonzalez, I., Jr. Species-specific tuning increases robustness to sampling bias in models of species distributions: An implementation with Maxent. *Ecol. Modell.* **2011**, *222*, 2796–2811. [CrossRef]
68. Hockings, M.; Stolton, S.; Leverington, F.; Dudley, N.; Courrau, J. *Evaluating Effectiveness: A Framework for Assessing Management Effectiveness of Protected Areas*, 2nd ed.; World Conservation Union: Gland, Switzerland, 2006.

diversity

MDPI

Article

The Influence of Ecological Engineering on Waterbird Diversity in Different Habitats within the Xianghai Nature Reserve

Dehao Li [1], Shiying Zhu [1], Jin Gao [1], Haibo Jiang [1,*], Guangyi Deng [1], Lianxi Sheng [1,*], Yingyue Cao [2], Lianshan Li [3] and Baoqing Lin [3]

[1] State Environmental Protection Key Laboratory of Wetland Ecology and Vegetation Restoration, Key Laboratory for Vegetation Ecology, Ministry of Education, Northeast Normal University, Changchun 130117, China

[2] Faculty of Engineering, Kyushu University, Fukuoka 819-0395, Japan

[3] Xianghai National Nature Reserve of Jilin, Baicheng 137215, China

* Correspondence: jianghb625@nenu.edu.cn (H.J.); shenglx@nenu.edu.cn (L.S.)

Abstract: The Xianghai National Nature Reserve of Jilin is located on the East Asian-Australian migration route of birds and plays an important role for migratory waterbirds. To explore the influence of restoration measures on migratory waterbird communities in Xianghai Reserve, we conducted a waterbird survey in the spring and autumn during waterbird migration from 2010 to 2020. The composition and spatial and temporal changes in waterbird communities were identified in the reserve and the diversity changes in waterbird communities were analyzed. The results showed that (1) the number of waterbirds increased significantly since restoration measures in 2012. Restoration measures has clearly promoted an improvement in the species and number of waterbirds. The number of waterbirds was approximately seven-times higher than that before restoration measures. (2) The number of rare waterbirds increased in the reserve, and the number of *Anseriform* waterbirds on the IUCN Red List of Threatened Species rapidly increased. (3) The diversity of the waterbird community has shown some recovery in the reserve, but the variation in water level during some periods led to a regular fluctuation in waterbird diversity. Restoration measures directly changed the water surface area and the water depth in the reserve, which are important variables that influence waterbird selection of wetlands as habitats. To effectively protect and manage important stopover sites on the migratory routes of waterbirds, we suggest to rationally regulate the supplementary water level of wetlands, optimize the hydrological pattern of wetlands, and strengthen the regulation of wetland water levels.

Keywords: waterbird; wetland; diversity; Xianghai Nature Reserve

Citation: Li, D.; Zhu, S.; Gao, J.; Jiang, H.; Deng, G.; Sheng, L.; Cao, Y.; Li, L.; Lin, B. The Influence of Ecological Engineering on Waterbird Diversity in Different Habitats within the Xianghai Nature Reserve. *Diversity* **2022**, *14*, 1016. https://doi.org/10.3390/d14121016

Academic Editor: Peng Hou

Received: 13 October 2022
Accepted: 19 November 2022
Published: 22 November 2022

1. Introduction

Waterbirds are flocking species that depend on wetlands for survival, and they are the most active component of wetland ecosystems [1]. Their community composition and diversity directly reflect changes in wetland ecosystems and monitor changes in the wetland environment as an objective biological indicator. [2]. Waterbirds have clear seasonal migration patterns. A series of available wetland stopover sites along the migration route are the basis for ensuring the success of waterbird migration [3,4]. These stopover sites are an important source of energy resupply for waterbirds during migration and provide food and habitat for different populations.

The Xianghai National Nature Reserve of Jilin is located in west of the Songnen Plain on the migration route of East Asian-Australian migratory birds. It is a breeding place for rare species such as *Grus japonensis* and *Ciconia boyciana* and a migration stopover site for endangered waterbirds such as *Grus leucogeranus*. At the beginning of this century, as a result of climate change and anthropogenic factors, wetland habitats were extensively lost in this region. This loss seriously affected the breeding and migration of waterbirds [5]. To better protect waterbirds and wetlands, since 2012, the River-Lake Linkage Project has been carried out to restore the hydrological conditions of lakes and marshes. However, there has been no study on the influence of ten years of restoration on the number and diversity of waterbirds in different wetland habitats. Therefore, the aim of this study was to analyze the effectiveness of the restoration measures for the conservation of waterbird species. We analyzed waterbird data for a total of 11 years before and after the restoration measures from 2010 to 2020. We explored the dynamic characteristics of waterbird diversity in the lake reservoir and marsh habitats in response to the restoration measures, which provide a basis for the conservation of migratory birds in the East Asia-Australia flyway and the scientific management of migratory stopover sites.

2. Materials and Methods

2.1. Study Area

The Xianghai Reserve is located in Tongyu of Jilin Province (Figure 1) (122°5′~122°31′ E, 44°55′~45°09′ N) and has a total area of 1050 km^2. This reserve was included in the first group of sites on the List of Wetlands of International Importance in China in 1992. The reserve is an important stopover site for migratory waterbirds and a breeding place for summer migratory birds in the Asia-Pacific region. The main objects of protection include birds (e.g., *Grus japonensis* and *Ciconia boyciana*) and their habitats [6]. The reserve is situated in a semiarid area with the continental monsoon climate of the northern temperate zone. Most rainfall occurs in summer. The average annual precipitation is approximately 370 mm, and the average annual evaporation is approximately 1890 mm [7].

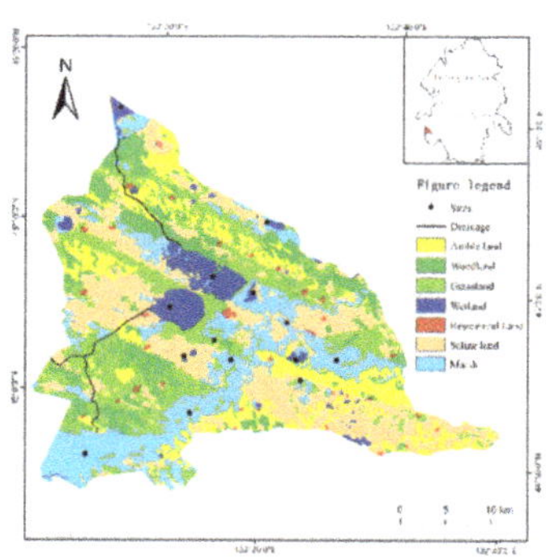

Figure 1. Location of the study area and sample points.

Before the restoration measures, due to the unreasonable exploitation of wetland resources, grassland resources, and water resources, a series of ecological and environmental problems had arisen in the reserve, mainly manifested in the decrease of grassland area, shrinkage of wetland area, increase of sandy and saline land area, etc. The restoration measures were carried out in this region in 2012. The lakes, reservoirs, and seasonal and dry marshes that were originally isolated were reconnected by the restoration measures in the Xianghai Reserve. The lakes and reservoirs in the region are supplied with water by the Tao'er River and the Huolin River and water to the surrounding marsh wetlands is supplied through their connected waterways. After the restoration measures, the area of water bodies and wetlands in the reserve increased, providing suitable habitat for the waterbird community in the Xianghai Reserve. At the beginning of the restoration measures, the annual water diversion in the region was approximately 4.00×10^7 m^3 from 2012 to 2014. The water diversion reduced to 1/3 of the initial diversion from 2015 to 2016. At the end of the restoration measures, water diversion is consistent with the level at the beginning of the restoration measures.

2.2. Bird Survey Methods

A total of 14 observation sites set up based on the historical stopover sites of waterbirds in the study area (Figure 1), including five reservoir habitats and nine marsh habitats [8]. The survey objects only include waterbirds that traditionally roost in or near water and did not include *Falconiformes*, *Accipitriformes*, or *Passeriformes* that depend on wetlands for survival. We classified the waterbirds in the area into four types based on their habitat preferences: dabbling birds, diving birds, large waders, and small waders. The dabbling birds mainly include *Anatidae*, *Laridae*, and *Rallidae*. The diving birds include *Anseriformes*, *Gavia*, *Podicipediformes*, and *Phalacrocoracidae*. The large waders mainly include *Ciconiiformes* and *Gruiformes*. The small waders mainly include most of the *Charadriiformes* and *Scolopacidae*.

Based on the migration pattern of waterbirds in the study area, waterbirds were observed from March to May and September to November each year from 2010 to 2020. The waterbird migration periods in spring and autumn were divided into three stages: premigration, mid-migration, and post-migration. There were two observations in each stage with an interval of 3~4 days. Binoculars (8 × 32x) and monoculars (20 × 60x) were used to scan the sample area clockwise during the survey to record the species and number of waterbirds. Direct counts were used for species with small numbers of individuals, and species with large numbers of individuals or flocks were counted by units of 10, 20, or 50 birds [9]. Species classification and identification were based on the List of the Classification and Distribution of Birds in China (3rd edition), edited by Zheng Guangmei [10]. The endangerment level of waterbird species in the study area was determined according to the Red List of the International Union for Conservation of Nature (IUCN).

2.3. Habitat Landscape in Remote Sensing Interpretation

Landsat 5 and Landsat 8 images with a path-row number of 120–029, which were captured on 2 July 2010, 8 July 2015, and 11 June 2020, were selected to interpret the marsh wetlands and water body for the area of the Xianghai Reserve. OLI images that were generated during the summer (from June to August) of each year were also selected to assist with interpretation, as multi-temporal features can help distinguish objects. The object-oriented segmentation method based on eCognition 9.3 was combined with a random forest model to extract land cover information. Water level and water diversion data of the study area were obtained from Xianghai Reservoir.

2.4. Data Analysis

Using a regression model, we examined the correlations between different physical features of the wetlands such as water body area, and water level, and the waterbird abundance, waterbird species richness, and diversity indices. We used the Shannon–Wiener diversity index and the G-F index to analyze the diversity of waterbird communities in the region [11].

Shannon's Diversity Index was calculated using the equation:

$$H' = -\sum_{i=1}^{S} p_i \, ln \, p_i \tag{1}$$

where p_i is the proportion of proportion of bird number of species i in the total bird abundance.

We used the Berger–Parker dominance index to judge the population size classes, which can be calculated as follows [12].

$$I = \frac{N_i}{N} \tag{2}$$

where N_i is the number of individuals of species i and N is the total number of all species in the community. When $I \geq 0.1$, species i is a dominant species, $0.01 \leq I \leq 0.1$ indicates that species i is a common species, and $I < 0.01$ indicates that species i is an occasional species.

3. Results

3.1. Waterbird Species and Numbers

3.1.1. Species and Population Changes in Regional Waterbirds

We recorded a total of 110 waterbird species from 2010 to 2020 in the study area, belonging to 10 orders, 17 families, and 50 genera.The main taxon was *Charadriiformes* (40 species), accounting for 36.36% of waterbird species in the study area. *Anseriformes*, *Pelecaniformes*, and *Gruiformes* had 36, 12, and 9 species, respectively, accounting for 32.72%, 10.91%, and 8.18% of the waterbird species in the study area. The proportions of the other six orders were all less than 5%. Among the recorded waterbirds, the dominant species were *Anser fabalis* and *Aythya ferina*. There were 17 common species (*Anser albifrons*, *Anas platyrhynchos*, *Grus leucogeranus*, *Grus monachal*, etc.), most of which belonged to the family *Anatidae*. The other species were occasional.

We analyzed the annual dynamics of the waterbird community in the Xianghai Reserve from 2010 to 2020 based on the number of species and individuals for 10 orders (Figure 2). The results showed that the number of waterbird species decreased in the study area from 2010 to 2012, increased from 2013, reached a maximum number of species in 2018, and then showed a decreasing trend. The number of waterbird individuals showed the same decreasing trend as the number of species from 2010 to 2012, increased annually from 2013 to 2016, decreased annually from 2016 to 2019, and suddenly reached a maximum in 2020. The change of regional water diversions influenced waterbird abundance and water body area in the reserve. At the beginning of the restoration measures, the water body area reached a maximum number, and then showed a decreasing trend year by year. We investigated the differences in water diversion on waterbird abundance and water body area through employing regression mode, which has a positive correlation with interannual water diversion. When the water diversion maintained at 5.00×10^7 m^3, the waterbird abundance was at a stable level.

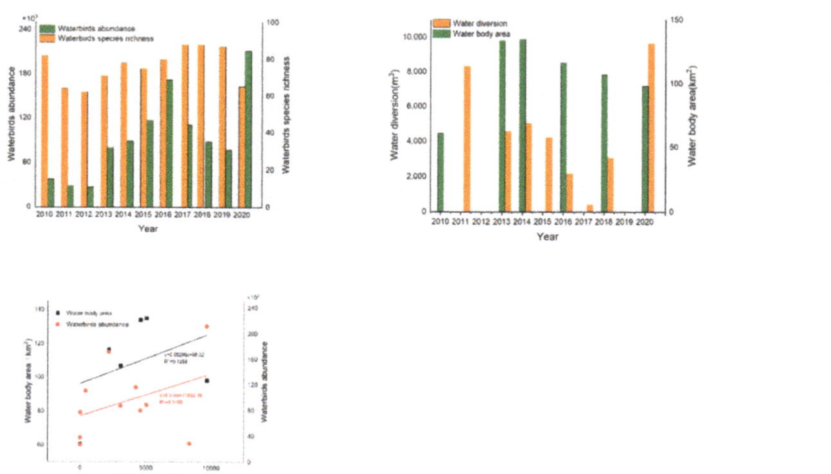

Figure 2. Annual dynamics of waterbird species and numbers in Xianghai Nature Reserve.

3.1.2. Changes in Waterbird Species and Numbers in Different Habitats

Anatidae waterbirds occupied the main part of the lake and marsh habitats in the Xianghai Reserve. The dominant species *Anser fabalis* and *Aythya ferina* were distributed in the marsh habitat (Table S1). The common species *Anas poecilorhyncha*, *Mergellus albellus*, and *Platalea leucorodia* were distributed in both habitats. *Gruiformes* waterbirds such as *Grus leucogeranus*, *Grus grus*, and *Fulica atra* were mainly distributed in the marsh habitat and less in the lake reservoir habitat.

The number of waterbird species in marsh habitats was consistently higher than that in the lake reservoir habitats from 2010 to 2020 in the Xianghai Reserve (Figure 3). Since 2012, restoration measures have been carried out in the study area. At the beginning of the restoration measures, the area of marsh habitat increased significantly. In the middle of the restoration measures, the area of marsh habitat showed a decreasing trend (Figure 4). With the increase in the marsh habitats, the proportion of waterbird species in marsh habitats continued to increase from 2010 to 2013 and reached a maximum (63%) in 2013. The proportion of species in marsh habitats decreased from 2016 to 2017 and from 2019 to 2020, which has a similar trend in marsh habitats change.

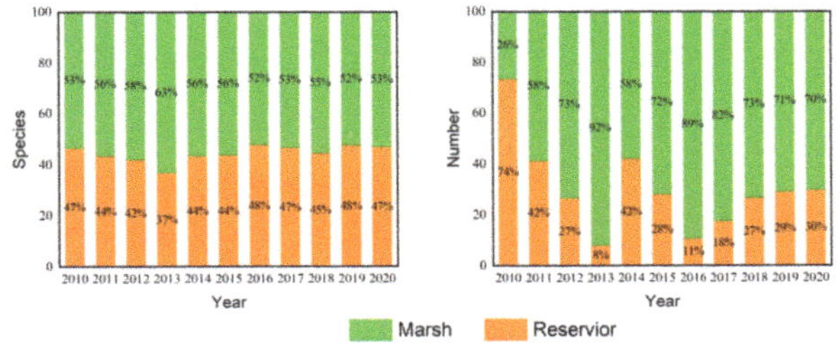

Figure 3. Proportion of species and number of waterbirds in different habitats in Xianghai Nature Reserve.

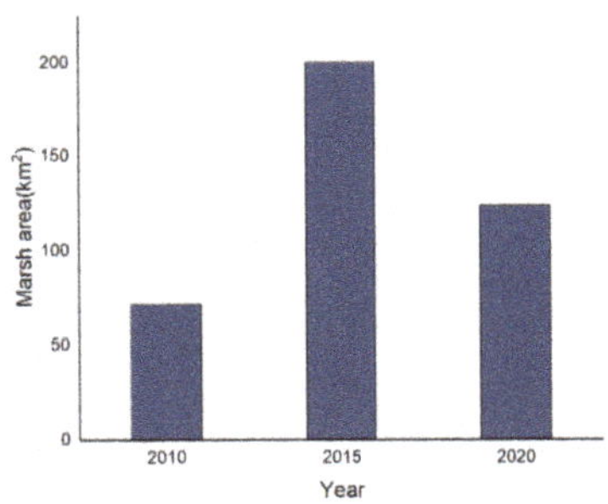

Figure 4. Changes in the area of marsh in the Xianghai Nature Resreve.

The number of waterbirds in the reservoir habitat was higher than that in the marsh habitat in 2010 (Figure 3). The proportion of waterbirds in the marsh habitat continued to increase from 2011 to 2013 and reached a maximum (92%) in 2013. The proportion of waterbirds in the marsh habitat fluctuated in other years, but overall, it was higher than that in the reservoir habitat.

3.2. *Rare Waterbirds*

Changes in Rare Waterbird Species and Numbers in the Region

Among the 110 species of waterbirds that are distributed in the Xianghai Reserve, 17 species are on the IUCN Red List of Threatened Species, accounting for 15.4% of the total waterbirds that are distributed in the reserve, which is much higher than the overall level of threatened waterbird species in China (10.6%) [13]. There are two species of critically endangered (CR) waterbirds (*Grus leucogeranus* and *Aythya baeri*), two species of endangered (EN) waterbirds (*Ciconia boyciana* and *Tringa guttifer*), eight species of vulnerable (VU) waterbirds (including *Grus japonensis* and *Grus vipio*), and six species of near threatened (NT) waterbirds (including *Aythya nyroca* and *Numenius arquata*).

We investigated the contact between the IUCN Red List of Threatened Species and time through employing regression models (Figure 5). The proportion of dependent variable (Y) impacts on independent variables (X) was explained by linear regression, which indicated the number of protected birds increased year by year after the restoration measures in 2012. In the Xianghai Nature Reserve, only the endangered waterbirds showed a decreasing trend. Critically endangered waterbirds, vulnerable waterbirds, and near threatened waterbirds showed the same trend, with their numbers increasing year by year. The number of vulnerable waterbirds changed more dramatically than the other two categories.

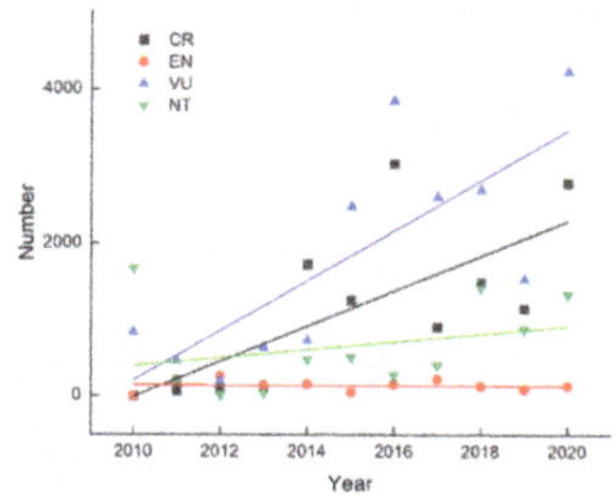

Figure 5. Bird numbers in relation to their IUCN red list status. Note: CR denotes critically endangered waterbirds; EN denotes endangered waterbirds; VU denotes vulnerable waterbirds; NT denotes near threatened waterbirds.

3.3. *Waterbird Diversity*

3.3.1. Changes in Waterbird Diversity in Different Habitats

The Shannon–Wiener diversity of waterbirds in the reservoir and marsh habitats of the Xianghai Reserve from 2010 to 2020 showed more dramatic interannual variation (Figure 6). The marsh showed a trend of regular fluctuation with the lowest (and similar) values in 2012, 2016, and 2020 (2.01 $\pm$ 0.07) and the highest values in 2010, 2014, and 2018 (2.60~2.84). The Shannon–Wiener diversity of the reservoir had a larger fluctuation with the highest value (3.12) in 2016 and the lowest value (1.83) in 2020. The change trend of the Shannon diversity index in the study area was similar to that of marsh habitat. After the restoration measures in 2012, there was an overall rising trend. The trend of reservoir habitats was opposite, showing a negative correlation.

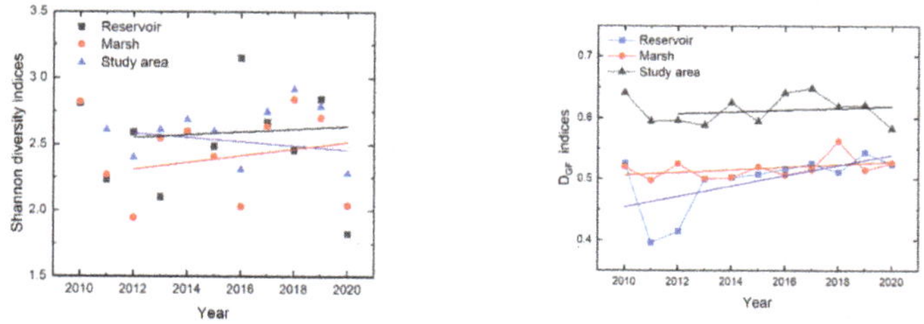

Figure 6. Annual variation of species diversity index of waterbirds in different habitats. Note: Study area denotes all data in the marsh and reservoir together.

The interannual variations of D_{G-F} in the reservoir and marsh habitats of the Xianghai Reserve from 2010 to 2020 are shown in Figure 6. In the marsh habitat, the indices of D_{G-F} showed overall increasing trends and reached their maximum values in 2018 (D_{G-F} = 0.56). In the reservoir habitat, the indices of D_{G-F} decreased from 2010 to 2012, showed an increasing trend from 2013, and reached their maximum value in 2019

(D_{G-F} = 0.54). In the study area, the index of D_{G-F} was generally stable and maintained at 0.58~0.65 in all years. After the restoration measures in 2012, the overall trend was rising.

3.3.2. Changes in Diversity for Different Types of Waterbirds

The correlations of waterbird abundance, waterbird species richness, and Shannon-Wiener indices of different types of waterbirds to water level fluctuations were analyzed during 2010–2020 (Figure 7). Waterbird abundance showed a positive correlation with the water level, and the wetland replenishment in the reserve promoted the increase of waterbird populations. Waterbird species richness is negatively correlated with the water level, and when the water level is too high it has a negative effect on waterbird species richness in the reserve. Waterbird abundance and species richness in the study area were maintained at relatively high levels when the water level was maintained at about 2 m.

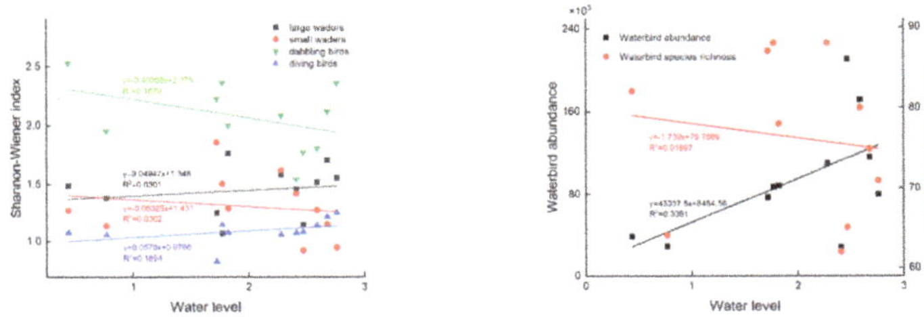

Figure 7. Effects of water level changes on waterbird abundance, waterbird species richness, and four waterbird guilds.

Dabbling birds and small waders showed a negative correlation with water level fluctuations, and the water level increasing would lead to a decreasing trend in waterbird diversity in these two waterbird guilds. Diving birds and large waders showed a positive correlation with water level fluctuations, and the water level increasing was conducive to promote the recovery of these two waterbird guilds. At the water level of about 1.5 m, the Shannon–Wiener diversity of four waterbird guilds was relatively high, and the water level at this time was conducive to the recovery of the waterbird population diversity.

4. Discussion

4.1. Waterbird Species and Numbers

The Xianghai Reserve is an important stopover site for migratory birds from East Asia-Australia during their north–south migration, and a large number of migrating waterbirds make a brief stopover here every spring and autumn. The dynamic changes in waterbird species and numbers directly reflect the utilization and quality of habitat. Water level change in wetlands is a key factor influencing waterbird foraging and habitat selection [14–16]. The species diversity of migratory waterbirds also changes with fluctuations in wetland water level [17]. A series of restoration measures have been carried out in the Xianghai Reserve since 2012. The water surface area increased and migrated to the north, and the water level and vegetation composition have changed [18,19]. The number of waterbird species decreased by 17 species before the implementation of restoration measures, and the number of waterbirds continued to decline from 2010 to 2013. After restoration measures in 2012, the number of species increased and remained stable, and the number of waterbirds significantly increased. In the marsh habitat, the waterbird community trends were the same as the overall trends in the study area, with significant recovery in waterbird abundance and species richness after the restoration measures. The number of waterbirds in 2016 was approximately seven times higher than that in 2012. The restoration measures significantly contributed to the enhancement of waterbird species and numbers.

The water level change significantly affected the number of waterbirds. The highest proportion of waterbirds was found in shallow water habitats [20]. The waterbirds in the Xianghai Reserve are mainly composed of *Anseriformes*, *Gruiformes*, and *Charadriiformes*, whose main habitats are shallow mudflats and grass flats [21]. These taxa are concentrated in the marsh habitats. During restoration measures, the change in water diversion can directly affected the water level of the marsh habitat. Changes in waterbird abundance and species richness in the study area mostly responded to fluctuating changes in water levels (Figure 7). If the water level is too high or too low, the number and species of waterbirds in the marsh habitat will be affected. When the water diversion decreased, and the hydrological conditions changed from 2017 to 2020. The number of dominant and common species such as *Anseriformes* and *Charadriiformes* greatly fluctuated and showed a downwards trend. The increase in water diversion caused the waterbird population to reach its peak since the restoration measures in 2020. The changes in the number of waterbird species concentrated in the fluctuation of *Charadriiformes* from 2017 to 2020, which are more sensitive to changes in water level and microhabitat than *Anseriformes* [22]. When the regional hydrological situation changed, the species and number of waterbirds also greatly changed.

4.2. Changes in Numbers of Rare Species

Waterbirds on the IUCN Red List of Threatened Species such as *Grus japonensis, Grus leucogeranus, Cygnus cygnus,* and *Anser erythropus* select plant tubers as their main food source in the Xianghai Reserve [23]. Birds require an appropriate water depth to meet their habitat needs [24]. Water depth is an important characteristic that affects the habitat of tuber-eating birds. The hydrological situation has fluctuated since the restoration measures that were carried out in Xianghai Reserve in 2012 (Figure 2). The increasing water level contributed to both waterbird abundance and large wader's diversity (Figure 7). The regional wetlands and vegetation composition have changed [25]. Changes in regional water levels will affect the composition of regional vegetation, which influence the availability of food sources for waterbirds. The traditional food source of *Grus leucogeranus*, the *Bolboschoenus planicalmis*, grows only in shallow water, and when the water depth is too high, the vegetation such as *Phragmites australis* and *Typha orientalis*, which are suitable for growing in deeper water, increases. This will lead to a decrease in the amount of food resources for *Grus leucogeranus* [26]. The increase in shallow marsh habitats (Figure 4) in the region has provided more food resources for tuber-feeding birds, resulting in an increase in the number of protected birds in the *Gruiformes*, *Anseriformes*, *Charadriiformes*, and *Pelecaniformes*. As *Anseriform* species have a wider ecological niche, their numbers rapidly increased.

Different groups of waterbirds have different habitat needs and make different habitat choices. Wild geese mainly inhabit grassy areas [27]. Ducks prefer open water and shallow marshes [28]. *Charadriiformes* have their main activity in mudflats. Compared with the waterbirds of *Anseriformes*, *Charadriiformes*, and *Pelecaniformes*, large wading waterbirds (e.g., *Grus japonensis* and *Ciconia boyciana*) prefer natural habitats without human disturbance and require larger wetland patches as habitats. Xianghai Reserve is a composite wetland and consists of natural wetlands (marshes) and artificial wetlands (reservoirs and canals). The area of marsh habitat in the region has increased, and the connectivity among wetland patches has been enhanced since the restoration measures in 2012 [29]. Waterbirds of national protection grade preferred shallow marshes and grassland habitats, and taxa such as *Anseriformes*, *Charadriiformes*, and *Gruiformes* have migrated from reservoir habitats to marshy habitats. The large wading birds (i.e., *Grus japonensis* and *Ciconia boyciana*) chose marsh habitats with less human disturbance as their habitats. Therefore, the interannual proportion of marsh habitats used was slightly higher than that of reservoirs.

4.3. Changes in Waterbird Diversity

Water level, water area, vegetation composition, and wetland size are important variables affecting waterbird habitat selection. The fluctuation of the water level in the Poyang Lake area that is affected by the Three Gorges Project directly influences the composition of vegetation in the area. The high water has a negative impact on the growth of aquatic vegetation and the number and diversity of wintering waterbird populations [30]. The interannual water diversions in the area have directly changed the water level of the wetland since the restoration measures. The high water will have a negative impact on the diversity of waterbird populations in the region. Therefore, the overall waterbird population diversity has recovered after the restoration of wetlands. However, the high water during some periods has caused regular fluctuations in waterbird diversity.

Large-scale water replenishment was carried out through a water diversion project for marsh habitats in the region from 2013 to 2015, resulting in changes in wetland area, patch connectivity, and water level [31]. The area of marsh wetlands in the central region of the reserve increased, and the stability of wetland patches increased from 2013 to 2015. With the improvement of wetland area, patch connectivity, and water levels, it had a positive effect on the conservation of waterbird communities in the study area. Waterbird abundance, waterbird species richness, and diversity all showed a significant increasing trend (Figures 2 and 6). Compared with unrestored wetlands, restored wetlands typically have higher waterbird abundance, waterbird species richness, and diversity [32]. At the beginning of restoration measures, the marsh habitat quality was restored, and waterbird diversity showed an increasing trend from 2013 to 2014. In the middle of restoration measures, the water level in the region increased from 2015 to 2016 due to a high level of water diversion. When the water level is higher, the area of the water–land transition zone is smaller [31]. The reduced waterbird habitat can lead to a decrease in diversity in the marsh habitat. Water replenishment to the marsh habitat stopped from 2016 to 2018, the water level of the wetlands began to decrease, and the Shannon diversity increased year by year. Compared with artificial wetlands, natural wetlands have higher waterbird diversity. Artificial wetlands can also be used as alternative habitats for waterbird populations [32]. Therefore, the reservoir habitat was an alternative to the marsh habitat before water replenishment in the study area. The diversity of waterbird species was relatively high. The waterbird populations shifted from the original reservoir habitat to the marsh habitat after the restoration measures. The diversity of the reservoir habitat showed a trend that was opposite to that of the marsh habitat.

When the diversity of biological communities is measured in a region, the G-F index can compensate for the deficiency of the Shannon–Wiener index and reflect the diversity among families and genera. The Shannon–Wiener diversity, D_{G-F} in the study area showed an overall increasing trend since 2012, indicating that the restoration measures had a positive effect on the conservation of regional waterbird populations. The Shannon index decreased and the indices of D_{G-F} increased in the marsh habitat. This reflects the relatively uniform distribution of waterbird species among families in marsh habitats. Therefore, D_{G-F} was relatively high. The decrease in the Shannon–Wiener diversity was due to a single dominant species of waterbird, and the number of single species populations in the region has been high since the restoration measures. The indices of D_{G-F} in the reservoir habitat showed an overall increasing trend after the restoration measure in 2013, indicating an overall increase in the diversity among families and genera. The water replenishment of the reservoir habitat by the restoration measures improved its quality and increased the number of waterbird species in that habitat [33].

The main objects of the reserve are birds such as *Grus japonensis* and *Ciconia boyciana* and their habitats. The Shannon diversity of large waders has been increasing annually since the restoration measures in 2012. The restoration measures played a positive role in the conservation of target species. The enhanced connectivity and aggregation of wetland patches provided suitable habitats for *Ciconiiformes* and *Gruiformes* [34]. However, large waders have declined in some years after restoration measures, which played a negative role in the conservation of target species (Figure 7). Diving birds showed an increasing trend at the same time. Diving birds are mainly dominated by *Podicipedidae*, which is a typical waterbird with a preference for deep water. Its numbers are positively correlated with water level in some areas, whereas the numbers of other waterbirds are negatively correlated with water level [20]. Therefore, the large number of *Podicipedidae* indicates a greater area of deep water. This water level is unfavorable for other waterbirds. Different waterbirds have different ecological niches due to differences in morphological characteristics and feeding habits. For example, wading birds are concentrated in areas with water depths of 15–20 cm [35]. The fluctuations in diversity indices of dabbling birds and wading birds reflect the changes in the hydrological situation in the region since the restoration measures. Therefore, based on the changes in the number and diversity of dabbling birds, regional water diversion can be regulated to provide a suitable wetland water level for the main protected species in the reserve and maintain their optimal habitat. The annual water diversion in the region was approximately 5.00×10^7 m^3 at the beginning of project construction, from 2013 to 2014. The water diversion was reduced to one third of the initial diversion from 2015 to 2016. When the water diversion was maintained at 4.00×10^7 m^3 at the beginning of the restoration measures, the overall water level in the study area was maintained at 1.5–2 m. This water level provided a suitable habitat for waterbird communities in the study area, and waterbird abundance, waterbird species richness, and diversity indices of the four waterbird guilds were all at high levels (Figure 7). Waterbirds in the study area were effectively protected at this time.

5. Conclusions

The Xianghai Reserve is an important stopover site on the East Asia-Australia migratory route. To better protect waterbird populations and wetlands, a series of restoration measures were constructed to restore the habitat quality of stopover sites. The number of waterbird species increased and remained at a relatively stable level in the reserve since the restoration measures. The number of waterbirds increased by approximately seven-fold compared with that before the restoration measures. The number of rare birds on the IUCN Red List of Threatened Species has significantly increased with the most rapid increase in *Anseriform* waterbirds. The diversity of waterbird populations has been restored. The Shannon–Wiener diversity reached its highest value of 2.84 in 2018. However, it has fluctuated at times. Changes in water diversions from the restoration measures will have an impact on the water body area, marsh habitat area, and water depth in the study area, ultimately leading to the changes in waterbird abundance, species richness, and diversity in the region. When the water diversion was maintained at 5.00×10^7 m^3, the increasing area of marsh habitat provided more suitable habitats for waterbird populations. The regulation of water diversion should consider rainfall and the water diversion of the previous year. Continuous and large amounts of water diversion can also affect the quality of waterbird habitat. Therefore, in the restoration measures, water diversion should reasonably regulate wetland replenishment, optimize wetland hydrological patterns, and strengthen wetland water level regulation to maintain suitable habitat quality for waterbirds.

Supplementary Materials: The following supporting information can be downloaded at: https://www.mdpi.com/article/10.3390/d14121016/s1. Table S1: Waterbirds species list in reservior habitats.

Author Contributions: Conceptualization, D.L. and H.J.; methodology, S.Z.; software, G.D. and J.G.; validation, H.J. and L.S.; investigation, J.G. and G.D.; resources, B.L. and L.L.; data curation, D.L.; writing—original draft preparation, D.L.; writing—review and editing, H.J. and Y.C. All authors have read and agreed to the published version of the manuscript.

Funding: This research was funded by the National Natural Science Foundation of China, grant numbers U19A2042 and 32271624; National Key Research and Development Program of China, grant numbers 2022YFF1300900.

Institutional Review Board Statement: Not applicable.

Informed Consent Statement: Not applicable.

Data Availability Statement: The data that are presented in this study are available on request from the corresponding author.

Conflicts of Interest: The authors declare no conflict of interest. The funders had no role in the design of the study; in the collection, analyses, or interpretation of data; in the writing of the manuscript, or in the decision to publish the results.

References

1. Ma, Z.J. Managing Wetland Habitats for Waterbirds: An International Perspective. *Wetlands* **2010**, *30*, 15–27. [CrossRef]
2. Gregory, R.D.; van Strien, A. Wild Bird Indicators: Using Composite Population Trends of Birds as Measures of Environmental Health. *Ornithol. Sci.* **2010**, *9*, 3–22. [CrossRef]
3. Birgita, D.H.; Peter, M.; Paul, M.; Richard, H.L. Long-term declines in multiple waterbird species in a tidal embayment, south-east Australia. *Austral. Ecol.* **2015**, *40*, 515–527.
4. Mu, T.; Cai, S.X.; Peng, H.B.; Hassell, C.J.; Boyle, A.; Zhang, Z.W.; Piersma, T.; Wilcove, D.S. Evaluating staging habitat quality to advance the conservation of a declining migratory shorebird, Red Knot Calidris canutus. *J. Appl. Ecol.* **2022**, *59*, 2084–2093. [CrossRef]
5. He, C.G.; Ishikawa, T.; Sheng, L.X.; Irie, M. Study on the hydrological conditions for the conservation of the nesting habitat of the Red-crowned Crane in Xianghai Wetlands, China. *Hydrol. Process.* **2009**, *23*, 612–622. [CrossRef]
6. Zhang, Y.H.; Deng, W.; Zhang, S.W. The spatial structure of the Red-crown crane's habitat in Xianghai National Nature Resreve based on RS and GIS techniques. *Acta Ecol. Sin.* **2006**, *11*, 3725–3731.
7. Qing, F.; Wang, S.; Cui, H.Y. Analysis on the characteristics of climate change in western Jilin Province in 1951–2019. *Anhui Agric. Sci. Bull.* **2021**, *18*, 177–179.
8. He, C.G.; Sheng, L.X.; Lang, H.Q.; Li, L.S.; Zou, L.F. Migration dynamics of Grus japonensis in recent years spring and conservation of its habitat in Xianghai Nature Reserve. *J. Appl. Ecol.* **2004**, *15*, 1523–1526.
9. Li, D.L.; Chen, S.H.; Guan, L.; Lloyd, H.; Liuc, Y.L.; Lv, J.Z.; Zhang, Z.W. Patterns of waterbird community composition across a natural and restored wetland landscape mosaic, Yellow River Delta, China. *Estuar. Coast. Shelf Sci.* **2011**, *91*, 325–332. [CrossRef]
10. Deng, G.M. *A Checklist on the Classification and Distribution of the Birds of China*, 3rd ed.; Science Press: Beijing, China, 2017.
11. Jiang, Z.G.; Ji, L.Q. Avain-mammalian species diversity in nine representative sites in China. *Biodivers. Sci.* **1999**, *7*, 220–225.
12. Zhang, Y.F.; Xu, Q.H.; Gao, S.P.; Li, D.M.; Wu, Y.F. Survey of bird diversity in Luanhekou wetland, Hebei Province. *Sichuan J. Zool.* **2010**, *29*, 244–248.
13. Zhang, Y.Y.; Zhang, Z.W.; Dong, L.; Ding, P.; Ding, C.Q.; Ma, Z.J.; Deng, G.M. Assessment of red list of birds in China. *Biodivers. Sci.* **2016**, *24*, 568–579. [CrossRef]
14. Oriane, W.; Mark, A.; Craig, R.; Rebecca, J. Waterbird Responses to Experimental Drawdown: Implications for the Multispecies Management of Wetland Mosaics. *J. Appl. Ecol.* **2002**, *39*, 987–1001.
15. Sándor, F.; Katalin, H. Effects of water level on waterbird abundance and diversity along the middle section of the Danube River. *Hydrobiologia* **2012**, *697*, 15–21.
16. Chang, Z.; Yu, J.Y.; Guang, M.Z.; Jie, L.; Sheng, L.G.; Lu, H.; Shan, S.H.; Hai, P.W.; Yuan, Z.; Hong, X.A.; et al. Influence of hydrological regime and climatic factor on waterbird abundance in Dongting Lake Wetland, China: Implications for biological conservation. *Ecol. Eng.* **2016**, *90*, 473–481.
17. Zhang, S.Q.; Zhang, P.Y.; Pan, B.H.; Zou, Y.A.; Xie, Y.H.; Zhu, F.; Chen, X.S.; Li, F.; Deng, Z.M.; Zhang, H.; et al. Wetland restoration in the East Dongting Lake effectively increased waterbird diversity by improving habitat quality. *Glob. Ecol. Conserv.* **2021**, *27*, 15–35. [CrossRef]
18. Ma, Y.M.; Guo, C.M.; Wang, Y.; Li, J.P.; Tang, X.L.; Chen, L.W. Remote sensing monitoring on area dynamic change of major water bodies in western jinlin province. *Bull. Soil Water Conserv.* **2018**, *38*, 249–255.
19. Ding, C.; Liu, J.P.; Ma, M.Y.; Ma, C.D.; Liu, Y.B. Plant species diversity of meadow-marsh ecosystem in Xianghai. *Wetl. Sci.* **2017**, *15*, 552–555.
20. Jia, Y.F.; Jiao, S.W.; Zhang, Y.M.; Zhou, Y.; Lei, G.C.; Liu, G.H. Diet Shift and Its Impact on Foraging Behavior of Siberian Crane (Grus Leucogeranus) in Poyang Lake. *PLoS ONE* **2013**, *8*, e65843. [CrossRef]
21. Elafri, A.; Belhamra, M.; Houhamdi, M. Comparing habitat preferences of a set of waterbird species wintering in coastal wetlands of North Africa: Implication for management. *Ekológia* **2017**, *36*, 158–171. [CrossRef]
22. Zhang, P.Y.; Zou, Y.A.; Xie, Y.H.; Zhang, H.; Liu, X.K.; Gao, D.L.; Yi, F.Y. Shifts in distribution of herbivorous geese relative to hydrological variation in East Dongting Lake wetland, China. *Sci. Total Environ.* **2018**, *636*, 30–38. [CrossRef] [PubMed]
23. Chatterjee, A.; Adhikari, S.; Pal, S.; Mukhopadhyay, S.K. Foraging guild structure and niche characteristics of waterbirds wintering in selected sub-Himalayan wetlands of India. *Ecol. Indic.* **2020**, *108*, 105693. [CrossRef]
24. Guan, L.; Lei, J.L.; Zuo, A.J.; Zhang, H.; Lei, G.C.; Wen, L. Optimizing the timing of water level recession for conservation of wintering geese in Dongting Lake, China. *Ecol. Eng.* **2016**, *88*, 90–98. [CrossRef]
25. Chen, L.W.; Zhang, G.X.; Xu, Y.J.; Chen, S.B.; Wu, Y.F.; Gao, Z.T.; Yu, H.Y. Human Activities and Climate Variability Affecting Inland Water Surface Area in a High Latitude River Basin. *Water* **2020**, *12*, 382. [CrossRef]

26. Romano, M.; Barberis, I.; Pagano, F.; Maidagan, J. Seasonal and interannual variation in waterbird abundance and species composition in the Melincue saline lake, Argentina. *Eur. J. Wildl. Res.* **2005**, *51*, 1–13. [CrossRef]
27. Sun, Z.D.; Huang, Q.; Opp, C.; Hennig, T.; Marold, U. Impacts and Implications of Major Changes Caused by the Three Gorges Dam in the Middle Reaches of the Yangtze River, China. *Water Resour. Manag.* **2012**, *26*, 3367–3378. [CrossRef]
28. Yu, Y.W.; Yi, F.J.; Lei, G.; Cai, L.; Guang, C.L.; Li, W.; Guan, H.L. Optimising hydrological conditions to sustain wintering waterbird populations in Poyang Lake National Natural Reserve: Implications for dam operations. *Freshw. Biol.* **2013**, *58*, 2366–2379.
29. Wu, H.B.; Zheng, B.H. Wetland area identification and waterbird protection management in consideration of lake topography and water level change. *Glob. Ecol. Conserv.* **2020**, *23*, e01056. [CrossRef]
30. Wang, C.; Liu, H.Y.; Li, Y.F.; Dong, B.; Qiu, C.Q.; Yang, J.L.; Zong, Y.; Chen, H.; Zhao, Y.Q.; Zhang, Y.A. Study on habitat suitability and environmental variable thresholds of rare waterbirds. *Sci. Total Environ.* **2021**, *785*, 147–316. [CrossRef]
31. Liu, J.P.; Ma, C.D. The spatial variation in the patch stability of marshes in Xianghai between 1985 and 2015. *Acta Ecol. Sin.* **2017**, *37*, 1261–1269.
32. Fan, J.; Wang, X.D.; Wu, W.; Chen, W.P.; Ma, Q.; Ma, Z.J. Function of restored wetlands for waterbird conservation in the Yellow Sea coast. *Sci. Total Environ.* **2021**, *756*, 144061. [CrossRef] [PubMed]
33. Lang, Z.H.; Xu, M.X.; Gao, W.H. Water environment situation and protection measures for Xianghai wetland in Jilin Province. *Yangtze River* **2011**, *42*, 34–38.
34. Cheng, W.; Gang, W.; Ling, J.D.; Hong, Y.L.; Yu, F.L.; Yi, Z.; Hao, C.; Bin, D.; Shi, C.L.; Yong, Q.Z. Diverse usage of waterbird habitats and spatial management in Yancheng coastal wetlands. *Ecol. Indic.* **2020**, *117*, 106–583.
35. Chris, S.; Lewis, W. Winter Management of Californian Rice Fields for Waterbirds. *J. Appl. Ecol.* **1998**, *35*, 95–108.

diversity

MDPI

Article

Impacts of China's Western Development and Protection Strategy: An Ecosystem Services Perspective of Western China

Yan Wang [1], Naifeng Lin [1], Jixi Gao [2,*], Changxin Zou [1,*] and Delin Xu [1,*]

1 Nanjing Institute of Environmental Science, Ministry of Ecology and Environment of the People's Republic of China, Nanjing 210042, China
2 Satellite Environment Center, Ministry of Ecology and Environment of the People's Republic of China, Beijing 100094, China
* Correspondence: gaoeco@163.com (J.G.); zcx@nies.org (C.Z.); xudelin11@163.com (D.X.)

Abstract: Policies directly affect land-use change, which in turn, affects ecosystem services. In parallel with the implementation of a full-scale development program for the western region, the Chinese government has introduced a series of ecological protection and restoration strategies for development and construction. This study conducted a quantitative spatial evaluation of the ecosystem service value (ESV) of national nature reserves in the western region under this dual policy of development and protection. On the basis of land-use data and related evaluation parameters, fluctuations in the valuation of ecosystem services during 2000–2010 were analyzed in response to land-use changes under the comprehensive policy. Results showed that the increases in the areas of forestland and water bodies led to an increase of CNY 74.1 billion in the ESV from 2000 to 2010, equivalent to 2.02%. Grassland with increased production capacity and water bodies were the main factors driving the total ESV dynamics. Values of all ES increased significantly. Therefore, the ecological conservation and restoration policy, along with the development policy, had a positive influence on ecosystem services in the nature reserves in western China.

Keywords: ecosystem service value; full-scale development policy; ecological protection policy; nature reserve; western China

Citation: Wang, Y.; Lin, N.; Gao, J.; Zou, C.; Xu, D. Impacts of China's Western Development and Protection Strategy: An Ecosystem Services Perspective of Western China. *Diversity* **2022**, *14*, 863. https://doi.org/10.3390/d14100863

Academic Editor: Panayiotis Dimitrakopoulos

Received: 22 August 2022
Accepted: 27 September 2022
Published: 12 October 2022

1. Introduction

For a long time, the human impact on ecosystems has been mostly negative, owing to an insufficient understanding of ecosystem services. The overexploitation of natural resources by humans and the substantial discharge of pollutants into the environment has led to continuous damage to, and weakening of, ecosystem services [1]. For one thing, the negative impact of human activities, such as urbanization construction, deforestation, and damaging forests to reclaim land, directly caused changes in land use, which in turn, directly weakened the ecosystem service [2,3]. Furthermore, human activities, such as long-term greenhouse gas emissions resulting in climate change, changing temperatures, and precipitation, have also directly affected ecosystem structure and functions [4]. According to the Fifth Assessment Report of the Intergovernmental Panel on Climate Change (AR5), it is likely (probability above 95%) that human activities have caused most of the global average surface temperature increase since the 1950s [5]. The trend of global warming has led to a continuous increase in high temperatures and changes in precipitation. According to the World Meteorological Organization, in 2018, extreme weather events such as high temperatures and drought occurred in many areas across the globe; for example, the temperature in the Arctic Circle reaching 32 °C, far higher than the average temperature of 10 °C in the same period of the previous year [6]. Drought and high temperatures have intensified wildfires in the northern hemisphere, and the extreme weather has had significant negative impacts on human health, agriculture, and ecosystems [6]. Climate change can, thus, affect ecosystem structure, composition, and function, and ultimately ecosystem service functions such as food production, water supply, flood regulation and storage, soil and water conservation, and biodiversity conservation [7–11].

National policies influence and determine the impact of human activities on the ecological environment. Since the 1950s, developed countries in Europe and North America have implemented policies of returning farmland to forest, such as the Roosevelt Engineering, Land Retirement, or Acreage Division and Conservation Reserve programs in the USA, and the policy of returning farmland to forest combined with reducing income in Britain [12–14]. These policy measures have transformed hundreds of thousands of hectares of arable land into parks, forests, and pastures [15,16], providing a reference for other countries to implement policies of returning farmland to forest. Additionally, some countries have begun to build large-scale green corridors, trans-boundary protected areas, or ecological networks between original protected areas to enhance biodiversity conservation. The most influential examples include the European Ecological Network and the Edmonton Ecological Network in Canada. The European Ecological Network also includes Natural 2000, the Emerald Network, the European Green Belt, the Pan-European Ecological Network, and other conservation planning projects, which have significantly improved the effectiveness of ecological protection in those countries [17–20].

China is the largest developing country in the world and economic development has long been the focus of its national policy. Over the past 40 years of reform and opening up, the proportion of China's total economic output to the world economy has increased from 1.8% in 1978 to approximately 15% in 2017, making China the world's second largest economy [21]. China's GDP growth rate (9.5%) was also significantly higher than the global average annual growth rate (2.9%) over the same period [18]. However, owing to the inherent vulnerability of China's ecological environment, social and economic pressures in the development process, and insufficient awareness of ecological and resource conservation, China's economic development came at the expense of abnormal consumption of resources and severe ecology deterioration [22,23]. During the rapid economic development, tremendous changes also occurred in the ecological environment. As a result of its increasing awareness of ecological protection, the Chinese government has begun to carry out various activities, such as ecological protection and restoration projects and ecosystem management, to protect natural resources and enhance ecosystem service functions [24–28].

Nonetheless, the ecosystems in western China are still relatively fragile, with poor natural conditions and low economic development. Owing to the fragile ecological environment, landlocked geographic location, as well as the non-preferential development policies of the central government in the past and the unplanned land-use practices of the local people, western China is currently underdeveloped and lags behind other regions of the country [29]. To promote the coordinated development of the regional economy and improve the ecological environment, the Chinese government implemented a strategy for developing the western region in 2000, covering 12 provinces, and also implemented a strategy to strengthen the protection and construction of the ecological environment at the same time, in terms of development and construction [30]. Driven by this national policy, projects such as the Natural Forest Protection Project, Grain for Green Program, Beijing–Tianjin Sand Source Control Project, and Returning Grazing Land to Grassland Project were implemented in the western region to improve the quality of the ecological environment by actively increasing and restoring forestland and grassland and reducing soil erosion and sandstorm hazards [31].

The first stage of the full-scale development policy targeting the western region has been completed (2000–2010), and the second stage is now being implemented (2010–2020). Under the comprehensive policies of full-scale western development and ecological protection construction, remarkable results in infrastructure and ecological construction were achieved in the region. Currently, the evaluation of policy effectiveness mainly focuses on economic growth, infrastructure construction, public health, and basic educations [32,33], while that of ecological protection construction is focused on the new forest and grassland area, soil erosion control area, the single ecological function of water conservation, nature conservation construction, and the ecosystem services of certain projects [34–38]. However, the changes in the overall ecosystem service status under the comprehensive policy are rarely studied.

To identify the ecosystem services changes in the western region during the first implementation stage of the full-scale development policy driven by the comprehensive policy of development and protection, the national nature reserves in the western development region were taken as a study area to evaluate ecosystem service value (ESV) and the changes during 2000–2010 based on remote sensing, meteorological, statistical, and other relevant data. The main objectives were as follows: (1) quantify land-use changes after the first implementation stage of the comprehensive development and protection policy in the western region, (2) determine the ESV changes during the analyzed period, and (3) discuss the ESV dynamics driven by the national policy.

2. Materials and Methods

2.1. Study Area

The study area (80°25′–123°11′ E, 21°14′–52°8′ N) was located in western China, including 144 national nature reserves in Shaanxi Province, Gansu Province, Ningxia Hui Autonomous Region, Xinjiang Uygur Autonomous Region, Inner Mongolia Autonomous Region, Sichuan Province, Chongqing Municipality, Guizhou Province, Yunnan Province, Guangxi Zhuang Autonomous Region, Qinghai Province, and Tibet Autonomous Region. The total area was 875,498.35 km^2, accounting for 13.01% of the total area of western China (Figure 1).

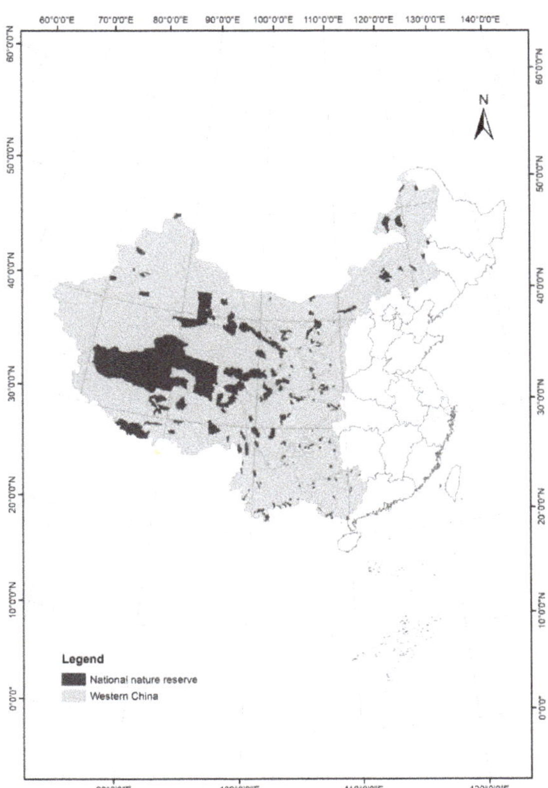

Figure 1. Location of the study area in China.

The region is vast in area, with a substantial span of latitude and longitude, and complex and diverse terrains and landforms. The natural conditions are harsh and varied and the topographic features are complex, with a high proportion of mountains and a fragile ecological environment. Land desertification is a serious issue in the study area. Natural resources, especially biodiversity resources, are abundant. The region is the source of several major rivers, such as the Yangtze River, Yellow River, and Lancang River. There are significant differences in the climatic conditions of the region. The northwest is dry with little rainfall, the southwest is warm and humid, and the Qinghai–Tibet Plateau is cold, while temperature varies greatly overall. The local ecological environment is relatively fragile and rather poor. Important ecologically fragile areas include the Loess Plateau with serious soil erosion and the Yunnan–Guizhou Plateau, which is one of the world's largest karst areas, with severe soil erosion.

2.2. *Data Sources and Processing*

The dataset of land use was extracted using ENVI 4.8 (Esri; Beijing, China) from Landsat Thematic Mapper imagery for 2000 and 2005 and from environmental satellite data for 2010 [39]. After completing the pretreatments, including band compositions, image fusion, image enhancement, and a unifying coordinate system, the maximum likelihood classifier of the supervised classification method was used to classify land use. The qualitative precision errors for deciphering the image in different years were controlled at the 90% level. The interpreted results were then compared with those of typical points from field survey results and early land-use data. The overall interpretation accuracies of the images from different years were over 90%. The total kappa coefficients were also above the minimum acceptable (0.7). ArcGIS 10.2 (Esri; Beijing, China) was used to process and analyze the land-use data with a background of raster images. According to the main geographical features of the study area, the region was divided into 28 land-use types, which were then classified under the seven primary land-use types listed in the resource and environmental database established by the Chinese Academy of Sciences (Table 1). The normalized difference vegetation index (NDVI) and meteorological data were obtained from the Goddard Space Flight Center (NASA) and National Meteorological Information Center [40,41], respectively. The dataset on vegetation was obtained from the GLC2000 database [42]. Data for the agricultural ecosystem net income from grain output per unit area were obtained from statistical data [43,44].

Table 1. Land-use types in the national nature reserves of western China.

Type	Concrete
Forest land	Arbor, broad-leaved forest, coniferous forest, mixed forest, and bush forest
Grassland	Herbaceous green space, meadow, scrub-grassland and steppe
Wetland	Herbaceous marshes, forest swamp, and shrub swamp
Farmland	Dry fields and paddy fields
Water body	Rivers, lakes, canal, ponds, reservoirs, glacier, permanent snow
Unused land	Bare land, bare rock, desert, and saline-alkali land
Construction land	Land used for mining, industry, residences, and transportation

2.3. *Methods*

2.3.1. Land-Use Changes

Land-use changes are crucial for ESV dynamics. The rate of land-use change was calculated as:

$$R_i = (U_b - U_a)/U_a \times 100\% \quad (1)$$

$$R_s = (U_b - U_a)/U_a \times 1/T \times 100\% \quad (2)$$

where R_i and R_s represent the range of land-use change and dynamic rate of land-use change (%), U_a and U_b represent the initial and final areas of a given land use (km^2), and T stands for the analyzed years, respectively.

2.3.2. Standard Equivalent

The standard equivalent refers to the economic value of the annual natural grain yield of farmland with per unit national average yield. The equivalent factor of other ecosystem services can be determined based on this equivalent combined with expert knowledge. It can characterize and quantify the potential contribution of different types of ecosystems to ecosystem services [45].

$$W = S_r \times F_r + S_w \times F_w + S_c \times F_c \tag{3}$$

where W represents the ESV of one standard equivalent (CNY·km^{-2}); and S_r, S_w, S_c represent the percentage of the sown area of rice, wheat, and maize in the total sown area of the three crops in 2010 (%); and F_r, F_w, F_c represent the national average net profit per unit area of rice, wheat, and maize in 2010 (CNY·km^{-2}), respectively.

2.3.3. Equivalent Value Factors

On the basis of Costanza and Folke's [46] theory and investigation by 500 Chinese ecologists [45], the equivalent value factors method was widely used to assess the ecosystem services [47–49]. However, the research of Costanza et al. focuses on the global scale and lacks pertinence for China, while the latter is greatly influenced by subjective opinions, and needs to be adjusted according to the conditions of China's ecosystems [45]. Some improvements were made to make unit ecosystem services value evaluation more practicable in China [39]. For instance, considering that ecosystem services generally have a positive correlation with biomass [45], it is feasible to overcome the spatial heterogeneity of the ecosystem services of a local area using a revision coefficient of net primary productivity (NPP) closely related to ecosystem services [18,45]. The corresponding equation of the equivalent value per unit area (E_i) is:

$$E_i = \frac{n_i}{N_i} \times E_0 \tag{4}$$

where E_i is the equivalent value per unit area in the study areas in year i, n_i and N_i are the average net primary productivity (NPP) values of the ecosystems in the study areas and the country in year i, and E_0 is the national average of equivalent value per unit area (Table 2), respectively. The Carnegie-Ames-Stanford approach (CASA) was used to evaluate the NPP value [50,51].

Because species resources in nature reserves are more abundant than in general areas, another revision coefficient was necessary to calculate the equivalent value of biodiversity conservation. The density of important species was adopted as the parameter for correction. A database of 3337 rare and endangered species in China was determined from the International Union for Conservation of Nature (IUCN) endangered species level 3.1, the Convention on International Trade in Endangered Species appendix, the IUCN Red List of Threatened Species, a list of China's endemic species, and lists of national key protected wildlife species. According to the information on species protection in 861 nature reserves in China, the distributions of 2157 rare and endangered species were established for the nature reserves. Given that protection in national nature reserves accounted for 96.2% of the objects and 99.7% of the area, the calculated value per unit area represented the important species well [18]. The equivalent value per unit area of biodiversity protection in nature reserves (E_b) was determined by Wang et al. [18] as:

$$E_b = \frac{\mathrm{d}}{D} \times Eb_0 \tag{5}$$

where E_b stands for the equivalent value per unit area of biodiversity protection in nature reserves, d and D stand for the density of important species in nature reserves and on a national scale (species·km^{-1}), and E_{b0} is the national average equivalent value per unit area of biodiversity conservation in China, respectively.

Table 2. Equivalent value per unit area of ecosystem services in China [45].

		Freshwater Supply	Gas Regulation	Climate Regulation	Environment Purification	Hydrology Regulation	Soil Conservation	Nutrient Circulation Maintaining	Biodiversity Conservation	Recreation and Culture
Forestland	Coniferous	0.27	1.7	5.07	1.49	3.34	2.06	0.16	1.88	0.82
	Mixed	0.37	2.35	7.03	1.99	3.51	2.86	0.22	2.6	1.14
	Broad-leaved	0.34	2.17	6.5	1.93	4.74	2.65	0.2	2.41	1.06
	Bush	0.22	1.41	4.23	1.28	3.35	1.72	0.13	1.57	0.69
Grassland	Steppe	0.08	0.51	1.34	0.44	0.98	0.62	0.05	0.56	0.25
	Scrub-grassland	0.31	1.97	5.21	1.72	3.82	2.4	0.18	2.18	0.96
	Meadow	0.18	1.14	3.02	1	2.21	1.39	0.11	1.27	0.56
Wetland	Wetland	2.59	1.9	3.6	3.6	24.23	2.31	0.18	7.87	4.73
Farmland	Dry	0.02	0.67	0.36	0.1	0.27	1.03	0.12	0.13	0.06
	Paddy	−2.63	1.11	0.57	0.17	2.72	0.01	0.19	0.21	0.09
Water body	Waterways	8.29	0.77	2.29	5.55	102.24	0.93	0.07	2.55	1.89
	Glacier and snow	2.16	0.18	0.54	0.16	7.13	0	0	0.01	0.09
Unused land	Desert	0.01	0.11	0.1	0.31	0.21	0.13	0.01	0.12	0.05
	Bare	0	0.02	0	0.1	0.03	0.02	0	0.02	0.01

2.3.4. ESV Evaluation

The ESV is mainly determined by land use and the equivalent value per unit area as [18]:

$$ESV_f = \sum_k U_k \times E_{ikf} \quad (6)$$

$$ESV_k = \sum_f U_k \times E_{ikf} \quad (7)$$

$$ESV = \sum_k \sum_f U_k \times E_{ikf} \quad (8)$$

where ESV_f, ESV_k, and ESV are the ESVs of service function f, land-use type k, and ecosystem in year i (CNY·km^{-1}), respectively; U_k is the area of land-use type k (km^{-2}); and E_{ikf} is the value coefficient for land-use type k with ES function type f (CNY·km^{-2}), where the value of ecosystem services for a standard equivalent value per unit area is determined by the agricultural ecosystem net income from grain output per unit area (CNY·km^{-2}) [43,44,52].

To assess the effect of ESV variation on land-use change, the contribution rate (C_{kt}) was estimated by Costanza et al. [38] as:

$$C_{kh} = \frac{|\Delta ESV_{kh}|}{\sum_{k=1}^{n} |\Delta ESV_{kh}|} \times 100\% \quad (9)$$

where C_{kt} refers to the percentage of the absolute value of ESV variation of land-use type k in the period h to the total amount of ESV variation of land-use type k in the period h (%), and ΔESV_{kh} refers to the ESV change of land-use type k in period h (CNY), respectively.

3. Results

3.1. Land-Use Dynamics

Since the implementation of the full-scale development policy targeting the western region, grassland has dominated the land use in the national nature reserves in western China, accounting for more than 3/5 of the total area, followed by unused land and forest land, accounting for more than 1/6 of the total area (Figures 2–4 and Table 3).

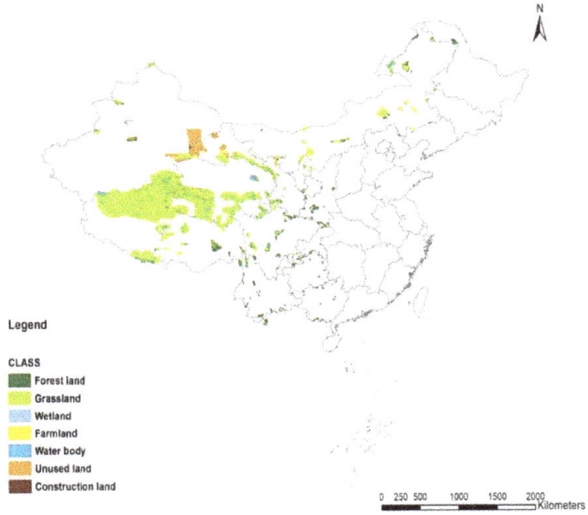

Figure 2. Land use in the national nature reserves of western China in 2000.

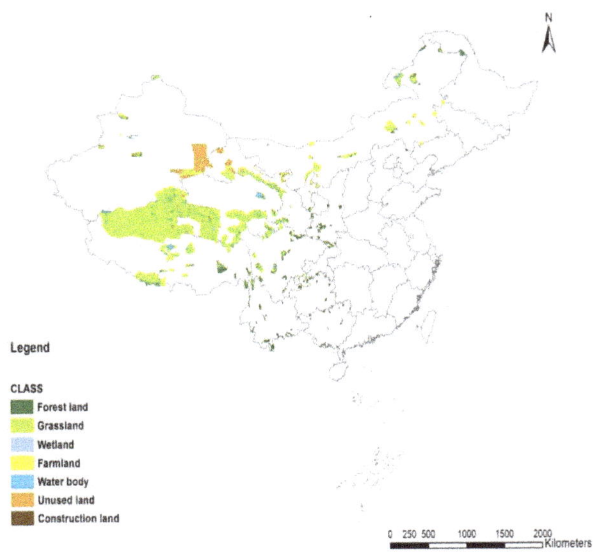

Figure 3. Land use in the national nature reserves of western China in 2005.

The most noticeable characteristics of land-use change in the nature reserves of western China in the whole study phase (2000–2010) were increases in water bodies and decreases in grassland. Especially in period I (2000–2005), the area of grassland declined by 1160.16 km^2, while the area of water bodies increased by 1474.98 km^2 (2.87%). In period II (2005–2010), other area changes were relatively stable except for construction land, which increased by 23.41%. The areas of forestland, water bodies, and construction land showed an increasing trend in both periods, while the areas of grassland, farmland, and unused land showed a continuously decreasing trend (Figure 5).

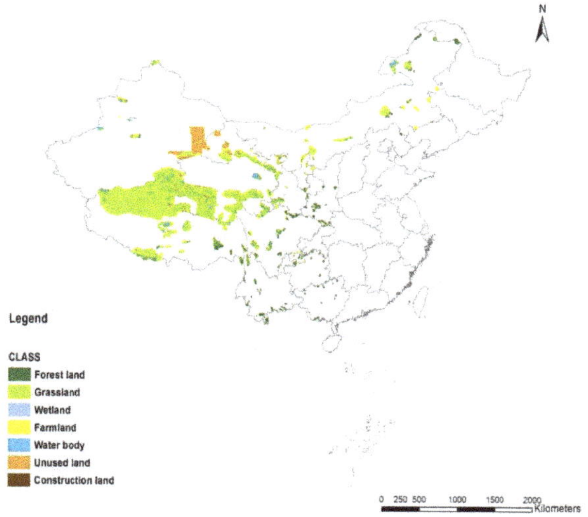

Figure 4. Land use in the national nature reserves of western China in 2010.

Table 3. Land-use dynamics in the national nature reserves of western China.

Land-Use Type	2000		2005		2010	
	Area (km^2)	Percentage (%)	Area (km^2)	Percentage (%)	Area (km^2)	Percentage (%)
Forestland	82,194.70	9.41	82,399.55	9.44	82,415.35	9.44
Grassland	565,771.61	64.79	564,611.45	64.66	564,418.54	64.64
Wetland	21,241.65	2.43	21,247.13	2.43	21,240.47	2.43
Farmland	5722.49	0.66	5614.51	0.64	5587.53	0.64
Water bodies	51,332.10	5.88	52,807.08	6.05	53,101.21	6.08
Unused land	146,264.00	16.75	145,773.76	16.69	145,520.99	16.67
Construction land	650.50	0.07	723.57	0.08	892.97	0.10
Total	873,177.05	100.00	873,177.05	100.00	873,177.05	100.00

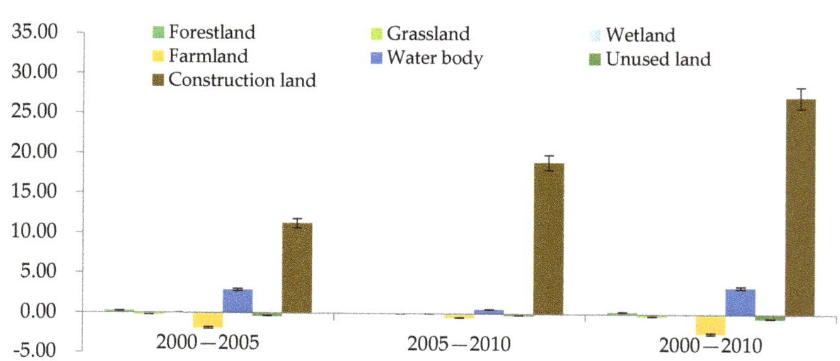

Figure 5. Dynamic rates of land-use types.

In 2000–2005, the areas of forestland, wetland, water bodies, and construction land increased by 0.25%, 0.03%, 2.79%, and 10.10%, respectively, while the areas of grassland, farmland, and unused land decreased by 0.21%, 1.92%, and 0.34%, respectively. In 2005–2010, the areas of forestland, water body, and construction land increased by 0.02%, 0.55%, and 18.97%, respectively, while the areas of grassland, wetland, farmland, and unused land decreased by 0.03%, 0.03%, 0.48%, and 0.17%, respectively. Furthermore, over the period of 2000–2010, the areas of forestland, water body, and construction land continued to increase by 0.27%, 3.33%, and 27.15%, respectively, while the areas of grassland, farmland, and unused land continued to decrease by 0.24%, 2.42%, and 0.51%, respectively.

3.2. ESV Dynamics

The results suggest that the total ESVs of the national nature reserves in western China were CNY 3661.9, 3655.2, and 3736.0 billion in 2000, 2005, and 2010, respectively (Table 4). Among the six ecosystem types, the ESVs of grassland were the highest because of the dominant area, accounting for 45.83%, 45.13%, and 45.35% of the total ESV, respectively. This was followed by water bodies and forestland. Although the area of forestland was approximately 1.6 times as large as that of water bodies, the ESVs of water bodies were approximately 1.2 times as large as the forestland because of the higher value coefficients (3.6 times higher than forestland for the average total value coefficients) (Tables 2–4). The ESVs of farmland were lowest because it occupied the smallest area. The ESVs of unused land were close to the value of farmland, but farmland had higher value coefficients (4.2 times higher than unused land for the average total value coefficients), although the area of unused land was approximately 26 times larger than that of farmland (Tables 3 and 4).

Table 4. Ecosystem services values of each ecosystem in the national nature reserves of western China.

Land-Use Type	2000		2005		2010	
	ESV (10^{10} CNY)	Percentage (%)	ESV (10^{10} CNY)	Percentage (%)	ESV (10^{10} CNY)	Percentage (%)
Forestland	73.43	20.05	73.20	20.03	73.74	19.74
Grassland	167.83	45.83	164.97	45.13	169.42	45.35
Wetland	37.90	10.35	37.92	10.37	38.69	10.36
Farmland	0.65	0.18	0.64	0.17	0.63	0.17
Water bodies	85.53	23.36	87.90	24.05	90.23	24.15
Unused land	0.86	0.23	0.89	0.24	0.89	0.24
Total	366.19	100.00	365.52	100.00	373.60	100.00

In 2000–2005, the total ESV in the national nature reserves of western China declined by CNY 6.7 billion (0.18%). The ESV declines in grassland, forestland, and farmland were the main contributing factors, with reductions of CNY 28.6, 2.3, and 0.1 billion, or 1.70%, 0.31%, and 1.54%, respectively, while the ESVs of water bodies, unused land, and wetland increased with CNY 23.7, 0.3, and 0.2 billion, or 2.77%, 3.49%, and 0.05%, respectively. In 2005–2010, the total ESV in the national nature reserves of western China increased by CNY 80.8 billion (2.21%). The ESV increases in grassland, water bodies, wetland, and forestland were the main contributing factors, with increases of CNY 44.5, 23.3, 7.7, and 5.4 billion, or 2.70%, 2.65%, 2.03%, and 0.74%, respectively, while the ESV of farmland decreased by CNY 0.1 billion, or 1.56%. The ESV of unused land remained stable.

During 2000–2010, the total ESV in the national nature reserves of western China increased by CNY 74.1 billion (2.02%). The ESV increases in water bodies, grassland, wetland, forestland, and unused land were the main contributing factors, with increases of CNY 47.0, 15.9, 7.9, 3.1, and 0.3 billion, or 5.50%, 0.95%, 2.08%, 0.42%, and 3.49%, respectively, while the ESV of farmland decreased by CNY 0.2 billion, or 3.08%. The ESVs of water bodies, wetland, and unused land increased in every period, while farmland continuously decreased.

Grassland and water bodies were the main factors determining the total ESV dynamics during 2000–2005, 2005–2010, and 2000–2010, with contribution rates of 94.70%, 83.77%, and 84.43%, respectively (Figure 6). Grassland was the dominant factor for the total ESV decline during 2000–2005 and increase during 2005–2010, accounting for 51.83% and 54.97% of the total variation, respectively. Water bodies were the main driving factor for total ESV increase in 2000–2010, accounting for 63.11% of the total variation. Farmland and unused land contributed least to the total ESV dynamics, together accounting for only 0.78%, 0.22%, and 0.82% in 2000–2005, 2005–2010, and 2000–2010, respectively.

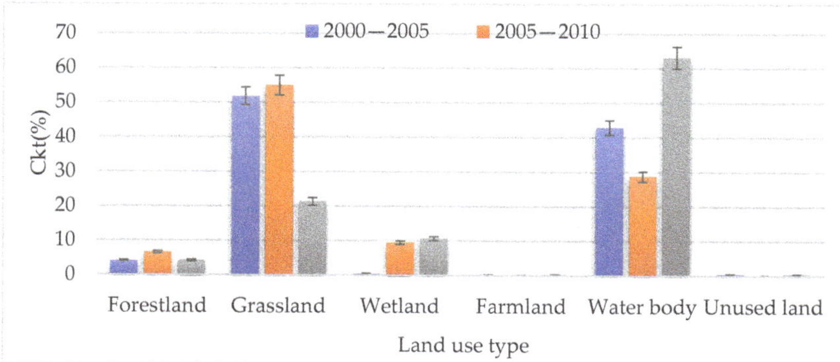

Figure 6. Contribution rates of land-use types.

The ESVs of each ecosystem service are shown in Figure 7. The ESVs of biodiversity conservation had the biggest proportion of the total ESVs in all periods (2000–2005, 2005–2010, and 2000–2010), accounting for 34.86%, 34.64%, and 34.62%, respectively. In other words, biodiversity conservation was the most prominent function in the national nature reserves, as expected, followed by hydrology regulation (accounting for 29.49%, 29.92%, and 29.98%, respectively) and climate regulation (accounting for 13.25%, 13.13%, and 13.11%, respectively). The ESVs of nutrient circulation maintenance (0.46%) and recreation and culture (accounting for 3.10%, 3.09%, and 3.08%, respectively) had the smallest percentages. During 2000–2005, the ESVs of freshwater supply and hydrology regulation increased, by 1.15% and 1.27%, respectively, while the other ESVs all decreased by 0.89% on average. During 2005–2010 and 2000–2010, the ESVs of every ecosystem increased, by 2.18% and 1.74% on average, respectively. Although the ESV for grassland registered a substantial reduction during 2000–2005, that of water bodies registered a substantial increase over 2005–2010, resulting in an overall ESV increase. The Wilcoxon signed-rank test in SPSS 21.0 indicated that the ESVs of each ecosystem service increased significantly from 2000 to 2010 ($p = 0.008 < 0.05$), showing that the total ESV registered a significant increase in the national nature reserves in western China during 2000–2010.

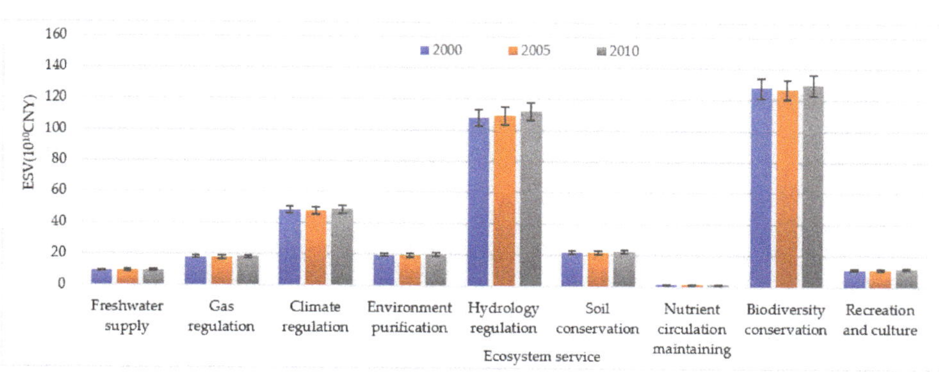

Figure 7. Ecosystem services value of each ecosystem service function in the national nature reserves in western China.

3.3. *ESV Spatial Variation*

Among the 12 provinces in the western region of China, the national nature reserves in Tibet and Qinghai Province had significantly higher total ESVs, accounting for approximately 36.79% and 25.73% of the total ESV of the national nature reserves in the western region, respectively (Figure 8). In contrast, the total ESVs in the national nature reserves in Ningxia and Guizhou Province were relatively low, accounting for only 0.31% and 1.31% of the total ESV. The ESV quantities were directly related to the non-construction land area. The highest proportion of non-construction land was in the national nature reserves in Tibet and Qinghai Province, accounting for 42.83% and 23.99% of the non-construction land in the national nature reserves in the western region, while the non-construction land area in the national nature reserves in Ningxia and Guizhou Province only account for 0.51% and 0.30%, respectively. From 2000 to 2010, the ESV in the national nature reserves in Guangxi increased the most significantly, with a growth rate of 40.36%, followed by Shaanxi and Gansu, with a growth rate of 7.74% and 7.41%, respectively (Figures 9–11).

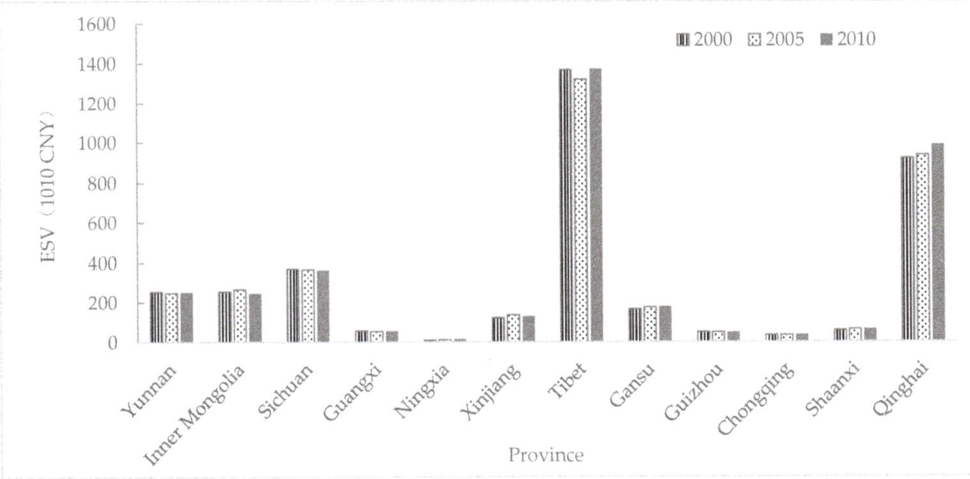

Figure 8. Ecosystem services values of the national nature reserves in each province of western China.

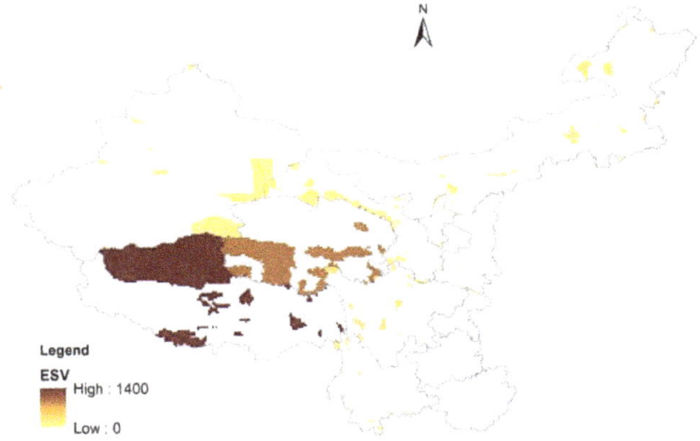

Figure 9. Ecosystem services values in the national nature reserves of western China in 2000.

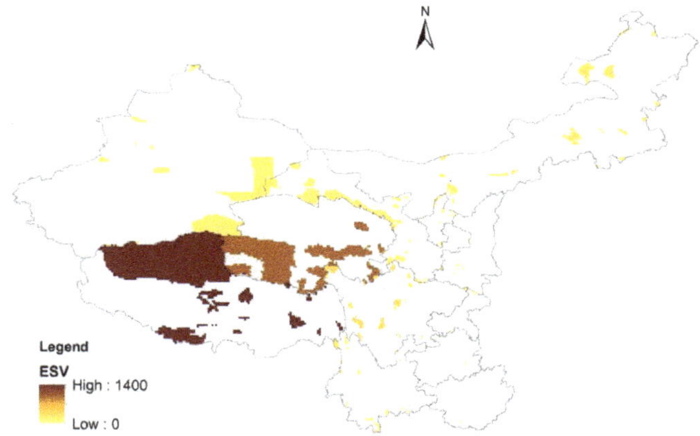

Figure 10. Ecosystem services values in the national nature reserves of western China in 2005.

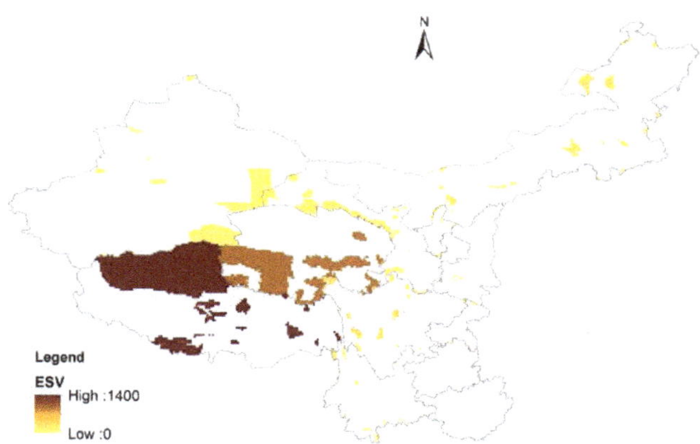

Figure 11. Ecosystem services values in the national nature reserves of western China in 2010.

4. Discussion

At present, ecosystem services value accounting can be divided into two main categories, namely, those based on the price of a unit service function and those based on the equivalent factor of a unit area value. The function value method is used to obtain the total value based on the amount of an ecosystem service function and the unit price of the function quantity. This method simulates the ecosystem service function of a small area by establishing the production equation between a single service function and local ecological environment variables. However, the approach has many input parameters, a complicated calculation process, and more importantly, it is difficult to unify the evaluation method and parameter standards of each service value. The equivalent factor method is based on quantifiable criteria used to construct the equivalent value of various service functions of different types of ecosystems on the basis of distinguishing different kinds of ecosystem service functions, and then evaluating them in relation to the distribution area of the ecosystems [45].

Compared with the function value approach, which estimates the ESV based on ecological processes with a series of predetermined functions or models, the equivalent value factor method is more convenient and appropriate for evaluating the dynamics of ESVs over large scales [49]. Furthermore, this method was widely used and, thus, continuously improved and optimized [18,45,46,53]. Generally, nature reserves show an advantage in species resources, particularly for rare species and biomass, and thus, a correction parameter is necessary. The management of nature reserves prohibits activities such as tree felling, grazing, fishing, and gathering of herbs, and only a small number of original residents can carry out a reduced number of production activities. The overall role of food and raw material production is not important in nature reserves, and it has only a slight influence on the total ESV. As such, we did not evaluate it. Compared with the national unit ESV (39.69 $\times$ 10^5 CNY/km^2) [49], the unit ESV in nature reserves was relatively high (42.799 $\times$ 10^5 CNY/km^2), showing that nature reserves provide more ecosystem services owing to their higher levels of biodiversity and biomass. Based on the results of a previous survey of biodiversity in nature reserves in China and the NPP data, the equivalent factors of biodiversity conservation values and other ecosystem services were adjusted to be more in line with the actual conservation of ecosystems and biodiversity in nature reserves. The biodiversity conservation value of each unit area of different ecosystems, namely, the value of other ecological services, was obtained through the value calculation method. The value of biodiversity and the value of other ecological services were combined with changes of land types to reflect the changes in the biodiversity conservation function and other ecosystem services.

The accurate construction of the equivalent factor table is the core of the equivalent factor method. Although the adopted equivalent factor table was improved through the method of meta-analysis, the original equivalent factor table is still somewhat subjective because it relies on expert opinion scores and cannot accurately and objectively reflect the situation of ecosystem services. In addition, owing to the complexity of the ecosystems and the impact of environmental and biological conditions, there are significant differences in the size and type of their service functions. Therefore, it is objectively necessary to distinguish the ecosystem types and service categories as finely as possible. However, the lack of relevant research resulted in a lack of relevant parameters and results for some types of ecosystem services. Thus, the secondary ecosystem classification had to be adjusted and merged—for example, forest types in different zones were not distinguished—and its impact on the evaluation results needs further research and analysis [45].

Implementing the strategy of full-scale western development and accelerating the development of the western region is of great significance for expanding domestic demand and promoting the coordinated development of China's regional economy. After the completion of the first stage, the GDP of the western region increased from CNY 1665.5 billion to CNY 6686.8 billion, giving an average annual increase of 11.9%. Development included the construction of 972,000 km of new highways, the reconstruction and expansion of 48 trunk/feeder line airports and 23 new airports, and the completion of large-scale water conservancy projects to solve the drinking water and safety problems of 94.37 million rural inhabitants, which greatly increased the economic and social development of the western region [52]. However, the fragile ecological environment in the western region should be considered in implementing the strategy of full-scale western development. Protecting and improving the ecological environment is essential for the full-scale, sustainable, and healthy development of the western region.

Western China is more economically undeveloped and ecologically fragile than central and eastern China. The environmental problems in western China are the result of the interactions between human and nature in the process of economic and social development. Even in historical times, the ecosystems in western China were destroyed by the pressure of population growth, resulting in the innate vulnerability of the ecosystems in western China. Beginning in 1953, Five-Year Plans have been China's most important programs for economic and social development, with the early Five-Year Plans focusing on the economic dimensions [54]. Large-scale economic construction was carried out in the western region, and many industrial enterprises such as steel, coal, and oil were built, which have consumed a substantial quantity of local natural resources. The exploitation of western ecosystems, especially substantial deforestation and destruction of vegetation, have directly had catastrophic consequences. A series of ecological and environmental problems continue to emerge, such as soil erosion, drying up of water sources, abnormal climate, sandstorms, and biodiversity destruction [54].

During the ninth Five-Year Plan period, the Chinese government began to implement the strategy of developing the western region. The western development strategy is divided into three main stages of implementation: (1) the foundation stage (2000–2010), which aims to adjust the structure and improve the infrastructure, ecological environment, science and technology education; establish and perfect the market system, fostering growth industries; make preliminary improvements to the investment environment in the western region; control ecological and environmental degradation, develop a virtuous cycle of economic operation and achieve a growth rate equivalent to the national average growth level; (2) the accelerated development stage (2011–2030), which builds on the foundation stage to enter a rapid stage of western development; cultivate characteristic industries; implement a comprehensive upgrade of economic industrialization, marketization, ecology and professional regional layout; and achieve substantial economic growth; and (3) the modernization stage (2031–2050), which strengthens some of the first development areas and integrates them into the modern domestic and international economic system for self-development. This stage will focus on accelerating the development of remote mountainous areas and underdeveloped agricultural and pastoral areas, generally improving the production and living standards of the people in western China and narrowing the gap in a comprehensive way [55].

With the continuous development and improvement of the concept of ecological protection, the Chinese government's ecological protection policy has made steady progress. The tenth Five-Year Plan (2001–2005) began to coordinate the development of the economy, society, and environment. During the eleventh Five-Year Plan (2006–2010), the government proposed "building ecological civilization", which attracted international attention. An ecological civilization is based on respecting and preserving nature; aims at harmonious coexistence between people, between people and nature, and between people and society; and aims to establish sustainable modes of production and consumption. It guides people onto a path of sustainable and harmonious development [56]. Priority areas for the conservation of biodiversity and key ecological functions were defined during the twelfth Five-Year Plan (2011–2015), leading to an unprecedented increase in the government's support for the conservation and improvement of the environment in the current thirteenth Five-Year Plan (2016–2020), which has further accelerated the pace of investments toward an ecological civilization [57].

In April 2022, the Chinese government included ecological civilization in the Constitution, the country's fundamental law, demonstrating China's determination to promote a new pattern of harmonious development between man and nature. China has implemented a wide range of national policies for ecological protection, including policies on afforestation, returning cultivated land to forest and grassland, forbidding grazing, zoning, ecological migration, which has improved land-use structure and the growth conditions of grassland vegetation, increased vegetation cover, total NPP, soil conservation, and water yield, and reduced soil erosion [34,58–60]. All these play important roles in nature conservation and restoring China's degraded ecosystems, especially in western China [61,62]. Over the past decade, the ESV in the study area has shown a trend of growth. On the basis of the current analysis, the main reason for this is the increase in the ecological land areas, such as forestland and water bodies, and the significant decrease in the farmland area, which is closely related to the conversion of farmland to forest and other policies.

The water body land use includes waterways and glaciers, and the increase in waterways is the main factor for the increase in water body area. The natural conditions mean that the western region is short of water resources and this shortage has become the main factor restricting regional development. To meet the conditions of ecological civilization construction and improve people's livelihood, the Chinese government strengthened key water source projects, water and soil conservation and ecological construction projects, and water resource allocation projects in the western region, including comprehensive watershed management projects such as the Heihe, the Tarim, and the Shiyang river basins; key soil and water conservation projects such as soil in northeast China, the Yangtze River, the middle and upper reaches of Yellow River; and water resources projects such as the first phase of the Tao Water Supply Project in Gansu Province, the Qingtongxia Irrigation area in Ningxia, the south bank trunk channel, and the irrigation area of the second phase of Qiapuqi Seawater Resources Project. These projects have greatly increased the water body area in the western region [63]. According to the ESV research in China, waterways have high ecosystem service functions, especially in hydrology regulation, freshwater supply, environmental purification, biodiversity conservation, climate regulation, and recreation and culture [45]. The expanded water body land use makes an important contribution to the ESV increase in western China.

However, the ESV in the national nature reserves in some provinces in the western region fluctuated and declined during the study period as a result of factors such as climate change and human activities. On the one hand, the increase in construction land was an important factor, such as in Sichuan, Yunnan, Guizhou. The area increase in construction land in the nature reserves was mainly due to the construction of infrastructures, such as corridors, roads, protection stations, posts, observatories, laboratories, and hydroelectric power stations. The planning of village construction and post-disaster reconstruction in nature reserves played another important role in the area increase in construction land. The older houses of the original inhabitants in the nature reserves were built of wood and were more vulnerable to the humid climate and frequent natural disasters (earthquakes, landslides, mud-rock flows, etc.). With social economic development and the growth of personal income, many rural houses were planned or rebuilt in nature reserves, resulting in an increase in the area of construction land [18]. On the other hand, the decline of ecosystem productivity caused by climate change and human activities also has a great impact on ESVs, such as in Tibet, Inner Mongolia, and Shaanxi. In the process of global warming, the plateau air and ground temperature increased significantly from 2000 to 2010, and NDVI decreased in the growing season, resulting in the decline in NPP [64,65].

With increasing global change such as climate warming and loss of biodiversity, achieving a harmonious coexistence between man and nature is a key issue that needs to be solved urgently. In the process of economic and social development, human beings should conform to the laws of nature, take measures in the light of local conditions to address important ecological and environmental problems, and transform the ecological environment from a vicious circle to a virtuous circle. Regions with an underdeveloped economy and fragile ecological environment must avoid the concept of restoration after destruction and adhere first to the practice of parallel economic development and environmental protection. Therefore, they should not exchange the ecological environment for a brief period of social and economic development, nor should they experience social and economic stagnation to promote ecological environment recovery and reconstruction. It is necessary to re-examine the rationality and appropriateness of human activities, actively adapt human activities to climate change, reduce the fragility of the ecological environment, and achieve a win–win situation of sustainable social and economic development under the premise of protecting and building the ecological environment. According to the results of the first phase of the western development strategy, the annual GDP growth rate of the western region was 11.9%, higher than the national growth rate in the same period. Breakthroughs were made in infrastructure development. A number of key projects were completed, including the Qinghai–Tibet Railway, west–east gas transmission, west–east power transmission, the western section of the main national highway, and large water conservancy projects, and the task of transmitting electricity to townships and oil roads to counties was completed. The substantial construction of transport infrastructure has changed the situation in the western region, improving logistics and allowing more convenient travel for people [66]. At the same time, the ecological environment protection measures adopted in the development and construction of the western region have greatly promoted the improvement of the ecological environment. A series of ecological projects, such as the Natural Forest Protection Project, the return of farmland to forest, and the return of grazing to grassland, have continued to expand the green area in the western region, the ecological environment was improved, and the western ecological barrier was strengthened. A total of 26.7 million ha of farmland was returned to forests, 92.7 million ha of forest resources was effectively protected by the Natural Forest Protection Project, 8.5 million ha of the Beijing–Tianjin sandstorm control project was managed, and atmospheric dust release was reduced. The number of people living in poverty fell from 55.53 million in 2001 to 26.48 million in 2008 [66]. China's western development strategy has achieved win–win results in economic and social development and ecological and environmental protection.

5. Conclusions

The current study assessed the ESV and the changes in the nature reserves of western China in 2000–2010. During this period, the area of forestland, water bodies, and construction land increased, while the area of grassland, farmland, and unused land decreased. The total ESV increased by CNY 74.1 billion (2.02%), to which water bodies, grassland, wetland, forestland, and unused land contributed the most. Grassland and water bodies were the main factors of the total ESV dynamics. The ESV of each ecosystem service increased significantly during 2000–2010 except farmland. Consequently, the ecological conservation and restoration measures achieved good results in the national nature reserves of western China after the first stage of western development.

Through the revision of the value equivalent factor method, this paper achieved the rapid assessment of the ESV of regional nature reserves. The revised method was easy to operate and can provide a reference for the evaluation of the effect of regional ecological policy implementation. Through the evaluation, the ESV changes of national nature reserves in western China after the first stage of the implementation of China's western development policy were analyzed, which has an important reference value for understanding changes in the national nature reserves in western China under the policy. However, limitations in the existing technology and methods resulted in a lack of relevant parameters and results for some ecosystem services. Therefore, the secondary ecosystem classification had to be adjusted and merged—for example, forest types in different zones were not distinguished—and its impact on the evaluation results needs further research and analysis.

Author Contributions: Conceptualization, Y.W. and J.G.; methodology, Y.W. and C.Z.; data analysis and processing, D.X. and N.L.; writing—original draft, Y.W.; writing—review and editing, J.G. and C.Z. All authors have read and agreed to the published version of the manuscript.

Funding: This research was funded by the National Key Research and Development Program of China (No.2020YFC1806305), Carbon Emission Peak and Carbon Neutrality Project of Jiangsu Province (No.BK20220021), and Youth Fund Project of Humanities and Social Sciences Research of the Ministry of Education(No.18YJC630216).

Institutional Review Board Statement: Not applicable.

Data Availability Statement: The data presented in this study are available in the article.

Acknowledgments: We are grateful to Lixia Wang for the assist with data processing.

Conflicts of Interest: The authors declare no conflict of interest.

References

1. Zheng, H.; Ouyang, Z.; Zhao, T.; Li, Z.; Xu, W. The impact of human activities on ecosystem services. *J. Nat. Res.* **2003**, *18*, 118–126.
2. Romanowicz, R.J. The impacts of changes in climate and land use on hydrological processes. *Acta Geophys.* **2017**, *65*, 785–787. [CrossRef]
3. Yu, F.; Wang, T.; Groen, T.A.; Skidmore, A.K.; Yang, X.; Ma, K.; Wu, Z. Climate and land use changes will degrade the distribution of Rhododendrons in China. *Sci. Total Environ.* **2019**, *659*, 515–528. [CrossRef] [PubMed]

4. Hopping, K.A.; Knapp, A.K.; Dorji, T.; Klein, J.A. Warming and land use change concurrently erode ecosystem services in Tibet. *Glob. Chang. Biol.* **2018**, *24*, 5534–5548. [CrossRef] [PubMed]
5. IPCC (International Panel on Climate Change). The IPCC Fifth Assessment Report 2014. 2014. Available online: http://www.ipcc.ch/report/ar5/ (accessed on 2 July 2021).
6. Xinhua. How Long Will the Four-Continent Fever Last? Global Weather Extremes are Expected to Increase in the Future. 2021. Available online: http://news.china.com.cn/2018-08/05/content_58101737.htm (accessed on 2 July 2021).
7. MEA (Millennium Ecosystem Assessment). Ecosystems and Human Well-Being: Synthesis. 2005. Available online: http://www.millenniumassessment.org/documents/document.356.aspx.pdf (accessed on 2 July 2021).
8. Trumbore, S.; Brando, P.; Hartmann, H. Forest health and global change. *Science* **2015**, *349*, 814–818. [CrossRef]
9. Scholes, R.J. Climate change and ecosystem services. *WIREs Clim. Chang.* **2016**, *7*, 537–550. [CrossRef]
10. Sample, J.E.; Baber, I.; Badger, R. A spatially distributed risk screening tool to assess climate and land use change impacts on water-related ecosystem services. *Environ. Model. Softw.* **2016**, *83*, 12–26. [CrossRef]
11. Bhattarai, U. Impacts of climate change on biodiversity and ecosystem services: Direction for future research. *Hydro Nepal J. Water Energy Environ.* **2017**, *20*, 41–48. [CrossRef]
12. Huggard, C.J.; Gómez, A.R. Struggle in an endangered empire: The search for total ecosystem management in the forests of Southern Utah, 1976–1999. In *Forests under Fire: A Century of Ecosystem Mismanagement in the Southwest*; University of Arizona Press: Tucson, AZ, USA, 2002; pp. 211–240.
13. Li, S.; Zhai, H. The Comparison Study on Forestry Ecological Projects in the World. *Acta Ecol. Sinic.* **2002**, *22*, 1976–1982.
14. Li, S. The ecological projects in the world. *World Environ.* **1993**, *4*, 31–33.
15. Li, S. Comparison on conversion of cropland to forest and grassland in the world. *World For. Res.* **2002**, *15*, 22–27.
16. Hao, T.; Xu, J.; Zhang, L. Chinese environmental economy policy and the preferential policy of reforestation and revegetation. *Chin. Agric. Sci. Bull.* **2005**, *21*, 349–363.
17. Yan, W; Li, H.; Meng, W. European and American ecological network planning enlightenment to China's. *Environ. Protect.* **2010**, *18*, 64–66.
18. Wang, Y.; Gao, J.; Wang, J.; Wu, Y.; Zou, C.; Tian, M.; Zheng, H. Evaluating losses in ecosystem services in nature reserves in Sichuan, China. *Int. J. Sustain. Dev. World Ecol.* **2014**, *21*, 290–301. [CrossRef]
19. Mu, S.; Zhou, K.; Fang, Y.; Zhu, C. The need and the prospects for developing large-scale green corridors to protect biodiversity. *Biodiv. Sci.* **2014**, *22*, 242–249.
20. Liu, G.; Yang, Z.; Chen, B.; Zhang, L.; Zhang, Y.; Su, M. An ecological network perspective in improving reserve design and connectivity: A case study of Wuyishan nature reserve in China. *Ecol. Model.* **2015**, *306*, 185–194. [CrossRef]
21. Xinhua. Statistics Represent a Historic Leap—The Statistics Showing China's Achievements in Economic and Social Development over the Past 40 Years of Reform and Opening-Up. 2018. Available online: http://www.xinhuanet.com/politics/2018-08/27/c_1123337404.htm (accessed on 2 July 2021).
22. Li, X.; Zhang, J.; Xu, L. An evaluation of ecological losses from hydropower development in Tibet. *Ecol. Eng.* **2015**, *76*, 178–185. [CrossRef]
23. Ameshева, I. Environmental degradation and economic development in China: An interrelated governance challenge. *Law Dev. Rev.* **2017**, *10*, 425–450. [CrossRef]
24. Ouyang, Z.; Zheng, H.; Xiao, Y.; Polasky, S.; Liu, J.; Xu, W.; Wang, Q.; Zhang, L.; Xiao, Y.; Rao, E.; et al. Improvements in ecosystem services from investments in natural capital. *Science* **2016**, *352*, 1455–1459. [CrossRef]
25. Zhang, L.; Fan, J.; Zhou, D.; Zhang, H. Ecological protection and restoration program reduced grazing pressure in the three-river headwaters region, China. *Rangel. Ecol. Manag.* **2017**, *70*, 540–548. [CrossRef]
26. Huang, L.; Shao, Q.; Liu, J.; Lu, Q. Improving ecological conservation and restoration through payment for ecosystem services in Northeastern Tibetan Plateau, China. *Ecosyst. Serv.* **2018**, *31*, 181–193. [CrossRef]
27. Liao, C.; Yue, Y.; Wang, K.; Fensholt, R.; Tong, X.; Brandt, M. Ecological restoration enhances ecosystem health in the karst regions of southwest China. *Ecol. Indic.* **2018**, *90*, 416–425. [CrossRef]
28. Sheng, W.; Zhen, L.; Xiao, Y.; Hu, Y. Ecological and socioeconomic effects of ecological restoration in China's Three Rivers Source Region. *Sci. Total Environ.* **2019**, *650*, 2307–2313. [CrossRef]
29. Wang, X.; Shen, Y.; Cong, R.; Lu, Q. Conflicts affecting sustainable development in West China since the start of China's Western development policy. *J. Res. Ecol.* **2012**, *3*, 202–208.
30. SC (The State Council of the People's Republic of China). Notice of the State Council on the Implementation of Policies and Measures for the Large-Scale Development of the Western Region. 2000. Available online: http://www.gov.cn/gongbao/content/2001/content_60854.htm6 (accessed on 2 July 2021).
31. Lu, F.; Hu, H.; Sun, W.; Zhu, J.; Liu, G.; Zhou, W.; Zhang, Q.; Shi, P.; Liu, X.; Wu, X.; et al. Effects of national ecological restoration projects on carbon sequestration in China from 2001 to 2010. *Proc. Natl. Acad. Sci. USA* **2018**, *115*, 4039–4044. [CrossRef] [PubMed]
32. WDDNDRC (West Development Department of National Development and Reform Commission). Achievement of western development in 10 years is brilliant. *New West* **2009**, *24*, 4–5.
33. Peng, X.; Chen, Z. Impact evaluation on China's western development policy. *China Pop. Res. Environ.* **2016**, *26*, 136–144.

34. Yang, B.; Wang, Q.; Xu, X. Evaluation of soil loss change after Grain for Green Project in the Loss Plateau: A case study of Yulin, China. *Environ. Earth Sci.* **2018**, *77*, 304. [CrossRef]
35. SC (The State Council of the People's Republic of China). Remarkable Achievements Have Been Made in Ecological Protection in the Western Region. 2006. Available online: http://www.gov.cn/ztzl/2006-01/01/content_145209.htm (accessed on 2 July 2021).
36. Renmin. Achievements in the Western Development Region in the Past 10 Years. 2009. Available online: http://finance.people.com.cn/GB/8215/174398/10547501.html (accessed on 2 July 2021).
37. Yang, X.; Xu, B.; Jin, Y.; Qin, Z.; Ma, H.; Li, J.; Zhao, F.; Chen, S.; Zhu, X. Remote sensing monitoring of grassland vegetation growth in the Beijing–Tianjin sandstorm source project area from 2000 to 2010. *Ecol. Indic.* **2014**, *51*, 244–251. [CrossRef]
38. Wang, J.; Peng, J.; Zhao, M.; Liu, Y.; Chen, Y. Significant trade-off for the impact of grain-for-green programme on ecosystem services in North-western Yunnan, China. *Sci. Total Environ.* **2017**, *574*, 57–64. [CrossRef]
39. MEE (Satellite Application Center for Ecology and Environment). Available online: http://www.secmep.cn/secPortal/portal/index.faces (accessed on 2 August 2020).
40. LAADS DAAC (Level-1 and Atmosphere Archive & Distribution System Distributed Active Archive Center). Available online: http://ladsweb.nascom.nasa.gov/data/search.html (accessed on 4 August 2020).
41. National Meteorological Science Data Center. Available online: http://data.cma.cn/ (accessed on 4 August 2020).
42. China Centre For Resources Satellite Data and Application. Available online: http://westdc.westgis.ac.cn (accessed on 9 August 2020).
43. NBS (National Bureau of Statistics). *China Statistical Yearbook*; China Statistics Press: Beijing, China, 2011.
44. PBNDRC (Price Bureau of National Development and Reform Commission). *Compilation of Data on Cost and Income of Agricultural Products Nationwide*; China Statistics Press: Beijing, China, 2011.
45. Costanza, R.; Folke, C. Valuing ecosystem services with efficiency, fairness, and sustainability as goals. In *Nature's Services: Societal Dependence on Natural Ecosystems*; Daily, G.C., Ed.; Island Press: Washington, DC, USA, 1997; pp. 49–68.
46. Xie, G.; Zhang, C.; Zhang, M.; Chen, W.; Li, S. Improvement of the evaluation method for ecosystem service value based on per unit area. *J. Nat. Res.* **2015**, *30*, 1243–1254.
47. Wu, K.-Y.; Ye, X.-Y.; Qi, Z.-F.; Zhang, H. Impacts of land use/land cover change and socioeconomic development on regional ecosystem services: The case of fast-growing Hangzhou metropolitan area, China. *Cities* **2013**, *31*, 276–284. [CrossRef]
48. Costanza, R.; de Groot, R.; Sutton, P.; van der Ploeg, S.; Anderson, S.J.; Kubiszewski, I.; Farber, S.; Turner, R.K. Changes in the global value of ecosystem services. *Glob. Environ. Chang.* **2014**, *26*, 152–158. [CrossRef]
49. Xie, G.; Zhang, C.; Zhen, L.; Zhang, L. Dynamic changes in the value of China's ecosystem services. *Ecosyst. Serv.* **2017**, *26*, 146–154. [CrossRef]
50. Potter, C.S.; Randerson, J.T.; Field, C.B.; Matson, P.A.; Vitousek, P.M.; Mooney, H.A.; Klooster, S.A. Terrestrial ecosystem production: A process model based on global satellite and surface data. *Glob. Biogeochem. Cycles* **1993**, *7*, 811–841. [CrossRef]
51. Monteith, J.L.; Moss, C.J. Climate and the efficiency of crop production in Britain. *Philos. Trans. R. Soc. B* **1977**, *281*, 277–294. [CrossRef]
52. SCIO (The State Council Information Office of the People's Republic of China). Remarkable Achievements in the Ten Years of the China Western Development. 2010. Available online: http://www.scio.gov.cn/ztk/xwfb/10/4/Document/695231/695231.htm (accessed on 2 July 2021).
53. Xie, G.; Zhen, L.; Lu, C.; Xiao, Y.; Chen, C. Expert knowledge based valuation method of ecosystem services in China. *J. Nat. Res.* **2008**, *23*, 911–919.
54. Qin, D.; Ding, Y.; Wang, S.; Wang, S.; Dong, G.; Lin, E.; Liu, C.; She, Z.; Sun, H.; Wang, S.; et al. A study of environment change and its impacts in western China. *Earth Sci. Front.* **2002**, *9*, 321–328.
55. HPRC(The History of the People's Republic of China). Large-Scale Development of China's Western Region. 2009. Available online: http://www.hprc.org.cn/gsgl/dsnb/zdsj/200908/t20090820_28292_2.html (accessed on 9 September 2022).
56. Xie, P. Natural arche of ecological civilization. *J. Lake Sci.* **2016**, *28*, 1–8.
57. Gao, J.; Wang, Y.; Zou, C.; Xu, D.; Lin, N.; Wang, L.; Zhang, K. China's ecological conservation redline: A solution for future nature conservation. *Ambio J. Hum. Environ.* **2020**, *49*, 1519–1529. [CrossRef]
58. Zhang, J.; Wang, T.; Ge, J. Assessing vegetation cover dynamics induced by policy-driven ecological restoration and implication to soil erosion in Southern China. *PLoS ONE* **2015**, *10*, e0131352. [CrossRef] [PubMed]
59. Dang, D.; Li, X.; Li, S.; Dou, H. Ecosystem services and their relationships in the grain-for-green programme—A case study of Duolun county in inner Mongolia, China. *Sustainability* **2018**, *10*, 4036. [CrossRef]
60. Ren, Y.; Kuuluvainen, J.; Toppinen, A.; Yao, S.; Berghäll, S.; Karppinen, H.; Xue, C.; Yang, L. The effect of China's new circular collective forest tenure reform on household non-timber forest product production in natural forest protection project regions. *Sustainability* **2018**, *10*, 1091. [CrossRef]
61. Cao, S. Why large-scale afforestation efforts in china have failed to solve the desertification problem. *Environ. Sci. Technol.* **2008**, *42*, 1826–1831. [CrossRef]
62. Wang, Y.; Gan, Y.; Fei, D.; Tai, F. Study on restoration effects of enclosed pasture in pastureland rehabilitation areas in northwestern Sichuan province. *Pratacult. Sci.* **2008**, *25*, 15–19.

63. China Daily. Major Water Conservancy Projects Carried out during the Decade of the Great Western Development. 2009. Available online: http://www.chinadaily.com.cn/dfpd/xibu10nian/2009-12/21/content_11176695.htm (accessed on 9 September 2022).
64. Wang, H.; Zhang, L.; Shi, X.; Li, D. Some new changes of the regional climate on the Tibetan Plateau since 2000. *Adv. Earth Sci.* **2021**, *36*, 785–796.
65. Zhang, X.; Lu, X.; Wang, X. The spatial and temporal variation of NDVI and its relationships to climatic factors in Northern Tibet over the period of 2000–2010. *Mt. Res.* **2014**, *32*, 475–480.
66. Xinhua. A Good Start and a Solid Foundation—A Review of the Achievements of the Great Western Development in the Past Decade. 2010. Available online: http://www.gov.cn/jrzg/2010-07/05/content_1646123.htm (accessed on 9 September 2022).

MDPI
St. Alban-Anlage 66
4052 Basel
Switzerland
Tel. +41 61 683 77 34
Fax +41 61 302 89 18
www.mdpi.com

Diversity Editorial Office
E-mail: diversity@mdpi.com
www.mdpi.com/journal/diversity

www.ingramcontent.com/pod-product-compliance
Lightning Source LLC
Chambersburg PA
CBHW040757250926
51850CB00007BA/1020
* 9 7 8 3 0 3 6 5 6 7 7 4 7 *